P9-DTI-573

Translational Neuroscience

Toward New Therapies

Strüngmann Forum Reports

Julia Lupp, series editor

The Ernst Strüngmann Forum is made possible through the generous support of the Ernst Strüngmann Foundation, inaugurated by Dr. Andreas and Dr. Thomas Strüngmann.

This Forum was supported by funds from the Deutsche Forschungsgemeinschaft (German Science Foundation)

Translational Neuroscience

Toward New Therapies

Edited by

Karoly Nikolich and Steven E. Hyman

Program Advisory Committee:

Steven E. Hyman, Julia Lupp, Robert C. Malenka,
Karoly Nikolich, Menelas N. Pangalos, and Bernd Sommer

The MIT Press

Cambridge, Massachusetts
London, England

This volume is the result of the 17th Ernst Strüngmann Forum,
held March 16–21, 2014, in Frankfurt am Main, Germany.

Series Editor: J. Lupp
Assistant Editor: M. Turner
Photographs: U. Dettmar
Lektorat: BerlinScienceWorks

MIT Press books may be purchased at special quantity discounts
for business or sales promotional use. For information, please email
special_sales@mitpress.mit.edu.

The book was set in TimesNewRoman and Arial.
Printed and bound in the United States of America.

Library of Congress Cataloging-in-Publication Data

Translational neuroscience : toward new therapies / edited by Karoly
Nikolich and Steven E. Hyman.
pages cm.— (Strüngmann forum reports)
Includes bibliographical references and index.
ISBN 978-0-262-02986-5 (hardcover : alk. paper)
1. Neuroscience—Research. 2. Nervous system—Diseases. 3. Drug development. I. Nikolich, Karoly. II. Hyman, Steven E.
RC337.T729 2015
616.80072—dc23

2015012872

10 9 8 7 6 5 4 3 2 1

Contents

The Ernst Strüngmann Forum

Founded on the tenets of scientific independence and the inquisitive nature of the human mind, the Ernst Strüngmann Forum is dedicated to the continual expansion of knowledge. Through its innovative communication process, the Ernst Strüngmann Forum provides a creative environment within which experts scrutinize high-priority issues from multiple vantage points.

This process begins with the identification of themes. By nature, a theme constitutes a problem area that transcends classic disciplinary boundaries. It is of high-priority interest, requiring concentrated, multidisciplinary input to address the issues. Proposals are received from leading scientists active in their field and selected by an independent Scientific Advisory Board. Once approved, a steering committee is convened to refine the scientific parameters of the proposal and select participants. Approximately one year later, a central gathering, or Forum, is held to which circa forty experts are invited. Expansive discourse is employed to approach the problem. Often, this necessitates reexamining long-established ideas and relinquishing conventional perspectives. When this is accomplished, however, new insights begin to emerge. As a final step, the resultant ideas and newly gained perspectives from the entire process are communicated to the scientific community for further consideration and potential implementation.

Preliminary discussion for this theme began in 2012, based on the recognition that translational neuroscience was at a critical juncture. Despite the impact that diseases of the nervous system have on individuals and society as a whole, available therapies are extremely limited and dated. The unique challenges and numerous failures to treat these complex and often slowly developing diseases have caused disinvestment by the private sector, just as breakthrough technologies have emerged to offer unprecedented insight into brain function and potential treatment possibilities. "What is to be done?" was the motivating call behind this Forum, which brought leading experts together to envision a conceptual roadmap—one that would create an effective, credible, and productive path emanating from the patient to the lab and back again. From February 22–24, 2013, the Program Advisory Committee (Steven E. Hyman, Julia Lupp, Robert C. Malenka, Karoly Nikolich, Menelas N. Pangalos, and Bernd Sommer) met to define the scientific framework for this Forum, which was held in Frankfurt am Main from March 16–21, 2014.

This volume communicates the synergy that emerged from a very diverse group of experts and is comprised of two types of contributions. Background information is provided on key aspects of the theme. These chapters, drafted before the Forum, have subsequently been reviewed and revised. In addition, Chapters 4, 7, 10, and 13 summarize the extensive discussions of the working groups. These chapters are not consensus documents nor are they proceedings; they transfer the essence of the multifaceted discourse, expose areas where opinions diverge, and highlight topics in need of future enquiry.

An endeavor of this kind creates its own unique group dynamics and puts demands on everyone who participates. Each invitee played an active role and embraced the process with a willingness to probe beyond that which is evident. For their efforts and commitment, I extend a word of gratitude to all. A special word of thanks goes to the Program Advisory Committee, to the authors and reviewers of the background papers, as well as to the moderators of the individual working groups (Mene Pangalos, Karoly Nikolich, Steve Hyman, Bernd Sommer, and Rob Malenka). The rapporteurs of the working groups (David Holtzmann, Stephan Heckers, Ilka Diester, and Gül Dölen) deserve special recognition, for to draft a report during the Forum and bring it to a final form in the months thereafter is no simple matter. Most importantly, I extend my sincere appreciation to Karoly Nikolich and Steve Hyman: as chairpersons of this 17th Ernst Strüngmann Forum, their dedication and guidance ensured a most vibrant intellectual gathering.

A communication process of this nature relies on institutional stability and an environment that encourages free thought. The generous support of the Ernst Strüngmann Foundation, established by Dr. Andreas and Dr. Thomas Strüngmann in honor of their father, enables the Ernst Strüngmann Forum to conduct its work in the service of science. In addition, the following valuable partnerships are gratefully acknowledged: the Scientific Advisory Board, which ensures the scientific independence of the Forum; the German Science Foundation, for its supplemental financial support; and the Frankfurt Institute for Advanced Studies, which shares its intellectual setting with the Forum.

Long-held views are never easy to put aside. Yet, when this is achieved, when the edges of the unknown begin to appear and the resulting gaps in knowledge are able to be identified, the act of formulating strategies to fill such gaps becomes a most invigorating activity. On behalf of everyone involved, I hope that this volume will convey a sense of this lively exercise, inspire future work in pathophysiology, and encourage the development of effective therapies to aid those who suffer from the devasting effects of neurodevelopmental and neurodegeneratives disorders.

Julia Lupp, Program Director Ernst Strüngmann Forum
Frankfurt Institute for Advanced Studies (FIAS)
Ruth-Moufang-Str. 1, 60438 Frankfurt am Main, Germany
http://esforum.de/

List of Contributors

Boeckers, Tobias M. Institute of Anatomy and Cell Biology, University of Ulm, 89081 Ulm, Germany

Bourgeron, Thomas Institut Pasteur, 75015 Paris, France

Broich, Karl Bundesinstitut für Arzneimittel und Medizinprodukte, 53175 Bonn, Germany

Brose, Nils Max Planck Institute for Experimental Medicine, 37075 Göttingen, Germany

Cuthbert, Bruce N. Director, Research Domain Criteria Unit, National Institute of Mental Health, Bethesda, MD 20892-9632, U.S.A.

Diester, Ilka Ernst Strüngmann Institute for Neuroscience, 60528 Frankfurt am Main, and Optophysiology – Optogenetics and Neurophysiology, Albert Ludwigs University, 79104 Freiburg, Germany

Dölen, Gül Department of Neuroscience, Johns Hopkins University, Baltimore, MD 21205, U.S.A.

Feng, Guoping Department of Brain and Cognitive Sciences, McGovern Institute for Brain Research, Cambridge, MA 02139, U.S.A.

Frackowiak, Richard Department of Clinical Neurosciences, CHUV, Lausanne University Hospital, 1011 Lausanne, Switzerland

Gur, Raquel E. Departments of Psychiatry, Neurology and Radiology, Hospital of the University of Pennsylvania, Philadelphia, PA 19104-4283, U.S.A.

Heckers, Stephan Department of Psychiatry, Vanderbilt University, Nashville, TN 37212, U.S.A.

Hefti, Franz Acumen Pharmaceuticals, Livermore, CA 94551, U.S.A.

Holtzman, David M. Department of Neurology, Washington University, St. Louis, MO 63110, U.S.A.

Hyman, Steven E. Stanley Center for Psychiatric Research, Broad Institute of MIT and Harvard, Cambridge, MA 02142, and Department of Stem Cell and Regenerative Biology, Harvard University, Cambridge, MA 02138, U.S.A.

Ip, Nancy Y. Division of Life Science, The Hong Kong University of Science and Technology, Kowloon, Hong Kong, China

Joyce, Cynthia MQ: Transforming Mental Health, London EC1M 5PA, U.K.

Kaiser, Tobias McGovern Institute for Brain Research, Department of Brain and Cognitive Sciences, Massachusetts Institute of Technology, Cambridge, MA 02139, U.S.A., and Molecular Medicine Graduate Program, University of Erlangen-Nuremberg, 91054 Erlangen, Germany

Koo, Edward H. Department of Neurosciences, University of California, San Diego, La Jolla, CA 92093-0691, U.S.A. and Departments of Medicine and Physiology, Yong Loo Ling School of Medicine, National University of Singapore, Singapore 119228

Koroshetz, Walter J. National Institute of Neurological Disorders and Stroke, National Institutes of Health, Bethesda, MD 20892, U.S.A.

Kroker, Katja S. Department of CNS Diseases Research, Boehringer Ingelheim Pharma GmbH & Co. KG, 88397 Biberach an der Riss, Germany

Malenka, Robert C. Director, Nancy Pritzker Laboratory, Department of Psychiatry and Behavioral Sciences, Deputy Director, Stanford Neurosciences Institute, Stanford University, Lorry Lokey Stem Cell Research Building, Stanford, CA 94305, U.S.A.

Mansuy, Isabelle Laboratory of Neuroepigenetics, Brain Research Institute, ETH Zurich, University of Zurich, 8057 Zurich, Switzerland

Masliah, Eliezer Neurosciences Department, University of California, San Diego, La Jolla, CA 92003-0624, U.S.A.

Mei, Yuan McGovern Institute for Brain Research, Department of Brain and Cognitive Sciences, Massachusetts Institute of Technology, Cambridge, MA 02139, U.S.A.

Meyer-Lindenberg, Andreas Central Institute of Mental Health, J5, 68159 Mannheim, Germany

Mucke, Lennart Gladstone Institute of Neurological Disease and University of California, San Francisco, CA 94158, U.S.A.

Nicotera, Pierluigi German Center for Neurodegenerative Diseases (DZNE), Bonn, Germany

Nikolich, Karoly Department of Psychiatry, Stanford University Medical School, CA 94305-5489, and Alkahest, Inc. San Carlos, CA 94070, U.S.A.

Owen, Michael J. MRC Centre for Neuropsychiatric Genetics and Genomics, Institute of Psychological Medicine and Clinical Neurosciences, Cardiff University School of Medicine, Cardiff CF24 4HQ, U.K.

Pangalos, Menelas N. EVP, Innovative Medicines and Early Development, AstraZeneca, Melbourn Science Park, Cambridge Road, Royston, Herts SG8 6EE, U.K.

Pascual-Leone, Alvaro Division for Cognitive Neurology, Berenson-Allen Center for Noninvasive Brain Stimulation, Beth Israel Deaconess Medical Center and Harvard Medical School, Boston, MA 02215, U.S.A.

Perlmutter, Joel S. Neurology, Radiology, Anatomy and Neurobiology, Occupational Therapy, and Physical Therapy, Washington University in St. Louis, School of Medicine, St. Louis, MO 63110-1093, U.S.A.

Robbins, Trevor W. Department of Psychology, University of Cambridge, Cambridge CB2 3EB, U.K.

Rubin, Lee L. Department of Stem Cell and Regenerative Biology, Harvard University, Cambridge, MA 02138, U.S.A.

Sawa, Akira Johns Hopkins Schizophrenia Center, Department of Psychiatry, Johns Hopkins University, Baltimore, MD 21287, U.S.A.

Schnaars, Mareike Max-Planck Research Group Neuroimmunology, Center of Advanced European Studies and Research (caesar), Bonn, Germany

Sommer, Bernd Division Research Germany, Boehringer Ingelheim Pharma GmbH & Co. KG, 88397 Biberach an der Riss, Germany

Spillantini, Maria Grazia Clifford Allbutt Building, Department of Clinical Neurosciences, University of Cambridge, Cambridge CB2 0SP, U.K.

State, Matthew W. Department of Psychiatry, University of California at San Francisco, San Francisco, CA 94143, U.S.A.

Wernig, Marius Institute for Stem Cell Biology and Regenerative Medicine, Stanford University School of Medicine, Stanford, CA 94305, U.S.A.

1

Introduction

Karoly Nikolich and Steven E. Hyman

Brain and nervous system disorders are the leading cause of disability worldwide and the primary overall cause of disease burden[1] in established market economies (Murray et al. 2012; Vos et al. 2012). Some brain disorders are important causes of death, most notably stroke and suicide (for which the far most potent risk factor is a mental illness). However, the overwhelming impact of brain disorders on societies and their economies results from their highly disabling effects (Bloom et al. 2011). After all, the brain is the organ of thought, emotion, and behavioral control; thus brain disorders not only cause suffering and stress in individuals and families, they also interfere with the ability of young people to learn in school, for adults to work, and, when severe, with the capacity for self-care. In addition, depression and other brain disorders significantly worsen the course of other chronic diseases, including heart disease and diabetes mellitus (Moussavi et al. 2007).

Developmental neuropsychiatric disorders, such as autism and schizophrenia, and mood and anxiety disorders have very high aggregate prevalence, begin early in life, and are either chronic or exhibit a relapsing and remitting course. With a growing proportion of the world's population living past the age of sixty years, the prevalence of highly disabling neurodegenerative disorders, such as Parkinson disease, Alzheimer disease, and other forms of dementia are projected to increase dramatically in upcoming decades and to entail staggering costs—not only direct medical costs, but also costs resulting from removal of caregivers from the workforce (Hebert et al. 2013).

Despite the severely negative impact of brain and nervous system disorders facing individuals, families, and societies, our current arsenal of effective therapies is extremely limited and remarkably dated. The molecular targets of the major classes of drugs to treat psychosis, depression, and anxiety have not changed from the prototypical drugs discovered serendipitously in the 1950s. Lithium, the therapeutic properties of which were surmised in 1949, is still a mainstay of treatment for bipolar disorder, as is L-DOPA, which was first used

[1] Disease burden is widely measured by disability adjusted life years (DALYs), the present value of future years of healthy life lost to premature mortality (YLL), and years lived with disability (YLD).

in 1961 for Parkinson disease. For many common brain disorders, including depression and schizophrenia, the efficacy of drug therapies plateaued more than a half century ago. Moreover, there are still no effective drug therapies for the core symptoms of autism, the cognitive symptoms of schizophrenia, or many individuals with epilepsy. While significant progress has been made in treating relapsing and remitting forms of multiple sclerosis, there are no disease-modifying therapies for Alzheimer disease, Parkinson disease, or amyotrophic lateral sclerosis (Hyman 2014).

Major pharmaceutical companies recognize the large and growing prevalence of brain disorders as well as the vast unmet medical need. At the same time, they see no clear path to successful drug discovery and development with the significant exception of Alzheimer disease, where several companies are in the midst of clinical trials, all based on the beta amyloid hypothesis. Given the lack of knowledge of fundamental disease mechanisms for most common brain disorders, a dearth of new molecular targets, the failure of most current animal models to predict drug efficacy, and a lack of biomarkers, many companies have de-emphasized or abandoned efforts in brain disorders and have applied, instead, their resources to such areas as cancer and immune system disorders (Pankevich et al. 2014).

Paradoxically, just as industry has begun to abandon brain and nervous system disorders, breakthrough tools and technologies have emerged that are beginning to produce unprecedented insights into pathogenesis of brain diseases. Leaps in genomic and computational technology, resulting from the Human Genome Project, have yielded remarkable progress in genetically complex brain and nervous system disorders, including autism (De Rubeis et al. 2014), schizophrenia (Schizophrenia Working Group of the Psychiatric Genomics Consortium 2014), and common forms of Alzheimer disease (Lambert et al. 2013). Findings from genetics studies are providing early insights into the molecular basis of these diseases, beginning to suggest new drug targets, and providing new hypotheses about potential biomarkers. The many hundreds of genetic loci that are emerging from genetics require relevant high throughput technologies to understand functional differences between alleles that confer disease risk and alleles of the same gene that do not. This has become possible over the last few years, with rapid development of methods to produce different neural cell types from stem cells, induced pluripotent cells, and directly from fibroblasts (Zhang et al. 2013). Different alleles can readily be engineered into and out of isogenic cell lines or patient-derived cell lines using genomic engineering tools (e.g., CRISPR) which have also emerged very recently.

These technologies have created the conditions for discovery and development of new drugs and therapeutic devices. However, because the technologies are new and results are at early stages, most companies are continuing their retreat from the brain, rather than incorporating new approaches. Indeed, approaches based on genetics differ greatly from prior drug discovery efforts for brain disorders—especially for neuropsychiatric disorders. In the past, much drug discovery was grounded in pharmacology and animal behavior. When drug discovery efforts were based on genetic findings, they were limited to rare monogenic forms of illness with the hope that if useful results were forthcoming, they might prove generalizable. An important question is whether, and under what circumstances, industry would participate in drug discovery for polygenic human brain disorders.

"What is to be done?" was the motivational cry that spurred this Ernst Strüngmann Forum. Behind the mere simplicity of this question stands immense power and potential, as evidenced almost a century ago by another who posed the same question, and whose resulting thesis, Што делать, provided the world with a roadmap that led to a very different type of revolution (Lenin 1917/1990).

The emergence of high-quality neuroscience over the last decade combined with the disappointing state of clinical treatment presents a unique paradox. Although the term "paradigm shift" is overused, there is, at a minimum, a vast discontinuity between recent central nervous system drug discovery efforts and the implications for current science. To utilize the scientific approaches that are emerging, academic and industrial efforts must be better connected, with more effective interfaces between the various stakeholders involved in drug discovery and development (i.e., academia, funding agencies, patient groups, biotech and pharmaceutical companies, and regulatory bodies).

Oncology provides an example of what can happen: over the last 20–30 years, based on genetics—most importantly, many somatic mutations that contribute to a large number of cancers—and steady advances in genomics and molecular biology, it has been possible to develop drugs targeted to the effects of specific mutations and to stratify patients in a manner that matches them with the best current treatments. Although advances in oncology provide an important set of aspirations, it differs greatly from the study of brain disorders in that oncology research has direct access to diseased tissue and many of the molecular mechanisms of disease are cell autonomous. In contrast, the human brain can generally be examined only indirectly in life, and many disease processes affect many cell types and are expressed through dysfunction in widely distributed neural circuits. These challenges for neuroscience can only be overcome with great ingenuity, perhaps with greater reliance on human neurons *in vitro*, artificial circuits, and organoids, as well as advances in human experimental biology.

This Forum was thus convened to mobilize and energize the efforts of experts involved in translational neuroscience. It addressed the complexity of the brain and the associated challenges of connecting levels of analysis: from molecules to cells, synapses, circuits, and thence to higher cognition, emotion regulation, and executive function. It also sought to identify (a) tools to stratify patient populations, critical "driver" pathways, and circuits that cause disease in these stratified patient populations; (b) tools to map and modulate critical "driver" pathways and circuits; (c) reliable markers that can quickly demonstrate target engagement and/or pathway modulation *in vivo* so that one can test and refine the hypothesis (and thus motivate a return to basic science); and (d) efficacy endpoints that do not require prohibitively long clinical trials. To do so, four working groups were formed to scrutinize the issues from the following perspectives:

1. *Neurodevelopmental disorders.* Heckers et al. (Chapter 4) begin by reviewing the current psychiatric classification of autism and schizophrenia, and show how this has impeded progress in understanding the underlying disease mechanisms of these disorders. They recommend that future research not be constrained by current nosology and explore alternative diagnostic approaches. Further, they review recent studies of prevention in the early stages of illness and the discovery of genetic and environmental risk factors. Both sets of observations can be used to guide future neuroscientific exploration of neurodevelopmental disorders unimpeded by current psychiatric nosology.
2. *Neurodegenerative diseases.* With the goal of developing more effective diagnostics and treatments over the next 5–20 years, Holtzman et al. (Chapter 7) identify areas where attention is needed to reveal a better understanding of these disorders: Major specific pathologies and circuit dysfunction in neurodegenerative diseases need to be identified during the life span along with dysfunctional circuits. In addition, genetically well-defined patient populations are likely to offer a better chance for therapeutic success. Therapies affecting neurotransmitter systems and signaling pathways should be further explored by utilizing defined patient populations and nodes of those affected by the disease. Better methods are needed to understand protein aggregation processes (from formation of misfolded proteins to the critical clearance pathways that regulate their levels and toxicity) as this could lead to novel therapeutics. A better understanding is needed of the role of apolipoprotein E (apoE), lipoproteins, and lipid biology under normal conditions as well as in neurodegenerative diseases. Attention must also be given to the blood-brain barrier, the neurovascular unit, and other barriers separating CNS from non-CNS compartments to facilitate both a better understanding of disease as well as drug/biological delivery. The role of the innate immune system and other immune mechanisms that contribute to progression of neurodegeneration remains to be elucidated. In addition, regardless of the upstream processes, it may be possible to activate neuroprotective mechanisms by defined factors, signaling pathways, or via cell-based methods. Given the current cost of these diseases to society and the expected increase in their prevalence, Holtzman et al. recommend that a major worldwide effort be put forth immediately and given the highest of priorities. With significant progress in each of these areas, they believe that substantial changes

could be made in the diagnosis and treatment of neurodegenerative diseases over the next twenty years.

3. *The pathophysiological toolkit: From genes to circuits*. Understanding the etiology and pathophysiology of neuropsychiatric disease requires the development of new tools, ranging from evolving diagnostic strategies to biomarkers of disease. Such tools must consider the unique challenges posed by neuropsychiatric disease, including the current lack of tractable interfaces between what can be learned in the clinic and the tools available using model systems in the laboratory. In their report, Dölen et al. (Chapter 10) outline some of these tools, addressing pitfalls and opportunities. They stress the importance of interfacing between model systems and acknowledge the iterative nature required to bridge the gaps between different levels of inquiry.
4. *Living systems: From models to patients*. Research tends to be done in a top-down or bottom-up manner, starting from symptoms or genes, respectively. While such unidirectional approaches work well in oncology and have the potential to advance understanding of monogenic neuropsychiatric diseases, successful application for complex, multifactorial disorders has resulted in a growing number of translational failures. Diester et al. (Chapter 13) investigate existing obstacles and explore options to overcome them. They find that it is essential to dissect complex diseases into measureable and manageable factors, which should then be investigated in a comparable and compatible assembly of model systems to test hypotheses, concepts, and ultimately drug candidates or other therapeutic interventions. Whereas some of these factors might be best investigated through a top-down approach, others might yield better results via a bottom-up approach. Both might be successful only up to a specific point in the experimental chain. Thus, Diester et al. consider it essential to define suitable bridging points between the two approaches, which they term a bidirectional approach. They stress that patients need to be included in this process, since disease-associated dysfunctions or symptoms are often behavioral in nature. In addition, to bridge behavioral readouts between models and humans, links to evolutionary conserved neural substrates need to be created. Some anchor points already exist that successfully bridge model systems to patients, and new promising ones (e.g., induced pluripotent stem cells) are emerging. Diester et al. argue that some recent developments could speed up translation of research into clinical applications (e.g., through faster drug screens in a patient-specific manner) and propose organizational structures that should permit faster transition from research to clinical applications. They also put forth the concept of a "third space" within which early proof of principle studies (Phase 0 and I) can be conducted more effectively.

As each group endeavored to identify existing "gaps" in knowledge and project potential solutions, it is important to note that consensus was not necessarily a goal. Ideas were pooled, as we sought to envision a conceptual roadmap that would create an effective, credible translational neuroscience leading from the patient to the lab and back again.

This volume sets forth the results of these collective efforts. It contains the state-of-the-art papers that served to ground our discussions as well as reports from each of the working groups (Chapters 4, 7, 10, and 13), which highlight, in particular, areas for future research. It concludes with the bold attempt to project the synergism that emerged into a concise conceptual roadmap. We hope that the ideas contained herein will fuel a radical change in translational neuroscience and initiate efforts to create new, effective therapies that will benefit the millions of people who suffer from the devastating impacts of brain and nervous system disorders.

Neurodevelopmental Disorders: A New Beginning

2

What Do We Know about Early Onset Neurodevelopmental Disorders?

Thomas Bourgeron

Abstract

Early onset neurodevelopmental disorders include a broad range of conditions that affect brain function in children: different diagnostic categories such as fetal alcohol syndrome, attention-deficit/hyperactivity disorder, intellectual disability, and autism spectrum disorders. Combined, they affect more than 10% of all children, and some disabilities are permanent throughout their lifetime. Causes are heterogeneous ranging from social deprivation, genetic and metabolic diseases, immune disorders, infectious diseases, nutritional factors, physical trauma, and toxic and environmental factors. Treatments often involve a combination of professional therapy, pharmaceuticals, as well as home- and school-based programs. This chapter briefly reviews the biological pathways associated with early onset neurodevelopmental disorders and provides useful links to progress in the field. Five main biological pathways are associated with autism spectrum disorders and intellectual disability: chromatin remodeling, cytoskeleton dynamics, mRNA translation, metabolism, and synapse formation/function. Three propositions to foster research, proposed by institutions, researchers, and the community of patients and families, will be discussed: (a) to use more dimensional and quantitative data than diagnostic categories; (b) to increase data sharing and research on genetic and brain diversity in human populations; and (c) to involve patients and relatives as participants in research. Finally, examples are provided of very stimulating initiatives toward a more inclusive world of individuals with neurodevelopmental disorders.

Introduction

Children with neurodevelopmental disorders can experience problems with language and speech, motor skills, behavior, memory, learning, or other neurological functions. These difficulties are also frequently associated with comorbidities such as sensorimotor, sleep, and gastrointestinal problems (Gillberg

2010). Symptoms of neurodevelopmental disorders often evolve and may improve as a child grows older, but many disabilities are permanent. Diagnosis and treatment of these disorders can be difficult; treatment often involves a combination of professional therapy, pharmaceuticals, and home- and school-based programs. With progress in neurobiology and genetics, the causes of early onset neurodevelopmental disorders are becoming better understood. In this chapter, I summarize current knowledge on the genetic causes and discuss propositions that have been suggested to improve research in this field.

Definition and Prevalence

Early onset neurodevelopmental disorders affect more than 10% of all children (Table 2.1) and often have consequences throughout their lives, in addition to significant effects on their families (Gillberg 2010; Developmental Disabilities Monitoring Network Surveillance Year Principal Investigators 2014; Perou et al. 2013). This grouping is diverse in terms of severity and pathophysiology: fetal alcohol syndrome, attention-deficit/hyperactivity disorder (ADHD), intellectual disability (ID), tic disorder, developmental coordination disorder, dyslexia, specific language disorders, and autism spectrum disorders (ASDs) (Flannick et al. 2014). Neuromuscular disorders, such as Becker or Duchenne muscular dystrophies, could also be included in neurodevelopmental disorders since they also affect cognition in a subset of patients; however, such disorders

Table 2.1 Prevalence and biological pathways associated with early onset neurodevelopmental disorders.

Neurodevelopmental disorders	Prevalence	Proteins or biological pathways
Learning disabilities	2–4%	Chromatin remodeling Metabolism Actin skeleton organization Channels Synaptogenesis Neurotransmission
Dyslexia	5–15%	Neuronal migration?
ADHD	1.7–9%	Synapses? Cortical maturation?
ASDs	0.6–1.2%	Chromatin remodeling Metabolism Actin skeleton organization Channels Synapses
Epilepsy	0.45–1%	Synapses Channels
Fetal alcohol syndrome	0.1–5%	—

are often considered as a separate cluster because of their predominant symptoms. Boys seem to be at elevated risk compared with girls for most neurodevelopmental disorders, suggesting gender-specific risk and protective factors.

The amount of funding and research dedicated to a disorder is often correlated to its prevalence and severity (Bishop 2010). Thus, it is noteworthy that the amount of research on ID is below the predicted level (Bishop 2010). The causes of early onset neurodevelopmental disorders range from severe social deprivation, genetic risk, metabolic diseases, immune disorders, infectious diseases, nutritional factors, physical trauma, as well as toxic and environmental factors. Among these, knowledge is increasing on genetic risk factors, which, in turn, is motivating new neurobiological research.

The Genetics of Early Onset Developmental Disorders

The list of genes that contribute to early onset developmental disorders numbers in the hundreds. This inherent complexity is exacerbated by the observation that each patient can carry a specific combination of alleles, with large and small effects, that occurs *de novo* or is inherited.

De Novo Mutations

De novo mutations include single base mutations, amplification of trinucleotide repeats, copy number variants (CNVs), large chromosomal rearrangements, and chromosomal aneuploidy (Gilissen et al. 2014). Chromosomal aneuploidy (an abnormal number of chromosomes) is observed in syndromic forms of neurodevelopmental disorders such as Down, Klinefelter, or Turner syndromes. Large chromosomal rearrangements and CNVs can be recurrent in some regions of the genome, such as on chromosome 22q11 (velocardiofacial syndrome), 15q (Angelman and Prader-Willi syndromes), or 17p (Smith-Magenis syndrome). However, in most cases, CNVs which affect from one to hundreds of genes are unique to each patient. A trinucleotide repeat expansion of CGG repeats is observed in fragile X syndrome. This expansion upstream of the *FMR1* gene impedes its expression resulting in increased translation at the synapse. Single nucleotide mutations are another example: X-linked genes, such as *MECP2*, can cause Rett syndrome or autosomal genes, such as *CDH8* or *SHANK3*, can cause ASDs.

Highly penetrant *de novo* mutations probably account for a significant fraction (15–50%) of severe early onset developmental disorders (Hoischen et al. 2014; O'Roak et al. 2011, 2012b; Sanders et al. 2011, 2012; Kong et al. 2012; Iossifov et al. 2012; Neale et al. 2012; Klei et al. 2012). This has been clearly demonstrated for intellectual disability (ID) (Gilissen et al. 2014) and autism spectrum disorders (ASDs) (O'Roak et al. 2012a; Sanders et al. 2011; Neale et al. 2012). Risk factors which increase the occurrence of *de novo* mutations, amplifications, deletions, or duplications are better understood (Campbell and Eichler 2013). For example, regions of the human genome flanked by large segmental duplications (such as on chromosome 15, 16p) are more prone to be deleted or duplicated through illegitimate recombination. Increased paternal age has also been shown to be a factor in *de novo* single base pair change. For ASD and ID, *de novo* chromosomal rearrangements and CNVs are more frequently observed in patients compared to controls. In contrast, patients and controls usually carry the same number of *de novo* single base mutations (on average 60–70 *de novo* mutations in each genome of 3 billions of base pairs and 1 in each exome of 60 millions base pair). However, in patients, there is a significant increase compared with controls of damaging (e.g., loss-of-function) mutations in evolutionarily constrained genes expressed in the brain (Figure 2.1) (Neale et al. 2012; O'Roak et al. 2012a; Sanders et al. 2012; Toro et al. 2010).

The vast majority of mutations reported in patients were identified using DNA isolated from their blood (or in some projects from saliva). Thus, *de novo* somatic mutations that occur in specific brain cell lineage were missed (Frank 2014; Poduri et al. 2013). Only studies using deep genomic sequencing and postmortem brain tissues of patients will be able to inform us as to whether somatic mutations in the brain are increased in early onset neurodevelopmental disorders.

Inherited Monogenic and Polygenic Forms

Among patients with early onset developmental disorders, inherited monogenic forms might account for a relatively significant fraction (> 10%) (Zhu et al. 2014). In ASDs, it is estimated that 3–6% of patients are "homozygous knockout" carriers of two loss-of-function mutations in the same gene (Lim et al. 2013; Yu et al. 2013b). In countries with higher consanguinity, the impact of recessive mutations is likely to be higher (Morrow et al. 2008).

Multiple hits in different regions of the genome might also contribute to a susceptibility to early onset neurodevelopmental disorders. Several studies have demonstrated the presence of more than one deleterious mutation in such patients (Girirajan et al. 2010, 2012; Leblond et al. 2012). In a large-scale study of 2,312 children known to carry a CNV associated with ID and congenital abnormalities, 10% carried a second large CNV in addition to the primary genetic lesion (Girirajan et al. 2012). Children who carried two large CNVs of unknown clinical significance were eight times more likely than controls to have developmental delay than controls. Among affected children, inherited CNVs tended to co-occur with a large second-site CNV. No parental bias was observed for the primary *de novo* or inherited site; in the second site, however, 72% of the second-site CNVs were inherited from the mother (Girirajan et al. 2012).

Other studies have supported a multiple-hits model in patients carrying a similar "first hit." In 42 carriers of a 16p11.2 microdeletion, 10 carried an

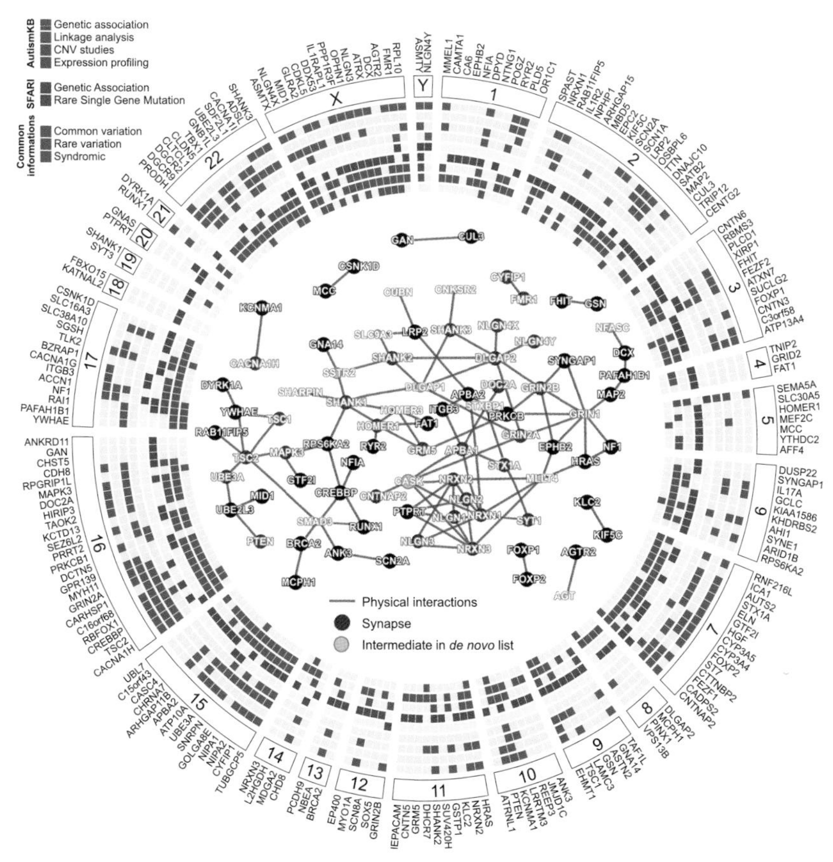

Figure 2.1 Circos plot of *de novo* mutations in autistic spectrum disorder. All coding-sequence variants and copy number variants present in AutismKB and SFARI Gene are shown. A GeneMANIA network analysis (Perou et al. 2013) highlights proteins with synaptic function: 36% of the proteins have at least one interaction with another protein, 61% are expressed in the brain, and 14% are known to be involved in synaptic function. Figure from Huguet et al. (2013), http://gnc.gu.se/digitalAssets/1455/1455513_rev-2cg-bourgerongnc-tb-_4_.pdf (accessed April 7, 2015).

additional large CNV: this is a significantly higher proportion when compared with controls conditional on a large first hit (10 of 42 cases, 21 of 471 controls; P = 0.000057, odds ratio = 6.6) (Girirajan et al. 2010). The clinical features of individuals with two mutations were distinct from and/or more severe than those of individuals carrying only the co-occurring mutation. Another study showed that three patients with ASD carrying a *de novo* SHANK2 deletion were also carriers of a second CNV at the 15q11 locus (Leblond et al. 2012). Two were carrying CNVs, including *CHRNA7* and *ARHGAP11B*; the third was carrying a mutation that removed *CYFIP1*, *NIPA1*, *NIPA2*, and *TUBGCP5*.

After this initial publication, another child with a neurodevelopmental disorder carrying a *SHANK2* translocation and a *CHRNA7* duplication was reported (Chilian et al. 2013).

In addition to *de novo* and inherited rare mutations, one of the current challenges for geneticists is to identify the myriad of frequent alleles across the genome, which in an additive manner increases the risk of developing a disorder. To date, genome-wide association studies have found few common sequence variants that contribute to risk of early onset neurodevelopmental disorders (Anney et al. 2012). However, quantitative genetics analyses (Yang et al. 2010) estimate that if each common variant has a very low effect, collectively they contribute to a relatively large proportion of the heritability of ASD (15–60%) and ADHD (25–30%) (Klei et al. 2012; Cross-Disorder Group of the Psychiatric Genomics et al. 2013). The same methodology was also used to estimate the contribution of genotyped single nucleotide polymorphisms in the heritability of IQ (> 40%) (Davies et al. 2011; Deary et al. 2012) and on the human brain anatomy (50%) (Toro et al. 2014). Based on these results, even if this genetic information is difficult to translate into clinical diagnostics, the identification of low risk alleles represents an important goal for understanding the genetic architecture of early onset neurodevelopmental disorders (Gratten et al. 2014). Moreover, even weak alleles that have been shown with confidence to influence disease risk are helpful to point to genes and pathways involved in pathogenesis.

Database of Associated Genes

Several genetic databases provide clinical and functional annotation of genes associated with early onset neurodevelopmental disorders. The Online Mendelian Inheritance in Man (OMIM) database[1] catalogues more than 5000 human genetic diseases. Decipher[2] and Database of Genomic Variants[3] are interactive web-based databases which incorporate a suite of tools designed to aid the interpretation of genomic variants. Two databases of genes associated with ASD are updated regularly: AutismKB[4] and SFARI Gene[5] (Huguet et al. 2013). A total of 197 genes are included in both databases, and 481 are additionally included in either one or the other (255 in AutismKB and 226 in SFARI Gene). The main difference between the two databases concerns the selection of the genes. AutismKB usually selects genes from linkage analyses, CNV studies, and genome-wide association studies, whereas SFARI Gene

[1] http://www.omim.org (accessed April 7, 2015)

[2] http://decipher.sanger.ac.uk (accessed April 7, 2015)

[3] http://dgv.tcag.ca/dgv/app/home (accessed April 7, 2015)

[4] http://autismkb.cbi.pku.edu.cn (accessed April 7, 2015)

[5] https://gene.sfari.org (accessed April 7, 2015)

usually selects genes from CNV studies, sequencing analyses of large cohorts, and case reports.

Biological Pathways Involved in Early Onset Developmental Disorders

Over the last ten years, tremendous progress has been made in our comprehension of early onset developmental disorders. Animal models (Peca et al. 2011; Bozdagi et al. 2010; Won et al. 2012; Schmeisser et al. 2012; Tabuchi et al. 2007; Jamain et al. 2008; Varoqueaux et al. 2006; Baudouin et al. 2012; Han et al. 2013a; Silverman et al. 2010; Ey et al. 2011) as well as induced pluripotent stem cells (Shcheglovitov et al. 2013; Boissart et al. 2013) have both contributed to better understanding of pathophysiology and new treatment suggestions. Understanding the symptoms and course for each individual as well as the biology (ranging from genetic and environmental risk factors to the neural circuits involved) remains a substantial challenge for geneticists and neurobiologists (Willsey et al. 2013; Parikshak et al. 2013; Gokhale et al. 2012).

Several pathway analyses have been performed using either genetic or transcriptome data to gain insight into the biological functions associated with ASD. Pinto et al. (2010) recently analyzed 2,446 ASD-affected families and confirmed an excess of genic deletions and duplications in affected versus control groups (1.41-fold, $p = 1.0 \times 10^{-5}$) and an increase in affected subjects carrying exonic pathogenic CNVs overlapping known loci associated with dominant or X-linked ASD and ID (odds ratio =12.62, $p = 2.7 \times 10^{-15}$, ~3% of ASD subjects). Consistent with hypothesized sex-specific modulators of risk, females with ASD were more likely to have highly penetrant CNVs ($p = 0.017$) and were also overrepresented among subjects with mutations in genes that encode fragile X syndrome protein targets ($p = 0.02$); this suggests that severe genetic lesions were required to overcome the lower liability to ASDs in girls. Genes affected by *de novo* CNVs and/or loss-of-function single nucleotide variants converged on networks related to neuronal signaling and development, synaptic function, and chromatin regulation. Voineagu et al. (2011) analyzed genes that are differentially expressed between two brain regions (frontal and temporal lobes) in patients with ASD and controls. Interestingly, the typical regional differences between the gene expression profiles of the frontal and temporal lobes were attenuated in patients. A first network module was related to interneurons and to genes involved in synaptic function, and was downregulated in brains from patients compared with those from controls. A second module was enriched for genes related to immunity and microglial activation, and was upregulated in brains from patients with ASD compared with those of controls.

To date, five main pathways have been identified as candidates for early onset neurodevelopmental disorders (Figure 2.2): chromatin remodeling,

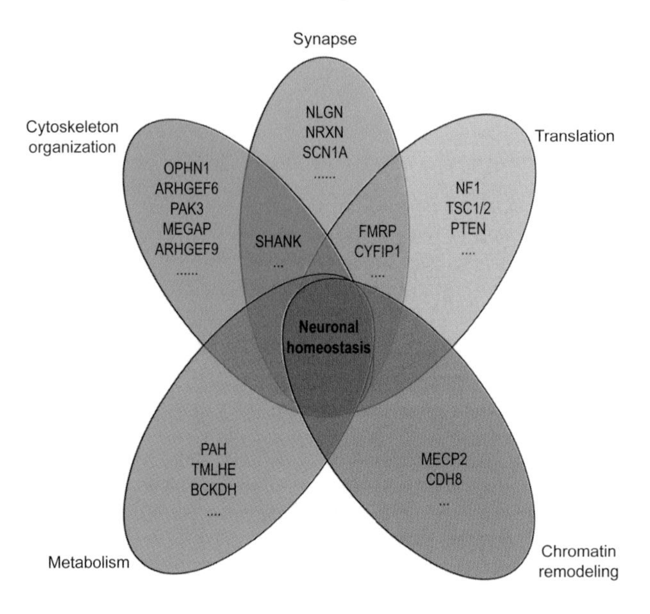

Figure 2.2 Five pathways are associated with early onset neurodevelopmental disorders. For each pathway, mutated genes are indicated.

cytoskeleton dynamics, mRNA translation, metabolism, and synapse formation/function. This list is, however, far from exhaustive.

The first pathway concerns chromatin remodeling and was suggested by reports of mutations in genes such as *MECP2* or *CDH8* in Rett syndrome and ASD, respectively (Amir et al. 1999; Neale et al. 2012; O'Roak et al. 2012a). A second pathway is related to metabolism and includes mutations in genes such as *PAH* in phenylketonuria, *BCKDH* in disorders of branched-chain amino acids, *TMLHE* in carnitine deficiency, or *AGAT* and *GAMT* in creatine-deficiency syndromes. Interestingly, patients with mild forms of inborn errors of metabolism may present with predominantly autistic symptoms (Yu et al. 2013b). Identifying such mutations is of clinical importance since treatments may already be available (Novarino et al. 2012). A third pathway is related to aberrant translation of mRNA-encoding synaptic proteins (Kelleher and Bear 2008) and includes mutations affecting several proteins that normally inhibit translation through the PI3K-mTOR signaling pathway (TSC1, TSC2, NF1, and PTEN) as well as mutations affecting proteins directly involved in inhibiting mRNA translation at the synapse (FMRP, CYFIP1, and EIF4E) (Kelleher and Bear 2008; Costa-Mattioli and Monteggia 2013). A fourth pathway concerns the actin cytoskeleton organization and includes mutations affecting OPHN1, ARHGEF6, PAK3, MEGAP, ARHGEF9, and the regulation of the RhoGTPase, the Ras, the Rab, the Arf, and the JNK pathways (Ba et al. 2013). While mutations affecting these pathways were mostly identified in patients with ID, they might account for a fraction of patients with ASDs (Pinto et al.

2010). Finally, a fifth pathway is involved in synapse formation and excitation/inhibition balance (Bourgeron 2009; Toro et al. 2010). Several genes associated with ASDs, such as *NLGN3/4X*, *NRXN*, and *SHANK1–3*, appear to be involved in the formation of excitatory and inhibitory synapses (Jamain et al. 2003; Durand et al. 2007; Szatmari et al. 2007). In addition, genes associated with epilepsy, such as *SCN1A*, which encodes a voltage-gated sodium channel, were also found mutated in patients with ASD (O'Roak et al. 2011).

While different, these pathways most likely affect neuronal homeostasis at the end point (Ramocki and Zoghbi 2008; Toro et al. 2010). Some suggest potential drug targets; indeed, some early clinical trials are ongoing to determine whether targeting some such proteins could improve the symptoms of patients (for reviews, see Spooren et al. 2012; Delorme et al. 2013).

Brain Regions and Functions Associated with Early Onset Neurodevelopmental Disorders

In some patients, a problem during cortical brain development can be clearly identified anatomically, such as presence of a double cortex (e.g., mutations in DCX), lissencephaly (e.g., LIS1), microcephaly (e.g., ASPM), or macrocephaly (e.g., PTEN). In other patients, a specific brain region seems to be more affected than others, such as the cerebellum in patients with mutation in the oligophrenin gene (Bergmann et al. 2003). However, the vast majority of neurodevelopmental disorders have less dramatic neuropathology, and thus key brain regions remain largely uncharacterized. In ASDs, the occurrence of a high rate of focal disruption of cortical laminar architecture in the cortex of young children could support dysregulation of layer formation and layer-specific neuronal differentiation at prenatal developmental stages as one mechanism (Stoner et al. 2014). Independently, two studies using ASD gene co-expression networks pointed at mid-fetal cortical glutamatergic neurons as a candidate brain region involved in ASD (Willsey et al. 2013; Parikshak et al. 2013). The interaction between genetics and brain imaging is at a nascent stage and progress in both statistical methods and the collection of large datasets is needed (Medland et al. 2014).

Three Propositions to Improve Research in the Field of Early Onset Developmental Disorders

Although tremendous progress has been made in understanding the causes of early onset neurodevelopmental disorders, several issues offer potential breaks for research in this field. Three propositions are listed below.

Proposition 1: Fewer Categories, More Dimensions

Recent advances in genomics has demonstrated that an identical genetic variant may increase the risk for a wide range of diagnoses formerly thought of as distinct (Cross-Disorder Group of the Psychiatric Genomics et al. 2013; Moreno-De-Luca et al. 2013; Kim and State 2014). These findings are contributing to an ongoing reconceptualization of current psychiatric nosology. The use of epidemiological samples, studies grouping individuals based first on genetic findings, and efforts to combine existing, categorical schema with dimensional phenotypes and biomarkers all promise important new insights into the etiology and classification of these disorders. DSM-5 now makes it easier to recognize overlap between different diagnostic categories; however, the existing narrow and rigid categories tend to disconnect researchers from the real phenotypes. Recently, initiatives have been undertaken to improve phenotype characterization using more dimensional approaches:

- The Research Domain Criteria (RDoC)[6] was launched by U.S. National Institutes of Medical Health (NIMH) to develop, for research purposes, new ways of classifying psychopathology based on dimensions of observable behavior and neurobiological and genetic measures. This effort is attempting to define basic dimensions of functioning related to known neural circuitry (e.g., fear circuitry or working memory) to be studied across multiple units of analysis, from genes to neural circuits to behaviors, cutting across disorders as traditionally defined. The intent is to translate rapid progress in basic neurobiological and behavioral research into an improved integrative understanding of psychopathology and the development of new and/or optimally matched treatments for mental disorders. A number of articles and editorials have been published describing and commenting on the project. NIMH anticipates that research grants employing this new experimental classification will represent an increasingly large share of its funding portfolio in coming years. NIMH has also announced that clinical trial funding will also be linked to RDoCs.
- The ESSENCE in child psychiatry: Christopher Gillberg (2010) coined the acronym ESSENCE, which stands for early symptomatic syndromes eliciting neurodevelopmental clinical examinations. ESSENCE aims to reflect the reality of children (and their parents) presenting in clinical settings with impairing child symptoms before the ages of 3–5 years, and it cover the areas of general development, communication and language, social interaction, motor coordination, attention, activity, behavior, mood, and/or sleep. Children with major difficulties in one or more (usually several) of these areas are referred

[6] For further description and links to articles, see http://www.nimh.nih.gov/research-priorities/rdoc/nimh-research-domain-criteria-rdoc.shtml (accessed April 9, 2015).

to health visitors, nurses, social workers, education specialists, pediatricians, general practitioners, speech and language therapists, child neurologists, child psychiatrists, psychologists, neurophysiologists, dentists, clinical geneticists, occupational therapists, and physiotherapists. In reality, however, children are usually seen by only one specialist, when in fact they have needed the input from two or more of these experts. Major problems in at least one ESSENCE domain before the age of 5 years often signals major problems in the same or overlapping domains years later.

In summary, progress in the comprehension of the risk factors for neurodevelopmental disorders will most likely come from dimensional and quantitative data that extends well beyond current psychiatric classification. To enable this, we need to gather information that is currently treated separately by DSM-5 diagnostic categories and located in different laboratories that do not effectively communicate results. Thus, there is a need for increased data sharing (discussed further below).

Proposition 2: More Research on Genetic and Brain Diversity in Human Populations and More Data Sharing

Based on current case-control design, there is a tendency for researchers to know the genotypes and phenotypes of the patients better than those of the controls. Indeed, in the vast majority of genetic studies, controls are often not investigated at the phenotypic level, and in phenotypic studies, controls are very limited in number as well as cultural and social economic status diversity (Manly 2008). As a consequence, early onset developmental disorders are considered binary traits ("affected" versus "non-affected") and this does not take into account the genetic and phenotypic diversity of both "affected" and "non-affected" individuals. The same is true for studies using transgenic mice: most of our knowledge is based on the effect of the mutations in C57BL6 mice. However, we know that mutations might produce a different phenotype in a different strain. The crucial role of the genetic background was very nicely illustrated in a recent paper that showed the phenotypic consequence of the scalloped mutation in different strains of *Drosophila melanogaster* (Chari and Dworkin 2013).

No progress could have been made in the genetics of neurodevelopmental disorders if thousands of genomes had not been sequenced to ascertain their genetic diversity. The same is true for human brains. The first initiatives of the Allen Institute for Brain Science[7] or the Sestan laboratory[8] are impressive in their description of human gene expression at very high resolution. However,

[7] http://www.brain-map.org (accessed April 9, 2015)

[8] http://medicine.yale.edu/lab/sestan/index.aspx (accessed April 9, 2015)

if we want to ascertain the links that exist between the variability of genomes and human brains, thousands of brains will need to be studied at the gene expression level as well as functional level, costly and difficult as this may be.

Integrating diversity in our experimental design will require increasing the sample size of our study populations. Indeed, risk factors for early onset neurodevelopmental disorders are either rare with large effect or frequent but with a small effect (McCarroll et al. 2014). In both situations, robust genotype-phenotype relationships are difficult to ascertain in small samples. One opportunity to increase sample size is to foster data sharing. There are many obstacles to efficient data sharing (Poline et al. 2012). There is, for example, an urgent need (a) to agree on an ethically informed consent for research subjects that will allow data sharing, (b) to agree on standardized measures, (c) to change the reward system regarding publications, and (d) to set up a system that will make data sharing both easy and secure.

There is an emerging community of researchers involved in data sharing. In neuroscience, for example, the Neuroscience Information Framework (NIF) and the International Neuroinformatics Coordinating Facility (INCF) have recently been launched. NIF[9] is a dynamic inventory of web-based neuroscience resources: data, materials, and tools accessible via any computer connected to the Internet. Its intent is to advance neuroscience research by enabling discovery and access to public research data and tools worldwide through an open source, networked environment. INCF[10] develops collaborative neuroinformatics infrastructure and promotes the sharing of data and computing resources to the international research community. Neuroinformatics integrates information across all levels and scales of neuroscience to help understand the brain and treat disease. In addition to increasing the sample size of studies, these data sharing initiatives may also lead to a reduction of publication bias in the field of early onset neurodevelopmental disorders (Joober et al. 2012).

Proposition 3: Patients and Relatives as Participants in Research

Many aspects related to the quality of life of patients and their relatives are not adequately addressed by researchers. For example, in ASDs, comorbidities such as gastrointestinal and sensory problems are underexplored.

The movement "no research about me, without me" is a call for patients and their relatives to be more involved in research design. In the United Kingdom, for example, the National Health Service (NHS) initiative INVOLVE[11] is a national advisory group that supports greater public involvement in NHS, public health, and social care research. Other examples include the James Lind Alliance,[12]

---

[9] http://www.neuinfo.org (accessed April 9, 2015)

[10] http://www.incf.org (accessed April 9, 2015)

[11] http://www.invo.org.uk (accessed April 9, 2015)

[12] http://www.lindalliance.org (accessed April 9, 2015)

the Patient-Centered Outcomes Research Institute,[13] and PatientsLikeMe[14] initiatives for patients that want to monitor their own health and chronic illness. Using such frameworks, patients can propose and conduct their own studies among members, and there is some success to report. For example, a trial of lithium for amyotrophic lateral sclerosis was completed faster than randomized control trials (Wicks et al. 2011). In this study, PatientsLikeMe reached exactly the same conclusion as previous randomized control trials, suggesting that data reported by patients over the Internet may be useful for accelerating clinical discovery and evaluating the effectiveness of drugs already in use. Another example is the initiative for cancer research at Sage Bionetworks. SAGE develops tools so that medical patients can keep their own data rather than storing such data in specified medical institutions. The aim is to offer predictive, personalized, preventive, and participatory (also known as P4) cancer medicine (Hood and Friend 2011). These types of initiatives require the creation of new types of strategic partnerships between patients, large clinical centers, consortia of clinical centers, and patient advocacy groups. For some clinical trials it will be necessary to recruit very large numbers of patients, and one powerful approach to this challenge is to utilize crowd-sourced recruitment of patients by bringing large clinical centers together with patient advocacy groups (Hood and Friend 2011).

Perspectives: Toward a More Inclusive World

For patients, the burden of neurodevelopmental disorders makes daily activities difficult and lowers the odds of living independently. Progress on the causes of neurodevelopmental disorders will hopefully lead to knowledge-based treatments aimed at improving quality of life for those affected. Nevertheless, in addition to improved medical care, innovative initiatives that call for a more inclusive world point toward other important advances. For example, Aspiritech, a nonprofit organization based in Highland Park, Illinois, places people who have autism (mainly Asperger syndrome) in jobs testing software.[15] The Danish company Specialisterne has helped more than 170 individuals with autism obtain jobs since 2004.[16] Its parent company, the Specialist People Foundation, aims to connect one million autistic people with meaningful work.[17] Laurent Mottron, a psychologist working in Montreal, has offered jobs for patients with ASDs in his group, and the perspectives gained have had a positive impact on his research into autism. In his words (Mottron 2011):

[13] http://www.pcori.org (accessed April 9, 2015)

[14] www.patientslikeme.com (accessed April 9, 2015)

[15] http://www.aspiritech.org (accessed April 9, 2015)

[16] http://dk.specialisterne.com (accessed April 9, 2015)

[17] http://www.specialistpeople.com (accessed April 9, 2015)

"The hallmark of an enlightened society is its inclusion of nondominant behaviors and phenotypes, such as homosexuality, ethnic differences and disabilities. Governments have spent time and money to accommodate people with visual and hearing impairments, helping them to navigate public places and find employment, for instance—we should take the same steps for autistics."

Acknowledgments

I wish to thank Steve Hyman for his helpful reading of the manuscript. This work was funded by the Institut Pasteur, the Bettencourt Schueller Foundation, Centre National de la Recherche Scientifique, University Paris Diderot, Agence Nationale de la Recherche (ANR-08-MNPS-037-01-SynGen), the Conny-Maeva Charitable Foundation, the Cognacq-Jay Foundation, the Orange Foundation, and the Fondation Fondamental.

3

Psychotic Disorders and the Neurodevelopmental Continuum

Michael J. Owen

Abstract

This chapter considers a related group of conditions: adult psychiatric disorders such as schizophrenia and bipolar disorder, which are believed to have their origins in disturbances of neurodevelopment, and childhood neurodevelopmental disorders, particularly autism spectrum disorder and intellectual disability. The primary focus is on the challenges faced in trying to understand etiology and pathogenesis as a prelude to developing more effective treatments.

Recent findings, particularly genetic studies that indicate extensive pleiotropy, reinforce the view that current categorical diagnostic systems do not map onto the underlying biology of psychiatric disorders. Unless these findings are integrated into research, understanding etiology and pathogenesis will most likely be impeded. A simple model is presented that integrates current knowledge. It hypothesizes that severe mental illnesses can be conceived as occupying a gradient with the syndromes ordered by decreasing relative contribution of neurodevelopmental impairment.

Emerging evidence from genetic studies indicates a convergence onto specific areas of synaptic biology, but the genetic architecture of psychiatric disorders is undeniably highly complex. This complexity poses both cultural and technical challenges to efforts to translate these findings into new mechanistic insights. In particular, networks rather than individual genes and proteins need to be studied, and experimental systems in which multiple variables can be manipulated and multiple endpoints studied need to be developed. The complexity and mutability of psychiatric phenotypes, and the shortcomings of current diagnostic criteria, pose additional challenges for clinical neuroscience, and require new ways of stratifying patients in neurobiologically meaningful ways, such as those specified by the Research Domain Criteria (RDoC) project. In both basic and clinical neurosciences, an increased focus must be allocated to large-scale experimentation, collaboration, statistical robustness, and reproducibility.

Introduction

In this chapter, focus is on the so-called "functional psychoses" (schizophrenia and bipolar disorder) and their relationship with "neurodevelopmental" disorders that occur in childhood, particularly autism spectrum disorder (ASD) and intellectual disability (ID). I will discuss the impact that recent research, particular genetics, has had on our conceptions of psychiatric diagnosis and classification, and the emerging implications that genetic and other findings have for future research on etiology and pathogenesis. A Gordian knot impedes progress in understanding psychiatric disorders: the complexity and inaccessibility of the brain has made it difficult to understand basic disease mechanisms, and this has meant that diagnoses cannot be validated in terms of etiology and pathogenesis. The lack of valid diagnostic criteria has, in turn, impeded progress in research on mechanisms. How we unpick this knot is the central challenge currently facing psychiatric research.

Current Approaches to Psychiatric Diagnosis

In the absence of a solid understanding of pathophysiology, psychiatric diagnoses are descriptive and largely syndromic in nature. Most Western clinicians use either the Diagnostic and Statistical Manual of the American Psychiatric Association (DSM) or the International Classification of Diseases (ICD). These operationalized, largely descriptive classifications were developed in the final third of the twentieth century to provide a reliable way of assigning a patient to a diagnostic category that represents one of the clinical syndromes recognized by clinicians. They were never intended, however, to define disease entities, and their uncertain validity was explicitly acknowledged. While the reliability of these systems has had many benefits on research and practice, the diagnostic categories have unfortunately become reified in the minds of many practitioners and researchers (Hyman 2010); they have been treated as if they are pathologically meaningful and relatively circumscribed disease entities rather than provisional and temporary clinical syndromes as originally intended.

This categorical and syndromic approach to diagnosis has been creaking under the weight of discrepant findings for some years (Craddock and Owen 2005, 2010). Substantial heterogeneity within diagnostic categories makes it possible for two patients with the same diagnosis to have few, if any, symptoms in common. It has been difficult to demonstrate clear boundaries (points of rarity) between diagnostic categories as well as between illness and wellness. In addition, patients often present with the features of more than one diagnostic category.

This has been addressed in two ways. First, in some instances interforms are recognized, the most prominent, and notorious, being schizoaffective disorder (Malaspina et al. 2013; Cardno and Owen 2014). This diagnosis, which is frequently used in clinical practice, reflects the fact that features of schizophrenia and severe mood disorder often occur in the same individual both concurrently or separately at different time points over the life span. Illnesses of this sort have posed challenges to classification systems over many years (Malaspina et al. 2013), and their relationship to the prototypical disorders continues to be debated (Cardno and Owen 2014). The second approach is to recognize so-called diagnostic "comorbidity"; that is, a patient can be diagnosed with more than one disorder. This may be clinically useful but it is often obscured in research through the use of diagnostic hierarchies or exclusions (Owen 2011).

Diagnostic categories in current use by psychiatry are extremely fuzzy, both in respect to other disorders as well as to wellness. Moreover, they are not underpinned by biological validity, and this no doubt explains why concerted efforts to develop diagnostic biomarkers have failed (Lawrie et al. 2011; Owen 2011). This clearly suggests that exclusive reliance on current diagnostic categories is likely to impede research into causes and mechanisms. These concerns have become impossible to ignore in the light of recent genetic findings.

Impact of Genetics on Classification

Genetics, in the form of family history and other genetic epidemiological data, has traditionally been regarded as a cornerstone of psychiatric nosology. It constitutes one of the three criteria proposed by Robins and Guze (1970)—the other two being clinical features and outcome—to justify nosological categories in the absence of a solid understanding of pathogenesis. Thus, the robust identification of genetic risk at the level of DNA variation has been eagerly sought for the insights it was expected to provide into the basic biological architecture of, and relationships between, psychiatric phenotypes, as well as for its contributions to understanding disease mechanisms.

Genetic epidemiology and population genetics suggest that a spectrum of allelic risk is likely to underlie complex traits, such as schizophrenia and other common diseases (Wang et al. 2005; Craddock et al. 2007). We should expect contributions from alleles that are common in the population but whose effect sizes will tend to be small due to the effects of natural selection, as well as from rare alleles, some of which might have a large effect on disease risk pending their removal from the population by selection. Although much of the genetic risk for psychiatric disorders remains unexplained, empirical data now support this general framework in psychiatric disorders (Sullivan et al. 2012). Of crucial importance to the present discussion, the findings to date point to extensive pleiotropy with respect to diagnostic outcome, though some examples, described below, suggest a degree of specificity.

Evidence for quite an extensive overlap in genetic risk between psychiatric disorders has come from the study of both common and rare risk alleles. In respect of common alleles, genome-wide association studies (GWAS) have found evidence for overlap at the level of individual genome-wide significant risk alleles (Green et al. 2010; O'Donovan et al. 2008; Williams et al. 2011), genes (Moskvina et al. 2009), and the *en masse* effects of multiple risk alleles (International Schizophrenia Consortium et al. 2009; Lee et al. 2012). A recent study (Cross-Disorder Group of the Psychiatric Genomics et al. 2013) found evidence for substantial sharing of the relatively common genetic risk variants that are tagged by the single nucleotide polymorphisms genotyped in GWAS between schizophrenia and bipolar disorder, bipolar disorder and major depressive disorder, schizophrenia and major depressive disorder, attention-deficit/hyperactivity disorder (ADHD) and major depressive disorder, and, to a lesser extent, between schizophrenia and ASD.

Perhaps more surprisingly, extensive pleiotropy has also been observed in the effects of rare risk alleles that individually confer much larger effects on risk than common alleles. A number of rare, but recurrent, chromosomal copy number variants (CNVs), which typically involve deletion or duplication of hundreds of thousands of bases of DNA sequence, have been found to confer risk of schizophrenia (Malhotra and Sebat 2012; Rees et al. 2014). These are also significantly associated with a range of childhood neurodevelopmental disorders: ASDs, ID, and ADHD as well as other phenotypes such as generalized epilepsy (Owen et al. 2011). Since pathogenic CNVs typically span multiple genes and are concentrated in a relatively small fraction of the genome, it is possible that this may not indicate cross-disorder effects at the level of specific genes. Although support for cross-disorder genetic effects emanates from family studies (Owen et al. 2011), the most compelling evidence comes from a recent large-scale study of small *de novo* mutations that affect one or a few nucleotides. Fromer et al. (2014) found evidence that genes with *de novo* mutations in schizophrenia overlapped those affected by *de novo* mutations in ASD and ID but not controls; loss-of-function (LoF) mutations were enriched even in the very small subset of genes (N = 7) with *recurrent* LoF *de novos* in ASD or ID. These findings demonstrate shared etiological overlap between schizophrenia, ASD, and ID at the resolution not just of loci or even individual genes, but of mutations with similar functional (LoF) impacts.

The finding of genetic overlap between schizophrenia and bipolar disorder was hardly surprising, given the frequent difficulties adult psychiatrists have in clinically distinguishing between these disorders (Craddock and Owen 2005). However, the genetic overlap between schizophrenia and childhood neurodevelopmental disorders posed a greater challenge to nosological orthodoxy. Schizophrenia has long been considered to have its origins in neurodevelopment (Weinberger 1987; Murray and Lewis 1987; Owen et al. 2011) but has, in recent years, come to be considered distinct from the neurodevelopmental disorders, which tend to have their clinical presentation in childhood (Rutter et al. 2006). This was largely due to the fact that schizophrenia, unlike the childhood disorders, was seen as a relapsing and remitting illness in accord with the time course of the psychotic symptoms that tend to result in presentation to clinical services. Many of the clinical features of schizophrenia are chronic, notably the negative and cognitive symptoms, and these often originate in childhood prior to the first psychotic break. There are, in fact, many similarities in phenotype between schizophrenia and the other neurodevelopmental syndromes (Owen et al. 2011). Importantly, all are associated with impairments of cognition. They also tend to be more common in males, and are frequently associated with varying degrees of developmental delay, neurological soft signs, and motor abnormalities. There is also substantial comorbidity among neurodevelopmental disorders, including schizophrenia (reviewed in Owen et al. 2011). As noted above, comorbidity is often obscured in research studies by the use of diagnostic hierarchies or exclusions, but it can also be concealed by developmental change in the predominant symptom type. Current service configurations also impose difficulties due to the administrative split between adult services and those directed toward the treatment of children and adolescents, as well as between psychiatric, ID, and, in the case of epilepsy, general medical services. These service splits have traditionally defined the purviews of researchers, with the result that much research in psychiatric and neurodevelopmental disorders has taken place in silos defined by the existing syndromic categories.

Support for shared neurodevelopmental etiology and pathogenesis across psychiatric disorders does not come solely from genetics. Obstetric complications and other factors (e.g., maternal infection, poor prenatal nutrition) associated with early cerebral insult have been consistently implicated as environmental risk factors for a range of neurodevelopmental disorders, including non-syndromal ID, ASD, ADHD, epilepsy as well as schizophrenia (Owen 2012a, b). The similarity between this range of outcomes and that seen in association with pathogenic CNVs is striking. In the 1950s, Pasamanick et al. (1956) proposed that a "continuum of reproductive causality" exists, consisting of brain damage incurred during pregnancy, or during or around birth, leading to a gradient of injury that extends from fetal and neonatal death through cerebral palsy, epilepsy, ID, and behavioral disorders, including schizophrenia. Given recent genetic findings, it seems reasonable to modify this concept to encompass a gradient of genetically *and* environmentally induced neurodevelopmental causality along which lie what we currently define as ID, epilepsy, ASD, ADHD, schizophrenia, and possibly the major affective disorders (Craddock and Owen 2010; Owen et al. 2011). This view recognizes the degree of etiological and symptomatic overlap between diagnostic groups as well as the lack of clear diagnostic boundaries. It also perceives the major clinical syndromes as reflecting, in part, the severity as well as the predominant pattern of abnormal brain development and the resulting functional abnormalities and modifying effects of other genetic and environmental factors (see Figure 3.1).

We have emphasized genetic findings that suggest common susceptibility across traditional diagnostic categories, implying that the underlying biology is not specific, at least at the level of current diagnoses. It is, however, important

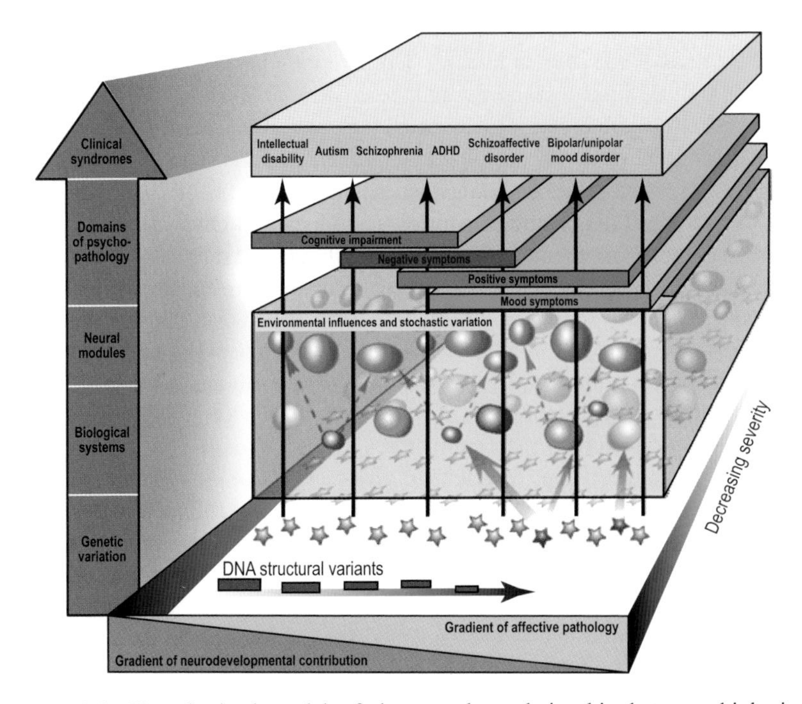

Figure 3.1 Hypothesized model of the complex relationship between biological variation and some major forms of psychopathology (adapted from (Craddock and Owen 2010). One axis depicts the relative contribution of type of pathology, and another (orthogonal) relates to clinical severity. The following description refers to the pathology axis. Starting at the level of genetic variation (lowest tier in figure), DNA structural variation (blue rectangles) has been represented as contributing particularly to neurodevelopmental disorders and associated particularly with enduring cognitive and functional impairment. Single gene variants, of which there are many, are shown as stars. In general, even single base-pair changes in a gene may influence multiple biological systems because genes typically have multiple functions and produce proteins that interact with multiple other proteins. For simplicity, we have shown one example of a variant that influences two biological systems and another that influences only one system (reddish stars and light blue arrows). Variation in the relevant biological systems (colored solid figures) is influenced by genotype at many genetic loci as well as by environmental exposures/experiences and random stochastic processes, both historically during development and currently, and this, in turn, influences the dynamic short-term state of the systems. The relevant biological systems influence the neural modules that comprise the key relevant functional elements of the brain (solid light blue figures). Typically, multiple biological systems influence each neural module. The (abnormal) functioning of the neural modules together influences the domains of psychopathology experienced and ultimately the clinical syndromes. Some important clinical syndromes are ordered along a single major axis with a gradient of decreasing proportional neurodevelopmental contribution to causation and reciprocal increasing gradient of proportion of episodic affective disturbance. The single axis is a simplifying device: there is substantial individual variation and it is not uncommon, for example, for individuals diagnosed with autism to experience substantial mood pathology. Other features of the model are described within the text.

to recognize that relatively nonspecific risk factors, which apply to a wide variety of cases, will be easier to identify than those associated with more specific outcomes. Both family and genomic studies provide evidence for risk alleles with differential effects on schizophrenia and bipolar disorder (Lichtenstein et al. 2009; Ruderfer et al. 2013) as well as for alleles that have a degree of specificity to more "refined" clinical phenotypes that do not necessarily relate well to traditional diagnostic categories (Craddock and Owen 2010). Another example comes from the mounting evidence that large, rare CNVs, which are more prevalent in schizophrenia, are actually underrepresented in bipolar disorder (Grozeva et al. 2010; Grozeva et al. 2013). Given that large CNVs have adverse consequences on brain development and cognition, these findings are consistent with the view that a diagnosis of bipolar disorder is unlikely to be made in the presence of cognitive and other sequelae of neurodevelopmental impairment, and that what we refer to as schizophrenia has a stronger neurodevelopmental component than bipolar disorder.

A Simple Model

A simple conception that integrates recent findings is that severe mental illnesses occupy a gradient with the syndromes, ordered by decreasing relative contribution of neurodevelopmental impairment as follows: ID, ASD, schizophrenia, ADHD, schizoaffective disorder, bipolar disorder (Craddock and Owen 2010). Elaborating on this view (Figure 3.1), the key variables are the particular constellation (i.e., number and nature) of disrupted neural circuits, which determine the type of syndrome that results (mental retardation, autism, or schizophrenia, etc.), and the severity (degree of disruption to individual circuits) of the syndrome (severity of ID, severity of ASD, severity of schizophrenia spectrum disorder, etc.). Of course this is a gross oversimplification. The precise nature and timing of critical events also plays a role as well as the modifying effects of genetic and environmental influences on factors such as the brain's capacity to buffer the effects of early damage, personality, and propensity to affective disturbances. Key features of the model can be described as follows:

- Dimensions or continua are preferred over categories to conceptualize the major clinical syndromes.
- There is broad organization along a major axis according to a gradient of increasing relative neurodevelopmental contribution to illness in one direction and increasing relative episodic affective contribution in the opposite direction, with syndromic severity represented orthogonally.
- Multiple domains or dimensions of psychopathology contribute to the major clinical syndromes in varying proportions, and these may relate more closely to dysfunctional brain systems than categorical diagnoses.

- States of relevant brain systems depend crucially on environmental influences and stochastic variation (both developmentally and dynamically over the short term).
- Brain systems are complex, interdependent, and modular in nature; modules are functionally discernible, not necessarily temporally or spatially stable subunits that are interconnected in complex, often multilayered networks of neuronal circuits.
- One-to-one relationships do not exist. The concept of "a gene for schizophrenia" or even "a gene for auditory hallucinations" is not plausible (Kendler 2006). Instead, sets of many one-to-one or one-to-many relationships are involved.

This simple model proposes an order for biological similarity of some of the major phenotypes, which can be tested empirically.

Recent Support for the Model

We have recently tested the predictions of this model by comparing *de novo* mutations in schizophrenia with those found in ID and ASD (Fromer et al. 2014). As noted already, we found evidence that genes with *de novo* mutations in schizophrenia overlapped with those affected by *de novo* mutations in ASD and ID, but not controls. We also observed that the genes hit by *de novo* mutations and the mutation sites themselves showed the highest degree of evolutionary conservation (a proxy measure of functional importance) in ID, then ASD, with schizophrenia least conserved. These findings suggest that highly disruptive mutations play a relatively lesser role in schizophrenia, and that the disorders differ by severity of functional impairment, consistent with the hypothesis of an underlying gradient of neurodevelopmental pathology indexed by cognitive impairment, with ID at one extreme.

Implications of Recent Findings for Basic and Clinical Neuroscience

Thus far, genetic studies have identified specific genetic risk variants that account for only a very small proportion of risk. However, it is now possible to discern the approximate shape of the genetic architecture of psychiatric disorders, and the potential complexity of the relationships between genetic risk and clinical endpoints. The implications for neuroscience are profound.

Genetic Complexity

The first key message is that a very large number of genes are involved. In schizophrenia, where the largest samples have been studied and our understanding is therefore greatest, over 100 distinct genetic loci harboring relatively common alleles of small effect (ORs < 1.3) have been identified to date at robust levels of statistical significance (Schizophrenia Working Group of the Psychiatric Genomics Consortium 2014). Furthermore, it is estimated that relatively common small effect alleles of this kind account for about 25% of total liability to schizophrenia (about 33% of genetic liability) (Lee et al. 2012) and that there are many thousands yet to be identified (Ripke et al. 2013). Presumably, the remaining genetic burden reflects alleles that are insufficiently frequent to be detected by GWAS, and some of these will have sufficiently large effects to be detected in realistically sized studies. As new generation sequencing (NGS) technology is applied, some of these rare variants are being identified at least at the level of specific gene sets (Fromer et al. 2014; Purcell et al. 2014). This complexity means that individuals, even those without the disorder, will carry a large number of risk alleles and that levels of genetic heterogeneity among those affected will be high: individuals with the same disorder, even those that are closely related, will be likely to have a different complement of risk alleles.

This genetic complexity poses challenges for basic neuroscientists as current methods only allow them readily to study the disruption of a single gene, or at most several at a time. Current translational paradigms aiming to understand the functional consequences of disease risk alleles were developed to study Mendelian disorders; however, we need to find new approaches if we are to translate findings from complex, polygenic genetic disorders into biological and therapeutic insights. Some traction from traditional methods might come from their application to rare high-penetrance risk alleles, and it is likely that more of these will be implicated as NGS technology is applied to larger samples (Fromer et al. 2014; Purcell et al. 2014). Clearly, multiple small effect risk alleles account for a substantial proportion of the heritability of psychiatric disorders, and thus it remains a major challenge to understand how to take these findings forward toward mechanistic insights. It is encouraging that the latest large GWAS data from schizophrenia have implicated a number of already known therapeutic targets, such as the dopamine D2 receptor and several proteins involved in glutamatergic neurotransmission (Schizophrenia Working Group of the Psychiatric Genomics Consortium 2014). This suggests that we can be optimistic that data of this kind will open up new therapeutic opportunities. However, it seems that it will be necessary to focus translational research on the function of gene and protein networks rather than on single genes. The identification of the key networks will require access to transcriptomic, methylomic, and proteomic data from the brain. Here, there is a need to develop resources initially across brain regions and developmental stages, but ultimately at a single cell level. The question of how to model network biology in experimentally tractable systems is beyond the scope of the present discussion, but it seems obvious, even to the nonspecialist, that this will require larger-scale, more quantitative approaches than are currently in general use. It should also not be forgotten that GWAS signals do not necessarily identify the key functional alleles, or even the gene or regulatory element that is affected. The research required to resolve these issues is labor intensive and often difficult, and there is an important question concerning the extent to which it will be necessary or even feasible to undertake this work for many hundreds of loci before undertaking the systems biological analyses that will come next.

The emerging evidence for a high degree of complexity also has important implications for clinical and basic neuroscience. The lesson from genomics is clear; the only way to render these complex disorders tractable is through adequately powered studies, and this requires both collaboration and methods that are scalable and standardized across many centers. Concerns have already been expressed about the low power and poor reproducibility of much neuroscience research (Button et al. 2013), and it seems inevitable that a culture change similar to that which took place in psychiatric genetics will be required in psychiatric neuroscience. Funders and journals will need to require adequately powered studies with robust statistical evaluation, data sharing, and an increased emphasis on replication.

Pleiotropy

The second key message is that there is a high degree of pleiotropy, at least at the diagnostic level, with even relatively high penetrance alleles and environmental risk factors being associated with a wide range of clinical outcomes. These findings reinforce the view that we should be seeking new ways to stratify patients for research that go beyond the descriptive clinical phenotypes inherent in DSM and ICD and that use markers of likely pathophysiological relevance (often termed endophenotypes, or intermediate phenotypes). These will include neurobehavioral, neuroimaging, and electrophysiological measures and should be based on current understanding of brain-behavior relationships rather than relying on clinical syndromes as enshrined in DSM-5 and ICD-10 (Craddock and Owen 2010). This philosophy is articulated, developed, and acted upon in the RDoC project of the U.S. National Institutes of Medical Health (Cuthbert and Insel 2013). The implications for clinical neuroscientists are clear: work should not be based on the assumption that current diagnostic categories delineate distinct disorders or even distinct groups of disorders with distinct pathogenic mechanisms. Instead, work should focus increasingly on identifying the mechanisms that underlie specific functional abnormalities, and we should expect these to cross current diagnostic boundaries. Researchers should stop organizing themselves by traditional phenotype and increase the level of interaction, discussion, and collaboration between clinical researchers and neuroscientists.

The requirements noted above for both adequately powered studies and new, more sophisticated ways of stratifying patients for neuroscience research brings a further challenge: if we are really serious about taking this agenda forward, then we need to study very large and deeply phenotyped cohorts. This, in turn, will require greater collaboration and data sharing between neuroscientists. However, some measures (e.g., neuroimaging measures like fMRI) will be difficult to apply to samples containing thousands of patients. Having said this, other behavioral, cognitive, symptomatic, or electrophysiological measures might well be sufficiently scalable. One explicit subgoal of RDoC is to foster new methods of measurement, and these could be tailored to the goal of feasible deeply phenotyped cohorts.

Finally, it is important to note an important fact that is often overlooked in clinical research on neurodevelopmental disorders: namely, these disorders manifest differently across the life span. Future research must factor this in and be aware of the possibility that key phenotypes and endophenotypes will change as the individual develops and ages. This will require research to characterize normative as well as pathological developmental trajectories.

Biological Convergence

The third key message is that, despite (a) the large number of genes implicated (a small proportion of the likely total) and (b) the divergent outcomes associated with individual alleles, there is already evidence for some convergence onto specific biological processes. Results from GWAS (Ripke et al. 2013), CNVs (Kirov et al. 2012), and sequencing studies (Fromer et al. 2014; Purcell et al. 2014) of schizophrenia point to highly functionally related sets of postsynaptic proteins involved in synaptic plasticity, learning, and memory. These include L-type calcium channels, postsynaptic scaffolding proteins involved in NMDA signal transduction, and proteins which interact with ARC (activity-regulated cytoskeleton-associated) protein, referred to as the ARC complex (Kirov et al. 2012), and brain-expressed genes that are repressed by fragile X mental retardation protein (FMRP).

A notable feature of these findings is not only their consistency across several studies, but also their convergence onto a coherent set of biological processes involved in the regulation of plasticity, particularly at glutamatergic synapses. These processes have major effects on plasticity in the postsynaptic region, although some implicated genes, including the L-type calcium channels and Neurexin-1 (an associated CNV locus; Kirov et al. 2009), also exert effects on plasticity presynaptically. These synaptic genes have also been implicated in cognition (Grant et al. 2005) as well as a range of neuropsychiatric conditions, including ASD and ID, and there is some support from the exome sequencing data for the notion that these disorders exist along a gradient of neurodevelopmental impairment (Fromer et al. 2014). However, the degree to which the associated pathways cross current diagnostic boundaries remains to be fully established. It is highly unlikely that this will represent the only set of biological processes implicated in the disorder, but the identification of at least one system involved in risk for schizophrenia and related disorders would pave the way for more detailed mechanistic studies and potentially stratified and novel therapeutic approaches.

While the degree of convergent support for glutamatergic synaptic processes is encouraging, it seems inconceivable that other processes are not involved. There is a highly convincing body of evidence implicating dopaminergic dysfunction in the genesis of psychotic symptoms, which occur commonly in schizophrenia and bipolar disorder but also in other neurodevelopmental disorders as well (Howes and Murray 2014). Indeed, the mechanism of action of antipsychotic drugs is understood to depend largely on the blockade of dopamine D2 receptors. Understanding the relationship between glutamatergic dysfunction, which is closely related to cognitive impairment, and abnormalities of dopamine signalling is likely to be the key focus in understanding how psychosis arises in schizophrenia and related disorders.

Further insights into possible mechanisms come from the observation that a wide variety of environmental mechanisms (e.g., traumatic, infective, inflammatory, toxic) converge on the same spectrum of neurodevelopmental outcomes as relatively disruptive, rare genetic events such as LoF mutations and CNVs. This suggests that genetic factors predisposing to adverse consequences of these environmental insults might play a role in these disorders. There may indeed be as many etiological mechanisms at work as there are ways of disrupting brain development, and it remains to be seen to what extent and how they converge on circumscribed pathogenic processes.

Phenotypic Complexity

A final important implication for future neuroscience research comes from the fact that neurodevelopmental disorders have multiple domains of function. Whether or not specific abnormalities are included in the definition of the disorder is largely arbitrary and a matter of convention. Kraepelin, for example, saw disorders of cognition as integral to schizophrenia, but over time there was an increasing diagnostic focus on psychotic symptoms and a de-emphasis on cognitive dysfunction. In recent years, as noted above, cognitive impairment has once again been viewed as central to the disorder, though it is possible to receive a diagnosis of schizophrenia without cognitive impairment. Movement disorders are also commonly seen in psychotic disorders but play little part in ICD and DSM definitions, whereas there is a whole tradition of nosology based on the work of Leonhard (e.g., Leonhard et al. 1999) in which psychotic disorders are central together with the course of disorder. In fact, studies of cognition and of brain imaging and electrophysiology in schizophrenia have found dysfunction to be widely distributed over multiple cognitive domains as well as brain regions and functions. This is not to say that all dysfunctions are present in all patients—far from it. There is widespread heterogeneity. Thus the complex constellation of symptoms and syndromes observed in individual patients likely reflects developmental and functional disturbances in a wide range of brain systems and psychological processes, and this is unlikely to be understandable in terms of a single pathway from pathology to diagnosis. Of course it is to be hoped that in time we will be able to decompose these into different groups of patients with specific constellations of symptoms and impairment, reflecting disturbances in different brain mechanisms and circuits. However, we need to recognize that it is possible for the underlying diathesis to involve the whole brain, or at least multiple, widely distributed systems, rather than reflecting dysfunction in specific brain regions or circuits. This accords with recent findings which implicate genetic abnormalities in the ubiquitous excitatory synapse as well as many hundreds of other risk loci, and the broad impact likely to result from many environmental risk factors.

How Do We Integrate Disease Classification with Neurobiology and Genetics?

The extensive shortcomings of current categorical criteria for research into etiology and pathogenesis have been addressed above and a number of possible ways forward discussed. Even the fiercest critics of DSM and ICD have to admit, however, that we are still some way off having sufficient new insights from genetics and neuroscience to replace current diagnostic approaches in the clinic. It has been argued that clinical practice would be improved by augmenting current categorical approaches with a number of cross-cutting dimensional measures; these, however, would introduce complexities into clinical practice and have not been welcomed sufficiently by practitioners. Attempts to introduce them into DSM-5 failed. Notwithstanding Robins and Guze (1970), there were good reasons for doubting whether genetics would allow us to carve nature at the joints and to map out clinical syndromes on the basis of nonoverlapping genetic etiology (Kendler 2006). This suggestion can probably be discarded once and for all on the basis of recent genetic findings. Given the challenges outlined above, it is probably too early to say what impacts neuroscience will have on classification and diagnosis. Current diagnostic categories are likely to remain clinically useful to the extent that they best inform management and prognosis, but these will require modification, as future research indicates closer relationships of specific phenotypes and endophenotypes to mechanism, and likely need to include both dimensional and categorical entities.

It seems likely, therefore, that the introduction of genetics and neurobiology into the clinic will occur piecemeal as new advances are made. There is probably already a good case to be made for the introduction of testing for rare CNVs in neurodevelopmental disorders, including schizophrenia. Although testing is unable to assist the identification of therapeutic options, it does provide patients and their families a partial explanation of the illness and has some implications for genetic counseling. It also has implications for prognosis, especially if testing is conducted in children, but the counseling

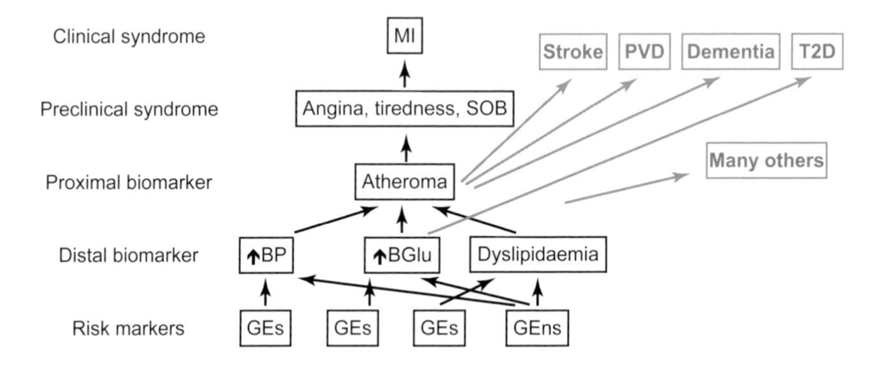

Figure 3.2 Schematic showing a simplified representation of the etiology and pathogenesis of myocardial infarction (MI). PVD: peripheral vascular disease; T2D: type 2 diabetes; SOB: shortness of breath; BP: blood pressure; BGlu: blood glucose; GEs: specific genetic and environmental factors; GEns: nonspecific GEs. Genetic and environmental factors are placed together for simplicity. Associated pleiotropic phenotypes are shown in gray. Diagnosis is multilevel and shown in box for exemplar case.

challenges should not be underestimated given the wide range of possible outcomes (Kirov et al. 2014).

If we wish to speculate how psychiatrists of the future, say in twenty years, will use genetic and neurobiological data in diagnosis and stratification, then it is instructive to consider how similar information is being used in other branches of medicine. Figure 3.2 illustrates how physicians currently approach a diagnosis in a case of myocardial infarction based on a combination of clinical syndromic presentation, biomarkers of underlying pathogenic mechanisms, and presence of environmental risk factors. It also indicates a potential role for genomic profiling for future individualized medicine. Figure 3.3 shows a similar, albeit highly speculative, schema of how psychiatrists of the future might approach a diagnosis of psychosis, integrating information across levels and using neurobiological and genetic markers to target therapies. It seems likely that clinical syndromes will remain useful but be augmented by biomarker and risk factor information in order to target therapies, devise secondary prevention, and inform prognosis. The extent to which a greater understanding of pathogenic mechanisms based on future advances in neuroscience and genetics will allow current clinical syndromes to be refined is unknown.

Conclusions

There seems little doubt that the widespread and uncritical use of current categorical diagnostic systems has impeded research into the etiology and pathogenesis of psychiatric disorders. Until recently it was nigh impossible to

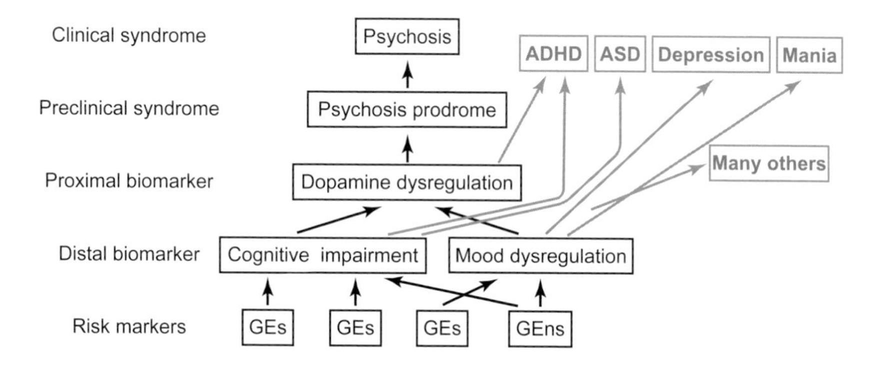

Figure 3.3 Schematic showing a simplified speculative representation of the etiology and pathogenesis of psychosis. ADHD: attention-deficit/hyperactivity disorder; ASD: autism spectrum disorder; GEs: specific genetic and environmental factors; GEns: non-specific GEs. Genetic and environmental factors are placed together for simplicity. Associated pleiotropic phenotypes are shown in gray. Diagnosis is multilevel and shown in box for exemplar case.

publish a paper or obtain grant funding without adhering to either DSM or ICD. Recent findings in psychiatric genetics have contributed to an increased realization that this situation must change, and there are encouraging signs (e.g., the RDoC project) that this is beginning to happen. However, the new findings pose a number of challenges. The genetic complexity of psychiatric disorders means that efforts to translate these findings into new mechanistic insights will require cultural as well as technological changes in the neurosciences. In particular, there will be a need to study networks rather than individual genes and proteins, and to develop experimental systems in which multiple variables can be manipulated and multiple endpoints studied. The complexity and mutability of psychiatric phenotypes also pose challenges for clinical neuroscience, and new ways of stratifying patients in neurobiologically meaningful ways are required, as is a move away from conducting research in diagnostic silos. In both basic and clinical neurosciences, there will need to be an increased focus on large-scale experimentation, collaboration, statistical robustness, and reproducibility.

A brief perusal of Figure 3.2 reminds us that psychiatric disorders are not unique in being genetically and phenotypically complex with extensive pleiotropy and multiple, related pathogenic mechanisms. These are features of common disorders in general. What sets psychiatry apart is the complexity and relative inaccessibility of the diseased organ. The continuing success of psychiatric genetics and the rapid development of new methods for studying the brain in humans and model systems are causes for optimism. The challenge now is to develop new approaches to identify pathogenic mechanisms and to identify the most suitable phenotypes to which these relate.

Acknowledgments

The neurodevelopmental gradient model was developed with Nick Craddock, and my work in psychiatric genetics in Cardiff has been conducted in collaboration with Nick Craddock, Mick O'Donovan, Peter Holmans, Valentina Moskvina, George Kirov, and Anita Thapar as well as many excellent junior colleagues. Many of the recent advances in psychiatric genetics have taken place as part of a global cooperative enterprise involving numerous scientists and clinicians from many countries.

First column (top to bottom): Stephan Heckers, Cynthia Joyce, Bruce Cuthbert, Raquel Gur, Andreas Meyer-Lindenberg, Bruce Cuthbert, Steve Hyman
Second column: Steve Hyman, Michael Owen, Steve Hyman, Matt State, Group discussion, Stephan Heckers, Matt State
Third column: Bruce Cuthbert, Thomas Bourgeron, Cynthia Joyce, Stephan Heckers, Raquel Gur, Thomas Bourgeron

4

Neurodevelopmental Disorders

What Is to Be Done?

Stephan Heckers, Steven E. Hyman, Thomas Bourgeron, Bruce N. Cuthbert, Raquel E. Gur, Cynthia Joyce, Andreas Meyer-Lindenberg, Michael J. Owen, and Matthew W. State

Abstract

Autism and schizophrenia are associated with abnormalities in brain development, the genetic and environmental risk factors of which are just now being discovered. How can these advances be translated into treatment? This chapter reviews the psychiatric classification of autism and schizophrenia, and discusses how this has impeded progress in understanding disease mechanisms. It recommends that future research not be constrained by current nosology and explores alternative diagnostic approaches. It discusses recent studies of prevention in the early stages of illness and discovery of genetic and environmental risk factors. Both sets of observations should be used to guide neuroscientific exploration of neurodevelopmental disorders unimpeded by current psychiatric nosology.

The Status Quo of Psychiatric Diagnosis

To appreciate the limitations of current psychiatric nosology, we need to review (a) the diagnostic encounter, during which a clinician collects data from a patient and assigns the patient to a diagnostic group; (b) the diagnostic rules that guide the clinician in the selection of clinical features and the interpretation of additional data (e.g., historical information and laboratory tests); and (c) the diagnostic paradigm, which asserts that diagnosis can be validated by clinical presentation, outcome, and disease mechanisms.

Typically, the diagnostic encounter is a cross-sectional assessment of psychopathology. The major focus has been on reliability: How can we make sure that a patient (seen by the same clinician or by several clinicians at various time points) receives the same diagnosis? The type of information collected in the psychiatric encounter is often constrained by schools of psychopathology and the training of the clinician. While it is well known that the reliability of the diagnostic assessment can be improved by collateral information (Ho et al. 2004), such information is often not collected. The psychiatric encounter occurs in a clinical setting and is thus unable to examine the person in the real world. Recent studies offer compelling examples that experience sampling can provide much richer phenotypic data (van Os et al. 2013). In addition, the social context of the person needs to be taken into consideration to understand the complexity and unique features of the individual struggling with a mental illness.

Diagnostic rules have been shaped primarily by the extensive literature on clinical psychopathology. For example, the current concept of schizophrenia is built upon the diagnostic concepts of Kraepelin, Bleuler, and Schneider. The International Pilot Study of Schizophrenia (Wing and Nixon 1975) and the U.S./U.K. study (Kendell et al. 1971) revealed wide discrepancies between the European and the American concepts of schizophrenia. This led to a much narrower definition of schizophrenia in the third edition of the Diagnostic and Statistical Manual of the American Psychiatric Association (DSM-III) (Andreasen 1989). Since then, only five domains (delusion, hallucination, thought disorder, catatonic behavior, and negative symptoms) have been included in the diagnostic criterion set for schizophrenia; others have been excluded (e.g., cognitive impairment, depression, mania). In addition, gatekeepers have been added to exclude (a) less severe forms of psychosis, (b) patients with prominent mood symptoms, and (c) patients with psychosis due to substance use or medical and neurological disorders. While these diagnostic criteria have increased the reliability of the schizophrenia diagnosis (intraclass kappa of 0.5 for DSM-5 diagnosis of schizophrenia) (Regier et al. 2013), they have threatened the validity of the concept (Andreasen 2007).

The diagnostic paradigm can be traced back to Kraepelin's concept of the natural disease unit (Hoff 1985). Kraepelin proposed that a psychiatric diagnosis can be validated through (a) clinical presentation, (b) disease course and outcome, and (c) the cause and mechanism of the illness. Implicit in this paradigm is the strong hypothesis that all validators converge on the diagnosis and reflect an underlying biology, for instance, the natural disease unit. Although there is little doubt that the validators' clinical presentation and course/outcome have much to offer in clinical practice, it is now clear that they often do not converge with the cause and mechanism of illness. In short, it is likely that several mechanisms and causes will be identified for one psychiatric disorder, and that shared mechanisms and causes will be identified for clinical features that traverse diagnostic categories (Heckers 2008). This challenges the current design of psychiatric research, for instance finding the etiology and pathophysiology of a diagnostic category defined by clinical presentation.

Where do we go from here? Do we need a paradigm change now or should we continue the current research strategy, with better methods and greater resources? For example, investigators could study patients with several diagnoses in one or multiple chapters of the DSM (e.g., psychotic disorders, mood disorders, anxiety disorders). In addition, the major domains of psychopathology could be redefined along dimensions (normal, mild, moderate, severe) rather than categories (Barch et al. 2013). We will review one proposal for a change in psychiatric nosology.

The Research Domain Criteria

The genomic and biomedical revolution of recent years has superseded the classical model of diagnosis based on symptoms, relatively gross measures of pathology, and standard clinical tests. In oncology, for instance, cancers are increasingly diagnosed by the genetic composition of the neoplasm rather than tissue of origin; although much treatment still relies on surgery and broad-spectrum chemotherapies, new approaches seek to determine optimal matches between the genetic signature and the particular chemotherapeutic agent. This has led to the development of large-scale adaptive trials that incorporate simultaneously a number of genetic factors; in breast cancer, for example, iSPY2. Even monogenic disorders are now known to comprise a number of different "diseases" at the genetic level. For instance, ivacaftor (Kalydeco™) is highly effective in treating patients with the clinical phenotype of cystic fibrosis who have the R117H mutation of the CFTR gene. R117H, seen in only 4% of cystic fibrosis patients, is one of over 1,000 different mutations that can produce the clinical syndrome, and subsequent clinical trials are going forward one mutation at a time.

While diagnostics in other areas of medicine have moved to increasingly sophisticated clinical and genetic tests, psychiatric diagnosis remains stuck in the practices of the past. The current DSM and ICD categories rely entirely upon presenting signs and symptoms to confer diagnoses. Further, they utilize an infectious disease model in which disorders are considered to be either present or absent, notwithstanding accumulating evidence that most symptoms can be arrayed along some dimension of severity and/or impairment.

This system largely suffices for current treatment practice, since both drug and behavioral modalities tend to be moderately effective across a range of mood/internalizing or psychotic spectrum disorders. However, modern genomics and neuroscience research increasingly indicates that the categories are a poor reflection of nature. Genetic risks appear across multiple disorders, such as unipolar and bipolar depression or bipolar disorder and schizophrenia, with a recent paper reporting commonalities across five different disorders (Cross-Disorder Group of the Psychiatric Genomics Consortium and Genetic Risk Outcome of Psychosis Consortium 2013). Further, neuroimaging and other studies indicate that common mechanisms are present across multiple disorders (e.g., emotion regulation or working memory disruptions), while almost all disorders are highly heterogeneous both in terms of symptom combinations and mechanisms of pathophysiology.

Unfortunately, the current symptom-based scheme has become the *de facto* standard for virtually all research into mental disorders, and it similarly constrains clinical trials, since the categories serve as the indications for new drug approvals by regulatory agencies. This situation severely hampers attempts to understand the fundamental genetics and pathophysiology of mental disorders and consequent attempts at mechanistically targeted treatment development. The paradox is thus that psychiatric research urgently needs studies which cut across diagnostic boundaries in order to provide the basis for developing an improved nosology that can usher the field into an era of precision medicine. Such a database cannot be developed, however, as long as research grants are funded solely in terms of DSM/ICD categories that obscure the relationships of genes, pathophysiology, and specific functional impairments.

This dilemma led the U.S. National Institutes of Mental Health (NIMH) to develop the Research Domain Criteria (RDoC) project, based on a goal in its 2008 Strategic Plan: "to develop, for research purposes, new ways of classifying mental disorders based on dimensions of observable behavior and neurobiological measures." RDoC is not a fully developed alternative system; it is a framework for research that is intended to provide a research literature that can support future revisions to the DSM and ICD that incorporate genetics and other areas of neuroscience. RDoC represents an attempt to free investigators from the shackles of research constrained to single DSM/ICD categories in several ways:

1. The system calls for research in fundamental functional dimensions (e.g., fear, reward, or working memory, and are formally termed "constructs") as studied along a full range of performance, from normal to abnormal. (It should be noted, however, that enriched sampling of subjects at the extremes of a distribution—that would correspond to DSM-level psychopathology—is envisioned for most studies, to maximize translation to clinical issues.)
2. Investigators are strongly encouraged to study functional dimensions in terms of multiple levels of measurement, including genes, circuit activity, behavioral measurements, and self-reports (at the extremes of a distribution, the latter measures would comprise symptoms).
3. The framework does not specify cutoffs for the presence of disorders, so as to focus attention on the dimensional relationships among the various measurements.
4. Researchers are free to devise sampling frames that may include patients drawn from multiple DSM categories (as well as a range of control subjects with subsyndromal pathology or normal functioning).

5. Studies of neurodevelopmental trajectories and environmental influences are encouraged for all areas of psychopathology to develop new ideas regarding pathogenesis.

Particularly in the early phases of the RDoC project, it is anticipated that research projects will provide maximum yield by including, in the sampling frame, a group of subjects drawn from related diagnostic groups (and milder forms of pathology along the same putative clinical spectrum; e.g., all of the anxiety disorders, unipolar or bipolar depression, or autism spectrum). In many cases, it may be advantageous to cover a range of clinical severity and include in the sampling frame all patients from a given chapter of the new DSM/ICD metastructure; for example, the new "Schizophrenia Spectrum" chapter contains not only schizophrenia but also schizophreniform disorder, acute and transient psychotic disorder, schizotypal disorder, etc.

Once a sampling frame is specified, the investigator is at liberty to implement a research design with independent and dependent variables appropriate to the research question. For instance, in a study with autism spectrum children, the independent variable could include genetic combinations, as identified by a computer algorithm, and dependent variables of social cognition and/or language facility. In a study with patients from multiple anxiety disorders (and controls), a hypothesis could be based on recent reports that patients with blunted fear-potentiated startle in an emotional imagery challenge (independent of anxiety disorder category) report higher symptomatic levels of overall distress and diffuse anxiety compared to patients with high levels of fear-potentiated startle (McTeague and Lang 2012); thus, the independent variable could be a magnitude of fear-potentiated startle, while the dependent variables would be scores on various measures of distress and nonspecific anxiety.

The goal of RDoC is to support research that contributes to the validation of intermediate phenotypes based on particular behavioral functions and the neural circuits that implement them. It is hoped that such constructs will provide an improved data set for relating findings from genomics and neuroscience to important clinical symptoms, such as excessive fear and anxiety, disrupted reward behavior (as in anhedonia or, at the opposite extreme, mania), or impaired cognitive functioning. RDoC is clearly intended to be a long-term effort, one that will take ten years or more to mature. However, it is also hoped that initial studies will yield near-term benefits, to inform the development of new ideas about psychopathology and new approaches to clinical trials. RDoC is frankly an experiment on a large scale, and its precise ramifications remain to be seen. We anticipate that it will inform necessary long-term efforts to incorporate genetics and pathophysiology research into new avenues for precision medicine in psychiatry.

The Early Stage of Neurodevelopmental Disorders: Detection, Intervention, and Prevention

The ultimate goal of our current research efforts is the early detection and prevention of neurodevelopmental disorders. How can we get there? One strategy involves the identification of genetic and environmental risk factors. If we can detect such factors in individuals before the disease phenotype becomes evident, this may provide an avenue for intervention and prevention. This strategy, however, creates ethical issues. For example, how do we weigh the benefits of diagnosis and treatment if we are unable to provide any treatment, or if there is the risk of adverse effects? Here we review some recent efforts to detect risk for neurodevelopmental disorders and to intervene before the full phenotype has developed.

It has been well established that untreated psychosis leads to poor outcome (Perkins 2006). This has led to intensive research of the early stages of psychosis, with the hope of being able to intervene early (Marshall and Rathbone 2011). More recently, researchers have moved further upstream in the disease process to study the psychosis high-risk state (also known as "ultra-high risk," "at-risk mental state," or "prodrome") (Fusar-Poli et al. 2013). The psychosis high-risk state can be defined as a genetic at-risk state (e.g., history of psychosis in parents) or as a subsyndromal clinical presentation, also known as attenuated psychotic symptoms or brief limited intermittent psychotic symptoms (BLIPS).

The North American Prodrome Longitudinal Study (NAPLS) reported five clinical variables that predicted later conversion to psychosis: genetic risk with functional decline, low social functioning, unusual thought content, paranoia, and history of substance abuse (Cannon et al. 2008). Additional studies provide preliminary evidence that the prediction of psychosis can be improved further by combining these clinical variables with family history of psychosis, neurocognitive measures, and advanced imaging studies (Fusar-Poli et al. 2013).

As cognitive deficits are core to schizophrenia syndrome, measures of cognition have been applied in the investigation of help-seeking people with prodromal features. Lacking are normative "growth charts" in well-phenotyped populations; these would enable integration of neurocognitive measures and psychosis features across development (Cannon 2014).

The Philadelphia Neurodevelopmental Cohort stratified youths (age 8–21 years) by degree of reported psychosis features assessed in a structured interview. All participants were administered a computerized neurocognitive battery, based on tasks acquired in functional neuroimaging studies. To compare performance-based growth charts, neurocognitive age was related to chronological age, established in typically developing participants (~1600). Across ages, the psychosis spectrum group (~1400) had a greater developmental lag than the group reporting mild psychotic symptoms (~900). Furthermore, neurocognitive delay was more prominent in youths endorsing psychotic symptoms than participants reporting nonpsychotic symptoms (~900). Combined clinical and neurocognitive assessment can facilitate early detection of individuals at risk for psychosis (Gur et al. 2014).

Early intervention of psychosis is a rapidly developing area of research (Marshall and Rathbone 2011). A small number of studies has found antipsychotic medication to be of little benefit in the prevention of psychosis. There is, however, compelling evidence that cognitive behavioral therapy (alone or in combination with antipsychotic medication) is helpful, and there is one intriguing study of omega-3 fatty acids reducing the conversion to psychosis (Amminger et al. 2010). These early studies provide hope that early intervention of psychosis is possible.

Not surprisingly, there is also considerable interest in both early detection and intervention in the service of secondary prevention of autism spectrum disorders (ASDs). There is currently a good deal of optimism regarding the value of reducing the age at diagnosis, which now is often several years after the initial signs are noted by parents, physicians, or caregivers, and in intervening as early as possible to alter developmental trajectories. Although some data supports this view, there remains a need for a substantial investment in prospective well-controlled longitudinal studies to address these issues definitively, including identifying the most efficient means of early detection and defining an evidence-based armamentarium for early behavioral and educational treatments (Kaale et al. 2014; Daniels et al. 2014).

At this stage, given the limitations of the available somatic treatments for ASD and the absence of therapeutics that target pathophysiological mechanisms, there is only a very small body of literature addressing early (as young as two years of age) pharmacological treatment (DeLong et al. 1998, 2002). However, as novel rational treatments are developed, and in light of the increasing ability to detect risk mutations at birth or before, the opportunities to test pharmacological therapies in the first weeks to months of life are likely to become conceivable, thus raising important clinical as well as ethical concerns.

From Genes to Treatment Targets in Neurodevelopmental Disorders

Recent successes in gene discovery in autism and schizophrenia are prompting a reassessment of established strategies for "bottom-up" translational neuroscience. The traditional approach of identifying a single gene of interest, creating one or more animal knockouts or knockins, and then elaborating molecular, cellular, and circuit level mechanisms has been tremendously fruitful in advancing basic understandings of the central nervous system and of rare monogenic brain disorders (e.g., FMRP, MECP2, and NF-1). The traditional approach has led to the identification of putative treatment targets and strategies. However, recent progress in the genome-scale genetics of neuropsychiatric disorders has revealed that, in general, risk of these disorders is polygenic. The scientific approaches that proved useful for the analysis of highly penetrant mutations that cause monogenic disorders are not likely to prove fruitful for the analysis of less penetrant genetic variants that act in diverse combinations to produce risk.

While the yield of recent gene discovery has been impressive, it has also revealed that both autism and schizophrenia (and other related neurodevelopmental disorders) reflect tremendous genetic heterogeneity. The genetic variants (alleles) involved tend to be biologically pleiotropic, and there is a surprising degree of phenotypic variability emerging from the same allele.

Given the early state of the research, the emerging picture of the genetic architecture of each of these syndromes remains unsettled. A dozen or more rare copy number variants (CNVs) have been found to increase risk substantially for both autism and schizophrenia.

Thus far there has been relatively greater success in schizophrenia in the search for common variation (polymorphisms) and somewhat greater success with regard to ASD in the discovery of rare *de novo* point mutations. It is not yet clear to what extent this reflects differences in the underlying genetic architecture that might be expected from the differences in severity and effects on fecundity versus differences in the predominance of different research strategies applied to the two disorders. However, this distribution of identified risk alleles has relevance for the design of translational strategies: common variation facilitates the identification of cohorts carrying alleles of interest, but individual alleles may be difficult to interpret *prima face* as they are often found in noncoding segments of the genome, and they typically carry small effects. In contrast, rare point mutations in ASD have been discovered through the analysis of *de novo* coding mutations. These are easier to interpret with regard to impact on protein function and, in some cases, may carry very large effect sizes; however, the population frequency of these alleles is low, with even the most common genes implicated by multiple independent mutations occurring in less than 1% of affected individuals. Moreover, none of the mutations identified to date are 100% penetrant and, for some, the range of possible outcomes has already been shown to involve a wide range of neurodevelopmental syndromes.

Together, these complexities make several aspects of the standard translational model more difficult:

- The selection of genes to study is complicated by their sheer numbers, and this will increase substantially with further genetic studies.
- Few of the identified mutations can be faithfully modeled by full knockouts, either for common or rare variations.
- Most disease-related variants identified so far are subtle variations that are present in either a predisposing or protective genetic background, which is a challenge for modeling the effects in inbred mouse strains.
- The means to model, in any system, multiple alleles of very small effect (either together or in parallel) are only just emerging and are currently

limited to a handful of risk alleles rather than the many hundreds likely to be implicated in neurodevelopmental disorders.

- The manipulation of biologically pleiotropic genes with alleles of varying effect sizes suggests that multiple functional (and potentially subtle) phenotypes are likely to emerge across development. To date there is little information regarding precisely where and when to look for a human specific pathology.

Despite the very recent nature of systematic and reliable gene discovery in complex neurodevelopmental disorders, multiple conceptual strategies have already started to emerge to begin to address these problems. Roughly, these can be characterized as follows:

- Continuation of the current traditional approaches: knocking out genes in animal models, which focus on recent findings, point to the most highly penetrant risk genes.
- Use of a systems biology approach to organize disparate risk alleles of varying effect sizes into molecular pathways as a first step to organize both the choice of which genes to model further and the functional phenotype on which to focus.
- Use of a systems biology approach to organize risk alleles into networks based on spatiotemporal expression properties to capture developmental as well as molecular contexts.
- Leveraging high throughput technologies to engineer mutations in *in vitro* systems, looking for convergence of functional phenotypes *post hoc* via high throughput, unbiased phenotyping approaches.
- Human genetics that searches for protective alleles as markers of treatment target development via large-scale population studies of at-risk cohorts.

While the traditional approach of deeply modeling a single gene knockout may not *a priori* address all of the complexity described above, the focus on highly penetrant alleles has already demonstrated an ability to elaborate critical biological and pathophysiological processes and to help identify questions that will need to be addressed in broader systems biological approaches. Next we review synaptic changes in autism as an example of the progress of this approach.

The Synapse in Neurodevelopmental Disorders

Human genetics, animal models, and iPSC studies have yielded compelling evidence that synaptic proteins are affected in neurodevelopmental disorders. The first evidence came with the identification of rare X-linked mutations affecting the cell adhesion molecules neuroligins (NLGN3-4X) in patients with autism or Asperger syndrome (Jamain et al. 2003; Laumonnier et al. 2004). Following these results, further support came from heterozygous *de novo* mutations in the scaffolding protein SHANK3 and the presynaptic cell adhesion molecule NRXN1, lending credence to the hypothesis of a gene dosage-sensitive synaptic pathway in ASD (Autism Genome Project Consortium et al. 2007; Durand et al. 2007). Since then, mutations in more than 50 genes related to synaptic functions (receptors, cell adhesion molecules, scaffolding proteins, ion channels) or regulating synaptic gene levels (chromatin modeling factors, transcription and translation factors, signaling pathways) have been reported in patients with ASD (Huguet et al. 2013). Unbiased pathway analyses also confirmed a significant enrichment of mutations in genes related to synapses and/or to the fragile X pathway (Steinberg and Webber 2013).

Remarkably, variation in synaptic genes such as NLGNs, NRXNs and SHANKs are not restricted to ASD, but have also been detected in patients with intellectual disability (ID), schizophrenia, attention-deficit/hyperactivity disorder (ADHD), and bipolar disorders (for a review, see Guilmatre et al. 2014). Among these synaptic genes, *SHANK1–3* genes code for large synaptic scaffold proteins of postsynaptic density (Grabrucker et al. 2011). Interestingly, deletions, duplications, and coding mutations in *SHANK* genes have been detected in the whole spectrum of autism, but with a gradient of severity in cognitive impairment. Mutations in *SHANK1* are present in males with normal IQ and autism; mutations in *SHANK2* are present in patients with ASD and mild intellectual disability; mutations in *SHANK3* are present in patients with ASD with moderate to profound ID (Berkel et al. 2010; Leblond et al. 2012; Sato et al. 2012).

SHANK3 is the most studied gene from the SHANK family since its haploinsufficiency has been identified in more than 900 patients affected with chromosome 22q13 deletion syndrome, known as Phelan-McDermid syndrome (Bonaglia et al. 2011). The genomic rearrangements observed in these patients are diverse, ranging from simple 22q13 deletions (72%), ring chromosomes (14%), and unbalanced translocations (7%) to interstitial deletions (9%), all resulting in haploinsufficiency of the *SHANK3* gene (Bonaglia et al. 2011). In more than 80% of the cases, autism or autistic-like behavior is present (Betancur and Buxbaum 2013).

Mice lacking any of the SHANK proteins display phenotypes relevant to ASD (Jiang and Ehlers 2013). *SHANK1* knockout mice show increased anxiety, decreased vocal communication, decreased locomotion and, remarkably, enhanced working memory but decreased long-term memory (Hung et al. 2008; Wohr et al. 2011). *SHANK2* knockout mice show hyperactivity, increased anxiety, repetitive grooming, as well as abnormalities in vocal and social behaviors (Schmeisser et al. 2012; Won et al. 2012). *SHANK3* knockout mice show self-injurious repetitive grooming and deficits in social interaction and communication (Peca et al. 2011; Yang et al. 2012b; Wang et al. 2011). Consistent with gene dosage sensitivity, a Shank3 transgenic mouse modeling a human *SHANK3* duplication exhibits manic-like behavior and seizures (Han et al. 2013a). Interestingly, the mood-stabilizing drug valproate, but not lithium, rescues the manic-like behavior of Shank3 transgenic mice.

Cortical neurons derived from induced pluripotent stem cells (iPSCs) have been generated from patients with *SHANK3* mutations (Shcheglovitov et al. 2013; Boissart et al. 2013). In accordance with the haploinsufficiency, neurons display lower *SHANK3* mRNA and protein levels. They also present defects in excitatory, but not inhibitory, synaptic transmission that could be restored by overexpression of SHANK3 or IGF1 treatment (Shcheglovitov et al. 2013).

The Glutamatergic Synapse in Schizophrenia

Pathway and network approaches offer a viable avenue to organize disparate genes and address the observed genetic heterogeneity. At present, methods development is a key issue and, as discussed below, the resources necessary to carry out rigorous unbiased systems biological pathway or network analyses are limited, with resulting limitations on the reliability, biological depth, and specificity of the output.

It is highly encouraging that, despite the complexity of genetic findings in schizophrenia and the realization that we are still at the beginning of a longer process of gene discovery, there is already evidence for some convergence onto specific and highly plausible biological processes. Results from genome-wide association studies (GWAS) (Ripke et al. 2013), CNVs (Kirov et al. 2012), and sequencing studies (Purcell et al. 2014; Fromer et al. 2014) point to the involvement of functionally related sets of postsynaptic proteins in synaptic plasticity, learning, and memory. These include L-type calcium channels, postsynaptic scaffolding proteins involved in NMDA signal transduction, and proteins that interact with activity-regulated cytoskeleton-associated (ARC) protein, referred to as the ARC complex (Kirov et al. 2012), and brain-expressed genes that are repressed by fragile X mental retardation protein (FMRP).

These findings are notable not only for their consistency across several studies, but also for their convergence into a coherent set of biological processes involved in the regulation of plasticity, particularly at glutamatergic synapses. These synaptic processes have been implicated in cognition (Grant et al. 2005) as well as in a range of neuropsychiatric conditions, including ASDs and ID (Fromer et al. 2014). While it is highly unlikely that this is the only set of biological processes implicated in the disorder, the identification of at least one system involved in risk for schizophrenia and related disorders paves the way for more detailed mechanistic studies and potentially to stratified and novel therapeutic approaches.

What Do We Need Now?

One strategy that has been suggested by many, albeit with clear downsides, is to differentiate fibroblasts, iPS cells, or hESC cells into desired neural types, and to purify them with a reporter gene that could be engineered into them. This would provide enough material for transcriptomics, proteomics, and epigenomics on a population of uniform cells. The downside, of course, is that cells *in vitro* are expected to be quite different from cells embedded in a circuit in a brain. Thus, to produce the best possible databases, it will be critical to iterate between postmortem human brain tissue (with all its problems) and *in vitro* cellular models.

High throughput cell-based assays, which leverage the recent development of iPSCs and genome editing, represent an exciting emerging possibility. The iPSC approach, like pathway and network analysis, seeks to leverage heterogeneity and phenotypic convergence, but does so based on the output of genetic manipulations. To achieve this goal, we need to develop assays that are reliable and replicable, address whether to model patient mutation in a natural genomic context or engineer known risk alleles into an isogenic background or both, and limit cell types that can be modeled and a set of phenotypes that might be reliably assayed using this approach.

Human genetics strategies that focus on protective alleles have recently been shown to be a productive avenue in identifying treatment targets subsequent to the identification of risk alleles (Flannick et al. 2014). These strategies would seem to be ideally suited at present to schizophrenia, given the highly productive efforts at common variant identification. In this case, a study would focus on a population sample, looking for individuals at high polygenic risk who do not have evidence of the phenotype. A similar strategy could be applied to rare loss-of-function (LoF) variants in ASD-associated genes, such as CHD8 in which LoF alleles have not yet been seen in large samples of unaffected individuals. Very large-scale research sequencing, or in-time, routine clinical sequencing, might conceivably capture a sufficiently large population to find first individuals with disruptive CHD8 mutations without ASD and then identify genetic moderators of that phenotype. Finally, we should not overlook the potential of high-throughput whole exome and genome sequencing in large samples to identify protective alleles at the population level by identifying rare alleles that are overrepresented in controls relative to cases. We note, however, that this will require the sequencing of large control populations.

These strategies all have considerable strengths as well as some limitations. In the final analysis, they represent important short-term options to begin to move forward with translational efforts in a broader context than is often employed at present. Of course, it is likely that complementarity among these approaches and very likely critical interactions will drive progress.

A key issue, however, is the lack of the foundational resources that are required if we are to exploit genetic discoveries through systems biology.

Understanding the functional effects of risk alleles requires an investigation of the biochemical pathways or protein networks in which they act. Genetic variation can impact directly on protein structure and function, but many of the small effect risk alleles identified in GWAS lie outside the protein coding sequence and presumably impact on regulatory regions. Put simply, the underlying idea is that the alleles contributing to polygenic disorders act by making small nudges in a protein network which either summate or reach a tipping point to alter its function. Thus, the identification of such networks provides a point of biological convergence to study pathogenesis. Moreover, the most promising drug target may be a protein within the network that is not itself implicated genetically.

If we are to move rapidly and efficiently from the implication of multiple disease risk alleles to mechanistic insights, we need access to large-scale data bases which can be interrogated to identify sets of functionally related risk alleles. For example, we wish to know whether specific gene expression networks or functionally related groups of proteins are enriched among the genes implicated in a specific disorder. The annotation of the genome and proteome is relatively advanced in some tissues, but there is a dearth of such data for the brain. This so-called annotation gap will limit progress as we seek to translate genetic findings into mechanistic insights and new models of disease pathogenesis. The field needs shared resources that will provide information about transcriptomes, proteomes, and epigenomes at different stages of development and about protein interactions. Initially a realistic short-term target is to obtain such data from several hundred brains in multiple brain regions over several different developmental periods. However, with time, the goal should be to have such data available for individual cell populations and types. The ability to isolate individual cells from the brain and to have enough material from which to perform single cell RNA sequencing or proteomics are still a challenge. For the former, there are encouraging experiments that single cell RNA sequence in neurons will be possible in the short term. However, we are a long way off from being able to perform proteomic studies in single cells.

In addition to these resources for genetic research, there are several other methodological advances for clinical and translational research:

- More complex phenotyping: experience sampling of behavior and mood states (van Os et al. 2013; van de Leemput et al. 2014) as well as more comprehensive documentation of clinical features not currently included in the diagnostic criteria for neurodevelopmental disorders, but which may be relevant as gene-behavior relationships are explored (e.g., sleep, circadian rhythms, gastrointestinal problems, anhedonia).
- Longitudinal studies of behavior and brain could provide growth curves for brain measures (volume, function) and behavior in healthy controls and patient cohorts: eye movement behavior in autism (Jones and Klin 2013) or cognitive function in schizophrenia (Cannon 2014).

- Studies of cognition as a continuous, quantitative measure to compare probands with noncarrier family members rather than a qualitative, dichotomous trait (Moreno-De-Luca et al. 2013).
- Studies of environmental risk factors (e.g., childhood maltreatment) and their effect on pathophysiological processes in neurodevelopmental disorders (Mehta et al. 2013).

Looking into the Future

We are making rapid progress in our study of neurodevelopmental disorders. At the same time, we face significant challenges in realizing the full potential of the opportunities given to us. Here we recommend three actionable steps for the research community.

First, patients and relatives need to become active participants in research. Translational neuroscience needs to learn from the experiences of the biotech and stem cell communities and work proactively to ensure that investigators stay connected with the public on research advances. The risk of negative backlash on research developments is high, given this group's focus on children, development, and intellectual/academic life achievement. In addition, public reaction to advances in understanding genetic and environmental risks will vary by country, culture, and generation, so anticipating communication issues will be helpful.

Patient-led initiatives to incorporate patient/public involvement (PPI) in the research process offer an important opportunity to inform the public and mitigate negative responses to research advances. Realistic steps include providing for PPI representation in research priority setting, consortia management, protocol development, study execution, and reporting. To be effective, PPI representatives need to have more than a general knowledge of science, statistics, and the translational research process. Training efforts have been incorporated into some PPI mechanisms, for example, in conjunction with the U.S. National Institutes of Health Councils, the INVOLVE program of the U.K. National Institute for Health Research, and programs sponsored by the U.S. Food and Drug Administration and the European Medicines Agency. Access to these programs, however, is limited and not necessarily focused on neuroscience. We recommend development of an integrated and internationally harmonized public education program to train patients/caregivers in the research process. The group of trained PPI representatives could be supplemented by seeking patient/caregiver representatives from within the ranks of neuroscience researchers. The Society of Neuroscience alone consists of over 33,000 international scientists, among whom mental health problems surely occur.

With best practices for PPI still in development, it is incumbent on research leadership to make the case for this approach. In the short term, PPI representation can have a positive effect on clinical research, shortening recruitment times and generally increasing satisfaction with research participation for the participants and investigators (Ennis and Wykes 2013). It is hoped that patient involvement will improve outcome assessments and trial design, thereby increasing the likelihood of successful studies. In the longer term, neuroscience must recognize the need for public education and take action to inform and advance research ethics discussions.

Second, research would be greatly enhanced by greater sharing of data and more stringent falsification of hypotheses. A fast-to-fail culture, well known in drug development, would speed up the neurobiological and genetic search strategies. This requires a change in the incentives for biomedical researchers. For example, researchers who need to compete for federal funding tend to produce more positive findings (Joober et al. 2012). NIMH has adopted the fast-to-fail study design for treatment trials, including a trial of kappa receptor antagonists for anhedonia (Reardon 2014).

Finally, we need to transform the education of mental health care providers and the public at large. We need to liberate ourselves from the constraints of diagnostic systems, to get closer to the individual as well as to the neural and genetic mechanisms of the mental disorders. We also need to develop a more positive relationship with regulatory agencies, so that the development of new treatments is not constrained by current diagnostic systems.

Similarly, we need to develop a core neuroscience curriculum for psychiatry residents and articulate a more differentiated view of genetic and environmental risk factors of neurodevelopmental disorders. For primary care, greater awareness of risk factors and attention to mental health assessments will be necessary for early identification and intervention strategies.

If we succeed with the research program as outlined in this chapter, what would success look like in ten years? We would have completed a full genetic screen of all risk alleles for neurodevelopmental disorders. We would have evidence for the causal mechanisms of environmental risk factors. We would have reasonable estimates of how much these genetic and environmental risk factors contribute to the development of a neurodevelopmental disorder in an individual. Finally, we would have made reasonable progress toward drug development, including the discovery and validation of new drug targets. Reaching these goals would greatly increase the return on research (see Wooding et al. 2013) and improve the lives of persons diagnosed with neurodevelopmental disorders.

Neurodegenerative Diseases: Coming Unstuck

5

Neurodegenerative Diseases

Where Are We Not Looking for Answers?

Walter J. Koroshetz and Lennart Mucke

Abstract

Preventing age-dependent neurodegeneration is a compelling yet elusive goal for neuroscientists, clinicians, affected individuals and their caregivers. As life expectancy increases and the global population ages, the burden of illness soars. A critical examination of what scientific areas require a fresh look is motivated by the growing magnitude of the health problem as well as by the numerous failed trials of agents once considered worth the investment of clinical trial resources in Alzheimer, Parkinson, and Huntington diseases. In so many of these cases it has become clear that the disease is more complicated than previously thought and that fatal flaws stemmed from assumptions based on imperfect or incomplete information from animal models. Renewed optimism that science will be able to provide relief derives from new knowledge that comes on a regular basis from bright, dedicated basic scientists. Lessons have been learned from failed experiments in the laboratory and clinic, and new tools are now available which allow questions to be addressed that were previously out of our reach.

By casting a wide net that tries to envelop the entire scope of the problem, this chapter considers the scientific gaps that stand in the way of effective treatments for neurodegeneration. Discoveries that present tantalizing therapeutic hypotheses are viewed as pieces of a puzzle so that the "magic" is stripped from the search for a "magic pill." At its core, neurodegeneration is a problem of cell health. Importantly, cells are interrelated within the brain, and thus limiting study to one type of cell is a flawed strategy. Neurodegenerative diseases must be considered as disorders of nervous system circuits. Patients' symptoms are the manifestations of neural circuit dysfunction. Accordingly, a successful treatment needs (a) to normalize the biology within individual cells as well as within the tissue and (b) to preserve or repair the information processing in important neural circuits. This approach places new biologic discoveries in perspective and promotes discussion of gaps that stand between discoveries and knowledge of neurodegeneration in overall tissues and circuit dysfunction that eventuates clinically.

Introduction

Neurodegenerative diseases represent a major public health challenge that threatens to reach crisis proportions with aging of the population. The science of neurodegenerative disorders has exploded over the last decades, yet treatments for patients have not been forthcoming. A number of clinical trials of potentially disease-modifying drugs have failed in Phase III. Critical reappraisal of these therapy-development efforts yielded ideas on the major hurdles that need to be overcome as well as how future research could be improved (Huang and Mucke 2012). At this Ernst Strüngmann Forum we asked, "Where are the major gaps in our understanding and capabilities?" and "Where are we not looking for answers?" A number of recent brainstorming efforts provided background for this forward-looking exercise. The National Institutes of Health convened three workshops: the Alzheimer's Disease Research Summit 2012,[1] a workshop on Alzheimer's Disease-Related Dementias: Research Challenges and Opportunities,[2] and a conference on research planning for Parkinson disease in 2014.[3] In addition the Alzheimer's Disease Summit: The Path to 2025,[4] convened in 2013 by the New York Academy of Sciences, addressed major public health issues and posited strategic recommendations for research and the Institute of Medicine Neuroscience Forum held a workshop in 2012 on cross-cutting themes in neurodegenerative disease research (for a summary, see Institute of Medicine 2013).

A number of common themes have come to the forefront as scientists, physicians, patients, caregivers, governments, and private organizations take on the growing public health crisis of age-related neurodegeneration. Two general classes encompass the major issues:

1. There are significant gaps in our knowledge of the basic biology that underlie aging and neurodegeneration. Although scientists have identified a number of important key features of human neurodegeneration, we are far from having a full picture and forced to rely on tremendous assumptions. Leading therapeutic hypotheses have, out of necessity, been based on assumptions rather than on solid information. It is thus critical to fill major scientific gaps in basic understandings of neurodegenerative disease if rational treatments are to be developed.
2. The nature of the public health problem and the personal suffering of affected individuals and their caregivers demand that we continue

[1] http://www.nia.nih.gov/newsroom/announcements/2012/05/alzheimers-disease-research-summit-offers-research-recommendations (accessed April 9, 2015)

[2] http://www.ninds.nih.gov/funding/areas/neurodegeneration/workshops/adrd2013/ (accessed April 9, 2015)

[3] http://www.ninds.nih.gov/research/parkinsonsweb/PD2014/index.htm (accessed April 9, 2015)

[4] http://www.nyas.org/pathto2025 (accessed April 9, 2015)

to test the most promising therapies, despite incomplete knowledge. Past failures call for smarter approaches if we are to gain significant knowledge, even from negative trials. It is essential, for example, to know that a drug candidate engages its target in the human brain and ideally that it modifies the relevant biological pathway at therapeutic doses prior to a long and expensive Phase III efficacy trial. Assuring target engagement and pathway modification requires the development of new biomarkers based on the pathophysiology that occurs within the specific subgroup of patients proposed to benefit. Well-designed clinical trials will likely be necessary to know definitively whether the proposed therapeutic target is actually involved in the disease process, as opposed to parallel or downstream processes which cause morphological or biochemical changes without functional relevance.

To maximize the chance for success, basic neuroscience must be able to explain neurodegenerative diseases at all levels: molecular and cellular, neural tissue, and neural circuit (where the patient's neurologic deficits and burden of illness are manifestations of the neural circuit abnormalities). In this chapter, we highlight areas in basic science where major gaps in knowledge exist, review new technologies, and discuss anticipated challenges to clinical research in neurodegeneration.

Causation on the Cellular and Molecular Level

What is the nature of the most upstream causative link between protein mutations and the neurodegeneration they cause?

Neurodegenerative Diseases as Proteinopathies

"Culprit" Proteins

The key discovery in neurodegenerative disorders over the past decades has been the evidence for "proteinopathies" in almost all disorders. Abnormal processing of specific proteins occurs in neurodegeneration due to uncommon mutations in both the genes for specific proteins as well as in the common "sporadic" disorders: for example, proteins identified in dominantly inherited and sporadic diseases, including amyloid beta (Aβ) in Alzheimer disease (AD), tau in frontotemporal dementia (FTD), TAR DNA-binding protein 43 (TDP-43) in FTD or amyotrophic lateral sclerosis (ALS), and synuclein in Parkinson disease (PD). This led to the plausible idea that neurodegenerative diseases are caused by pathogenic effects of specific "culprit" proteins. Indeed, in each of these cases, the suspect protein has been found to misfold and aggregate, giving rise to a pathologic "signature protein" for each disease. Notably, most of the proteins involved in neurodegenerative disorders can exist in a dazzling array of conformations and assembly states; however, the most pathogenic effects of these conformations and states remain to be identified. This lack of knowledge has hindered both the development of treatments aimed specifically at the most pernicious culprits and the assessment of whether any of the treatments which have undergone clinical trials actually affected their levels in relevant brain regions.

Gain-in-function mutations in the genes coding for signature proteins are causal in several inherited forms of neurodegenerative diseases. This seems to offer an inherently important pathogenetic clue and enables experimental models to explore the molecular biology of neurodegenerative diseases using modern genetic techniques. A variety of cell function abnormalities have been identified in cellular and animal models expressing the mutant gene: some are related to abnormalities in pathways regulated by the normal protein, some to pathways unique to the mutant protein, and some to the transport and cellular location of the culprit protein. In general, though, the function of the normal protein is incompletely understood. The difficulty lies in determining the most upstream consequence of production of the abnormal protein versus the many downstream consequences that are expected in a diseased cell. This distinction is important as therapies aimed at a single upstream target may be effective, whereas a therapy aimed at only one of many downstream targets may not.

The temporal sequence of pathology may also be important. For instance, can early stages of pathology be driven by the abnormal protein but later stages of pathology be less dependent on the initial trigger? The idea of temporal stages has been advanced to explain the early failures of immunotherapy for AD patients. Is the disease in such a late stage in persons with dementia that removal of Aβ is no longer able to attenuate disease progression? Given this concern, there is a move toward earlier treatment and studies of asymptomatic persons with AD biology; that is, people known to have inherited amyloid precursor protein (APP) or presenilin mutations or who have been identified through positron emission tomography (PET) and cerebrospinal fluid (CSF) studies to be at high risk for developing AD. Amyloid PET scans, which show evidence of extracellular Aβ plaques before cognitive changes, may enable clinical trials in presymptomatic populations. Tau PET scans, which show evidence of intracellular tau aggregates, should enable detection and subsequent monitoring of the spread of neurodegeneration in AD as well as in the other tauopathies: progressive supranuclear palsy, FTD-tau, and chronic traumatic encephalopathy. The NIH planning groups for PD have recommended that new efforts be made to identify synuclein pathology years before onset of motor diagnosis. Major studies (PHAROS and PREDICT) identified premotor symptoms and imaging markers in Huntington disease to enable clinical trials that might prevent disease onset (Paulsen et al. 2014).

Toxicity of Misfolded Proteins

The second intriguing idea dominating much of the current research follows from the evidence that in each of the neurodegenerative diseases, the culprit protein—whether in monogenic or sporadic disease—misfolds and forms aggregates within the cell (tau, TDP-43, synuclein) or outside the cell (Aβ). This has motivated the question, "Are neurodegenerative diseases disorders of protein misfolding and subsequent aggregation?" Considerable work is ongoing to determine if misfolded proteins are toxic to the cell, either as oligomers or full-fledged aggregates. Some studies suggest that the inclusion bodies filled with protein aggregates are a protective mechanism that walls off the toxic effects of monomers or oligomers. What are the important targets of toxic oligomers? Do they impair mitochondrial function, as has been purported to occur in PD? Biomarkers that allow measurement of these signature protein aggregates in humans can make profound contributions for diagnosis, the detection of presymptomatic onset, and the following of spread and propagation; they also can serve as targets for therapies aimed at decreasing production or increasing clearance. However, the identity of the most pathogenic conformation or assembly state is uncertain for most of the proteins causally involved in neurodegenerative diseases.

Impaired Proteostasis

The study of protein misfolding, due either to mutations in their coding sequence or simply overproduction, intersects with the cell biology of protein processing and metabolism (Tian and Finley 2012; Chen et al. 2011; Hamasaki et al. 2013; Hara et al. 2006; Nishida et al. 2009; Park and Cuervo 2013). Molecular chaperones shepherd normal protein folding in the endoplasmic reticulum and cytosol (Voisine et al. 2010). Does the production of mutant proteins or overproduction of "signature" proteins overwhelm the chaperone system, thus creating and resulting in widespread aberrant misfolding of the culprit protein (Komatsu et al. 2006; Koga et al. 2011; Orenstein et al. 2013)? Does this also lead to misfolding of other essential proteins in the cell? A number of mutations in protein-processing genes have also been causally linked to ALS and FTD: ubiquilin 2, vasolin-containing peptide, optineurin, charge multivesicular body protein 2, and vesicle-associated membrane protein-associated protein 2. The downstream effects of the production of misfolded proteins are an area of intense study. The molecular chaperone system preserves normal protein folding in stress conditions that promote denaturation. However, proteins that exceed the capability of the chaperones and become misfolded are metabolized by the ubiquitin proteasome system (UPS) as well as by chaperone-mediated autophagy. Larger aggregates are substrates for micro- or macro-autophagy into lysosomes. The fact that the culprit proteins form intracellular aggregates suggests that normal protein-processing mechanisms are not sufficient. Investigators are pursuing questions such as: Do culprit proteins in neurodegenerative diseases overwhelm these systems to the detriment of the cell? Do they create unsustainable endoplasmic reticulum stress and affect normal protein processing, lipid metabolism, and autophagy of other organelles (i.e., mitophagy of mitochondria)? The intersection of protein misfolding of culprit proteins and their effects on protein-processing systems also allows for consideration of a variety of other cellular circumstances to contribute to neurodegeneration. A disorder in protein processing might also explain the common co-occurrence of multiple "signature protein" aggregates in the same brain (i.e., Lewy bodies and AD pathology, TDP-43 and AD biology, the interaction of Aβ and tau in AD biology, and TDP-43 and tau in chronic traumatic encephalopathy). Indeed, one unanswered question is why these specific sets of proteins are so prominent among the entire set of intracellular proteins.

Banking on the possibility that deficiencies in protein metabolism may be common to neurodegenerative disorders, groups are testing whether agents which upregulate the capability of a cell's chaperone, UPS, or autophagic systems can attenuate the toxic effects of the "proteinopathies" in cell and animal models (Lee et al. 2010). New high-resolution confocal microscopy methods now allow investigators to follow fluorescently tagged proteins and organelles from their synthesis to degradation in cell models. Biomarkers that reflect the activity of autophagy or mitophagy in humans would enable testing this strategy in patients with a variety of neurodegenerative disorders.

Transmission and Propagation of Protein Misfolding

Over the past decade, evidence has accumulated that points to the propagation of misfolded proteins in the nervous system (Clavaguera et al. 2014a). In most neurodegenerative diseases there is spread of pathology from regions affected early (i.e., hippocampus in AD, vagal motor nucleus and olfactory bulb in PD, monolimb motor neurons in ALS, and dorsolateral tail of caudate in Huntington disease) to widespread involvement at late stages of disease. All brain regions do not degenerate synchronously. Spread and propagation of neurodegeneration was first identified in prion diseases. Recently, fibrillar aggregates of tau and synuclein have been seen to be taken up into otherwise normal cells and give rise to misfolding. Injections of fibrillar aggregates purified from human brain have caused aggregation of the corresponding mouse protein (Clavaguera et al. 2013; Watts et al. 2013). Local injection of mutated tau in viral vectors has been seen to cause tau aggregation in synaptically connected neurons that do not contain the virus, suggesting cell-to-cell transmission and propagation of misfolding (de Calignon et al. 2012). How are these proteins transferred from cell to cell? Some evidence suggests they are released from degenerating axons into the synaptic space. How are they transported in the cell to seed further misfolding? What processes are necessary for propagation of misfolding after entry of misfolded proteins? Are the current propagation models relevant to the human condition? Does the spread of misfolded proteins cause neuronal dysfunction or degeneration? Can transmission or propagation be blocked for therapeutic benefit (Clavaguera et al. 2014a)? Where in the nervous system is the site(s) of initiation? What is the initiating factor? Is it intrinsic to the neural cell? Is it triggered by some stressor coming from the vasculature, the nasal cavity, oral cavity, the gut, or the lung (Clavaguera et al. 2014b)? For example, in PD, synuclein fibrils may be seen in the fibers of the enteric nervous system that innervate the gut and parasympathetics that innervate the submandibular glands. Does synuclein aggregation begin in these peripheral nerve fibers and spread into the central nervous system?

Neurodegenerative Disorders as Errors involving RNA

Recent discovery of mutations in RNA-binding proteins in neurodegeneration (TDP-43, FUS/TLS, C9orf72, spinocerebellar ataxias) have shifted the focus to errors in RNA (Birman 2008; Ling et al. 2013; Ramaswami et al. 2013). Toxic RNAs have been postulated as causal features in ALS, FTD, and fragile X tremor ataxia. Some spinocerebellar disorders are due to mutations in noncoding RNA. In most cases of ALS and FTD, ubiquitinated TDP-43 is seen in neurons and glia, and it is common in AD as well. FUS inclusions have been found in Huntington disease, and spinocerebellar ataxia 1 and 2. Cytoplasmic and nuclear aggregates of RNA-binding proteins are seen in a number of neurodegenerative disorders, which may affect RNA homeostasis, translation of mRNA, sequester important RNAs and RNA-binding proteins, and affect transport with unhealthy consequences for the cell. Aggregates of RNA-binding proteins may also add stress to the cell's protein-processing capacity, thereby accounting for the co-occurrence of RNA-binding protein aggregates with aggregates of the "signature proteins." Abnormal gene regulation has been identified in a number of other disorders, including Huntington disease. The causative RNA mechanism in these disorders has not been definitively defined but includes affecting RNA processing and metabolism, gene transcription, spliceosome function, miRNA processing, and transport of RNA to specific cellular regions (synapse, axon). The role of noncoding micro RNAs in neurodegenerative disease has only recently started to be explored.

Neurodegenerative Disorders as Maladaptive Aging

As aging is an inherent component of all the disorders, any full picture of neurodegeneration must account for its role. Changes in the capability of cells, over time, to manage the stress response and process misfolded proteins in the UPS or autophagy systems, especially in nonregenerative, long-lived neurons, could provide the biologic link (Cuervo and Wong 2013). Aging might also be defined as an integral part of the cell's lifetime of protein stress. What are the ramifications for maintaining normal protein homeostasis for decades in the face of (a) a mutant protein, (b) overproduction of protein, (c) protein misfolding due to activation of Ca^{2+}-dependent proteases related to excitatory neurotransmission, (d) oxidative stress due to mitochondrial uncoupling and free radical production, (e) mechanical trauma, and (f) other yet unknown cell stressors? Aging science has identified a number of molecular and cellular processes of senescence, which should be considered as contributors to the "ticking clock" of neurodegeneration. In addition to aging processes in the neuron, aging effects at the tissue (i.e., vascular, glial, cerebrospinal fluid flow, or even whole body system) level (immune, metabolic) could contribute to the occurrence of dementia with age.

Mitochondrial Dysfunction

Accumulations of mitochondrial mutations and decreased mitochondrial function have been identified as contributing to neurodegeneration in most disorders as well as in aging itself. In some models, mitochondrial dysfunction is a consequence of the signature proteinopathies. Mitochondrial dysfunction is likely central to a number of neurodegenerative disorders as it is causally linked in Pink1 and Parkin mutations in PD (Burchell et al. 2013; Hertz et al. 2013), and superoxide dismutase mutations in ALS. The mitochondrial hypothesis has led to clinical trials of a variety of agents to improve energy metabolism and/or decrease oxidative stress in sporadic AD, ALS, PD, and Huntington disease. None of these trials (with the possible exception of CoQ10 effects on elevated brain lactate in Huntington disease) were accompanied by direct evidence that the drugs had any effect on energy metabolism in patients. Improved means of assessing mitochondrial function in brain are necessary to test the mitochondrial hypothesis.

Epigenetics in Neurodegenerative Disease

A variety of epigenetic changes with aging of the cell may result in changes in gene transcription that contribute to the age dependence of neurodegenerative disease. A relatively new idea of possible relevance is the finding that transposable elements in the genome become free to move as cells age and are able to reinsert into the genome with age (Perrat et al. 2013; Reilly et al. 2013). Possibly related to this notion are observations that Aβ accumulation increases the level of neuronal activity-induced DNA double-strand breaks, thus delaying their repair.

Compensatory Cellular Processes

Somewhat related to the study of additive effects of cellular stressors, it is known that a variety of "trophic" factors attenuate the effects of stress on cell health. Early studies sought to leverage the protective properties of single "growth factors" without full knowledge of their effects on the cells affected by neurodegeneration, and they did not meet with success. After acute brain injuries, there is almost always functional recovery over time associated with immune responses to injury, enhanced neurogenesis, movement of endogenous neuroprogenitors into the injured area, axonal and dendritic sprouting, and new synapse formation. What is the spectrum of compensatory processes in neurodegeneration? How are compensatory pathways affected by the proteinopathies? Do such compensatory processes attenuate the onset of pathology? Do compensatory processes deteriorate with age and does their rundown contribute to the requirement for aging in neurodegenerative diseases? Would enhancement of these compensatory processes be expected to attenuate disease progression? On the other hand, does the likelihood that evolution has already optimized compensatory processes limit the therapeutic potential of strategies aimed at their enhancement? Furthermore, could these compensatory processes go array, as occurs in posttraumatic epilepsy, and result in epileptiform activity that worsens neurodegeneration?

Neurodegeneration and Neural Cellular Dysfunction: Neurons and Glia

Cell death is common to the neurodegenerative diseases, but cell dysfunction is expected to occur long before a cell dies. Less studied than the molecular/organelle dysfunction in neurodegeneration is the function and structure of the neuronal components—such as the axon (Morfini et al. 2009; Yue et al. 2008), dendrite (Stephan et al. 2013), and synapse—or how, under what circumstances, the molecular abnormalities affect the neuron's ability to function in its network. At what stage do the molecular abnormalities underlying neurodegenerative diseases affect the neuron's ability to function in its network? How does this affect the circuit function? Aggregates are seen in many neurodegenerative diseases in glial cells (tau in tufted astrocytes in progressive supranuclear palsy, synuclein in oligodendrocytes in multiple system atrophy) but little is known about their effects on glial cell biology or glial functions. This gap in knowledge is highlighted by the finding that mutations confined to microglial cells or glial cells in genetic mouse models can have profound neuropathological effects. In a neuroncentric way of thinking, these effects are called non-cell autonomous.

Genomics of Neurodegeneration

Although there has been considerable progress in understanding the genetic underpinnings of neurodegenerative disease, much more can be learned with the newer techniques that are now available. Genomic studies on brain tissue have become especially important to study epigenetic signatures of neurodegenerative diseases. A new area of study, understanding whether mosaicism contributes to neurodegenerative disease, has been ushered in by the discovery that somatic mutations, especially copy number variants, are enriched in the brain. Whole exome sequencing also allows the fairly rapid discovery of *de novo* mutations in studies of trios (mutations in affected individuals not found in unaffected parents). Genomic studies of brain tissue also allow the tracking of insertions by large volumes of transposable elements in a cell's DNA, which may escape epigenetic processes that hold these in check, whether due to age-related changes in histone modification or epigenetic events.

Extrinsic Triggers of Cellular Degeneration

Advances in genomics have opened a great new knowledge base in neurodegenerative disease. Much less is known, however, about the causal influences that are extrinsic to the cell. For instance, excitotoxicity was long studied as a common stressor that might lead to neurodegeneration. Epileptiform activity has recently been identified in AD as a potential contributory factor. The identification of environmental influences that affect the development of neurodegeneration remains a knowledge wasteland. There is some evidence that pesticide exposure links to PD; ALS was seen at slightly higher frequency in military personnel after the Gulf War; and repetitive concussion has been linked to a neurodegenerative tauopathy as well as to motor neuron disease. It has been suggested that concussion is a risk factor for AD, but that risk has not been clearly defined. The interaction between vascular disease and AD is complex, since they co-occur commonly in persons with dementia, and vascular risk factors are highly correlated with risk for dementia. A variety of research has tied stressors or "second hits" to neurodegenerative disease, but whether they are one of thousands or an important influence by themselves is not clear. Of interest, some protective factors have been identified from epidemiologic studies (e.g., caffeine, brain penetrant calcium channel blockers, and high urate levels) to decrease risk of PD. Because techniques are not available to measure lifelong exposures to chemicals, microbes, oxidative stress, metabolic stress, mechanical insults, or even brain circuit activity, the environmental and extrinsic influences which likely contribute to development of neurodegenerative disease may remain largely unknown and difficult to study.

Systems Biology of Neurodegeneration

Some investigators highlight the need for a systems approach that interrogates many, if not all, molecular pathways within the cell to avoid the pitfalls of focusing on a single pathway and not appreciating how it is integrated into tens or hundreds of others. A cell's state may be defined by networks of interconnected proteomic, metabolomic, and transcriptomic patterns. No single pathway aberration is sufficient to cause neurodegeneration, but specific "network states" may underlie neurodegeneration. The therapeutic target, then, becomes normalization of the cell's abnormal network "state," as opposed to intervening in a single pathway.

Tissue Level Dysfunction in Neurodegenerative Disease

Neurovascular Unit and Glymphatics

Some argue that the current and past approaches to neurodegenerative disease have been overly neuroncentric. The neuron functions within a dense cellular tissue made up of different types of glia, resident inflammatory cells, hematopoietic inflammatory cells, various cellular, and matrix components of the vasculature. A recent term from vascular biology is the "neurovascular unit." In addition to the coupling of blood flow with neural activity, control of movement of cells, and molecules across the blood-brain barrier, another system (called the glymphatic system) has been identified which parallels the lymphatic system in other tissues (Iliff et al. 2012) The fluid movement in the interstitial space is dynamic, as is the movement of cerebrospinal fluid from production in the choroid plexus to absorption into the blood stream via the arachnoid villi. Most recently, the flow of interstitial fluid through the glymphatic system has been ascribed a role in the "flushing" of proteins such as Aβ from the brain, especially during sleep. Interstitial fluid is described to enter from the cerebrospinal fluid (CSF) space along penetrating arterioles, pass through glia, and exit along perivenous channels back to the CSF. Aβ clearance from brain has been identified as a potential contributor to amyloid plaque burden. Unexplored remain how glymphatic function changes with age, the distortion to brain structure that occurs with aging, hypertension, brain atrophy, and changes in CSF flow (e.g., enlarging ventricles, function of meningeal tissue, geometry and flow in widened subarachnoid space, arachnoid granulations affecting CSF absorption, choroid plexus aging and CSF production).

Vascular Disease and Alzheimer Biology

In persons dying with dementia, a combination of vascular diseases (gross infarcts, micro infarcts, diffuse white matter disease) commonly occurs with AD pathology. Indeed some data suggests that their combined burden of pathology leads to the most common forms of dementia. Vascular risk factors, most prominently the presence of infarcts, are one of the strongest risk factors for dementia of the AD type. Of note, the predominant site of infarction is in deep white matter and basal ganglia nuclei, not cortex. The nature of the association of mixed dementia with diffuse white matter disease (a mix of infarction, wide periarterial spaces, rarefaction of myelin) is also unknown. New data suggests that metabolic syndrome (obesity, type 2 diabetes, sedentary lifestyle) is associated with dementia, but whether this is via effects on vascular disease or some hormonal effect of the abdominal fat pad or a combination of factors is not clear.

Vascular function and the biology of AD are intertwined. Aβ clearance into the blood has been shown to rely on normal function of the blood-brain barrier which may be disordered with age and hypertensive vascular disease. Aβ actually appears to be toxic to pericytes, which are important regulators of the blood-brain barrier. Aβ amyloid deposits in the blood vessel wall in a large percentage of persons with AD and affects blood-brain barrier function, even giving rise to multiple hemorrhages in many (amyloid angiopathy). How dysregulation of blood flow, oxygen, glucose delivery to tissue, and blood-brain barrier function contribute to AD pathology is not entirely clear.

Functional Consequences of Gliosis and Inflammation in Neurodegenerative Disease

Gliosis is ubiquitous in neurodegenerative disease yet little is known about glial function or how it may contribute to progression in these disorders. In some neurodegenerative disorders, the signature protein misfolding occurs in glial cells (multiple system atrophy, chronic traumatic encephalopathy). In progressive supranuclear palsy, tau fibrils are seen in oligodendroglia; how that affects axonal transmission of action potentials is, however, not known. Much more is becoming known about the normal biologic interactions between glia and neurons that might now be explored in disease. Inflammation is ubiquitous in most neurodegenerative diseases. The role of inflammatory cells, however, is now known to be complex (Glass et al. 2010) and has recently been identified as essential to the process of normal synaptic pruning (Stephan et al. 2013). They are also involved in regulating the movement of neuroprogenitor cells along vascular pathways. Neuroprogenitor cells in the dentate, periventricular stream, and nasal mucosa may have a role in repair in the nervous system, but we know little of their role in combatting neurodegeneration. Cellular inflammation and cytokine secretion occurs in ischemic vascular disease and may be one of the links between cerebrovascular disease and neurodegeneration. The tissue processes that contribute to or combat cell-to-cell transmission of culprit misfolded proteins is an acute area in need of research.

Reparative Tissue Responses in Neurodegeneration

Reparative processes at the tissue level are also mainly unexplored. Pathologic studies, in most cases, indicate that the brain tissue structure is abnormal by the time symptoms occur. Whether this hysteresis is related to some threshold of pathology required to disturb neural circuits or whether there is some reparative process that temporarily compensates for the neurodegeneration is currently not known. Some of the early cellular and molecular changes may be related to compensatory changes at the tissue level, as opposed to contributors to pathology. For example, protein aggregates have been shown in some cases to have protective attributes by preventing the harmful activity of toxic oligomers. Microglia have important normal functions in pruning ineffective synapses. The protective effects of gliosis are not clear. In general, we know little about the tissue function level in neurodegenerative disease.

Circuit Level Dysfunction in Neurodegenerative Diseases

Behavior and neurological deficits which characterize neurodegenerative diseases are considered manifestations of neural circuit dysfunction. In the past, correlation of neurologic deficits with pathology pointed to the circuit dysfunction in neurodegenerative diseases (i.e., semantic memory dysfunction with pathology in hippocampal and cholinergic regions, progressive aphasia with focal cortical atrophy in language cortex, loss of visuospatial skills with right parietal atrophy, bradykinesia with ventral substantia nigra). Most recently, new technologies are allowing dynamic assessment of neural circuit function, such as resting state magnetic resonance imaging (MRI), graphic analysis of electroencephalography (EEG), magnetoencephalography, or in response to specific tasks. These techniques have provided evidence for signatures of circuit dysfunction for specific disorders and are being validated as a means to measure progression of circuit dysfunction over time. This work is at an early stage. As its final product, a comprehensive predictive model of neurodegenerative disease would explain neural circuit abnormalities and exactly how these translate to neurologic dysfunction. If known, then tracking changes in circuit function would enable dynamic monitoring of progression and provide targets for therapy that are tied closely to the features important to patients. In addition, circuit tracking might enable answers to an important conundrum in neurodegenerative disease: the lack of perfect correlation between burden of pathology and burden of illness.

One explanation is that we cannot actually see or measure the full burden of pathology. Another is that individual variation exists in how neural circuit function is affected by neurodegenerative changes at the tissue level. Studying this problem might lead to an understanding of the mechanisms underlying "resilience." Studies of the healthy "very old," for instance, show good cognitive function despite presence of AD pathology. Is this due to compensation at the molecular or tissue level, which preserves circuit function, or are other circuits called in to compensate for dysfunction in the usual disturbed circuits? Is it possible to alter neural circuits in neurodegenerative disease to improve function? Mental activity, physical exercise, and educational level have been suggested to attenuate or delay neurodegenerative disease. Suppression of network hyperactivity by treatment with an antiepileptic drug has been shown to improve cognitive functions in a mouse model of AD and patients with amnestic mild cognitive impairment. Are these interventions altering circuit function in some beneficial manner related to or independent of the pathobiology at the tissue level? Deep brain stimulation certainly produces profound benefits for motor function in PD, but has had no effect on disease progression.

New Technologies

Many of the questions posed above are not novel and have been pursued in some manner for decades. For instance, transmission and propagation of misfolded proteins was tested by Carleton Gadjusek in the last century. New technologies often lead to breakthroughs in testing key hypotheses. New technologies also allow measurements of things never seen before, which in turn opens up entirely new hypotheses. Certainly, many of the advances over the past decade come from impressive new capabilities in human brain imaging, computer analysis of large data sets, genetics and genomics, molecular imaging, and cellular imaging. Looking to the future, one might expect continued major improvements and the introduction of new technologies. One objective of the Brain Research through Advancing Innovative Neurotechnologies (BRAIN) initiative in the United States is to expand the power of these types of tools and develop others for interrogation of neural circuits.

Interrogation of Neural Circuits

In circuit analysis, intracranial recordings have been of value in understanding abnormal circuit function in small, defined areas in patients with epilepsy, and advances will allow greater resolution and coverage of brain. Genetically engineered light emission technologies linked to changes in intracellular calcium and membrane voltage now allow simultaneous recording of thousands of neurons. This technology may be expanded to monitor a variety of other signaling processes in animal and cellular models (e.g., light emission linked to enzyme activity, organelle function, channel activation, protein aggregation). A new technique to extract lipid to make volumes of brain accessible to light microscopy will allow rapid three-dimensional analysis of molecules and structure. Human brain circuit analysis is under intense study using the BOLD (blood oxygenation level dependent) technique, EEG and evoked potentials, and magnetoencephalography. Each of these has inherent limitations in terms of temporal or spatial resolution and the accessibility of deep brain structures, which completely new technologies will be required to overcome. It may occur that exquisitely sensitive and specific intracerebral techniques will be developed to record circuit behavior in animals and animal models of disease; however, their invasive nature does not allow use in humans. Developing safer means of recording from inside the skull in humans may rise as a major challenge to translate the advances made in animals.

Fiber Tracking in Neurodegenerative Disease

White matter atrophy in neurodegenerative diseases is known to be quite marked, but little is known about the specifics of the loss or the effects on neural circuits. Efforts, such as the NIH connectome project,[5] in charting the white matter connections of the human brain in thousands of individuals using diffusion MRI will likely enable a new field of exploration in neurodegenerative diseases that parallels current and past studies of regional gray matter loss. Though studies of white matter structure in neurodegenerative diseases may expand, there is still only crude technology to interrogate white matter function (i.e., focal transcranial magnetic stimulation coupled to EEG or fMRI to detect target activation).

Induced Pluripotent Stem Cells

Scientific progress is often dependent on the available experimental models and their ability to dissect mechanisms important in disease. Transgenic animal models have transformed the field over the last few decades but as yet have not predicted therapeutic success in human trials. The advances in induced pluripotent stem cell (iPSC) technologies, which are only a few years old, allows experimentation to move to human cells (Bilican et al. 2012, 2013; Ryan et al. 2013; Serio et al. 2013). In just a short time frame, iPSC technology has provided the opportunity to establish specific phenotypes in cells from individuals with genetic causes of neurodegeneration (with the abnormal gene corrected in clonal iPSCs as controls) and to study the mechanisms underlying those phenotypes. The key issue is to discern which phenotype, discovered in iPSC cultures, proves to be relevant to disease pathogenesis. Though initially studied *in vitro,* the transplantation of iPSCs into an *in vivo* environment in animals may allow important observations about their interaction within the tissue and neural circuits and the effects of aging. Whether iPSCs from patients with sporadic disease will also manifest valuable specific phenotypes is yet to be definitively determined. An event that occurs in a spinal motor neuron or enteric nerve fiber due to an environmental stressor may not be shared by a skin fibroblast. Replacement therapy in human patients is an ultimate goal and may be on the horizon in focal atrophies, such as dopaminergic cell replacement in PD. Whether iPSCs themselves can be used to deliver trophic factors to attenuate neurodegeneration widely in the brain is not clear.

Manipulation of Genes and mRNA

Though genetic techniques are prominent in cellular and animal studies, gene therapy in human neurodegenerative diseases is still in its infancy. Delivery

[5] http://www.humanconnectomeproject.org/ (accessed April 9, 2015)

problems are being worked on using a variety of viral vectors and direct delivery into brain or CSF of "protected" iRNA and antisense. If toxicity can be managed and delivery to the necessary intracellular sites is efficient, it is conceivable that knockdown of gain-in-function mutations will effectively treat the autosomal dominant neurodegenerations. A variety of other genetic manipulations then become possible in non-Mendelian neurodegenerations, including decreases in synthesis of "culprit proteins," and upregulation of protein clearance mechanisms such as autophagy-lysosome systems or proteasome-ubiquitin systems. Similar issues face antibody therapy, which is currently being tested to improve clearance of extracellular accumulation of misfolded proteins. Antibody therapy has been postulated as a potential treatment to prevent cell-to-cell transmission of protein misfolding.

Advanced Genomics and Proteomics

Molecular techniques to assay large numbers of genes, noncoding DNA, histone modifications, and mRNA are becoming more sophisticated at a tremendous rate and should enable the study of gene networks, epigenetics, mosaicism, and the movement of transposable elements in a cell's genome using sophisticated computational techniques. iRNA screens can be used to identify the complete set of interrelated genes. Proteomic technologies that couple mass spectroscopy to antibodies, DNA, or protein aptamers are also able to detect femtomolar quantities of thousands of known proteins. These will allow the study of protein networks in models of neurodegenerative disease, such as iPSCs, but can also be applied to measure important signatures of disease or pharmacological targets in cerebrospinal fluid. This field could be advanced by a concerted effort to characterize the normal CSF proteome across the life span.

PET Radioligand Development

PET tracer development is an intense area of investigation, and even commercialization now, to detect and follow protein aggregation in certain neurodegenerative diseases. Improved tracers promise to transform diagnosis of disease in the presymptomatic stages and provide biomarkers of disease progression. Radioligands that bind to extracellular amyloid and intracellular tau aggregates as well as inflammatory cells have been incorporated into ongoing clinical studies. Efforts to find radioligands for synuclein imaging are underway.

Computational Techniques and Big Data

A number of groups have questioned whether science could be advanced by a data repository for clinical and basic research data. Would such a large database lead to discoveries from the application of analytic tools to "big data"? The density of data accumulated in genomic, proteomic, neuroimaging, and neurophysiology studies requires sophisticated analytic techniques to identify important relationships. Certainly the lack of foresight necessary to combine data from the many large and expensive ongoing studies around the world appears wasteful.

Anticipated and Current Challenges to Clinical Research in Neurodegeneration

How to Approach the Environmental Determinants of Neurodegeneration

Probably the most difficult challenge in understanding and preventing neurodegenerative diseases is to identify those factors across the life span of the individual that either increase or decrease risk (Rodriguez-Navarro et al. 2012). It is likely that we will know the genetic load within decades, but even then we will struggle to account for the environmental load that leads to non-Mendelian neurodegeneration, which affects the majority of individuals. Though genetics has advanced enormously, the ability to identify causal "environmental" risk factors to measure their effects in brain and to develop methods to mitigate their effects over time is almost nonexistent. Epidemiologic studies sometimes highlight particular correlations: traumatic brain injury and vascular risk factors in dementia, pesticides in PD, and protective effects of caffeine in PD. They suffer, however, from lack of precision in measurement, abundance of confounders, and purely correlative relationships to disease. The fact that these exposures may occur decades before disease onset makes study very difficult. It may be possible and insightful to approach the gene-environment problem by studying persons with defined genetic risk and attempt to uncover the environmental factors that affect penetrance (i.e., LRRK2 PD or apoE4 homozygotes).

Bioethical Issues Associated with Ultra-Early Detection of Neurodegenerative Biology

Current thinking in neurodegenerative diseases also places importance on early detection, before symptoms of disease are evident by neurologic examination. This is now possible in AD with technologies (amyloid PET imaging and CSF Aβ/P-tau ratios) to identify and measure the pathology that precedes symptoms. Efforts are underway to replicate this in other neurodegenerative disorders. This new technology raises significant ethical questions which must be resolved culturally to proceed: How should investigators ethically recruit and inform individuals at risk for neurodegeneration in the future? How should these individuals be supported by researchers and the medical system? How is privacy of "risk" to be maintained? In designing interventions, what ratio of risk/individual benefit is appropriate in trials of asymptomatic individuals, especially with variable degrees of penetrance, variable confidence in estimates of risk, etc.? How should we ethically manage expectations in trials of asymptomatic individuals?

Biomarkers for Therapy Development in Neurodegenerative Diseases

Perhaps the most acute challenge to neurodegenerative disease research, as well as therapy development in other CNS disorders, is the failure of preclinical models to predict success in clinical trials. This may be a factor of the limitations of the models but it may also be an indictment of the clinical trial strategies. It has been difficult to demonstrate that a particular treatment has the intended molecular effect in humans at the implemented dose and duration plan. Current thinking puts highest priority on developing the means to test for target engagement in human brain, proof of principle that a therapy has the expected biologic effect at feasible doses and delivery strategies. The science of discovery and validation of useful measurement of target engagement in neurodegenerative diseases requires renewed effort. Particular gaps exist at the validation stage of a biomarker of disease pathophysiology. Numerous reports of measurements that differ in small numbers of patients versus some control group occur in the literature, but very few ever become useful for research. Lacking is the standardization of the measurement, study of confounders, assessment of variance in the measure from the same individual over time (test-retest variability), measurements from different populations with the same disease (site to site variability), and understanding changes that occur with changes in techniques, reagents, or different scanners, etc. These problems plague current techniques such as task-dependent fMRI, resting state MRI, diffusion imaging of white matter tracks, CSF proteomic studies and multiple genomic studies. Academic investigators are unlikely to solve these problems by themselves. Since useful biomarkers for neurodegenerative disease research are necessary, partnerships between academics, industry, and government will be required.

6

Preclinical and Clinical Understanding

Major Gaps in Our Understanding and Capabilities

Edward H. Koo

Abstract

Age-associated diseases are an inevitable, costly, and burdensome outcome to societies as life expectancy increases. All countries face this growing problem even as there are notable successes in treating some of these diseases. For example, effective control of hypertension has been accompanied by a noteworthy reduction in the incidence of cerebrovascular disease. Unfortunately, the same cannot be said for neurodegenerative diseases, such as Alzheimer, Parkinson, motor neuron, and Huntington diseases, despite the tremendous recent progress that has been made in understanding the molecular pathogenesis of this group of brain disorders. Currently there is no effective treatment that delays the onset or slows the natural progression of these diseases. Even the very effective pharmacologic and surgical therapies for Parkinson disease are directed solely at motor symptoms; neither the nonmotor symptoms nor the gradual deterioration can be effectively treated or prevented. This chapter discusses some of the difficulties faced in discovering and developing effective treatments for these diseases, from both the preclinical and clinical perspectives. Some of the gaps in our knowledge and critical questions that must be addressed in the near future are described, with an emphasis on Alzheimer disease. The challenges are many and include incomplete understanding of disease pathophysiology, deficiencies in animal models, and inefficiencies in translating new genetic, molecular, cellular, and neurobiological insights into the clinical arena. There is increasing consensus that treatments which target the potential disease triggers will likely be more effective when given before the onset of clinical symptoms. This, however, brings additional challenges—from the need for new biomarkers to improvements in the design and execution of clinical trials—which the research community must quickly address.

Introduction

In many ways, diseases of the brain represent the "final frontier" in our quest to understand disease causation and to discover effective treatments. For those working on diseases of the nervous system, the brain is wondrously elegant and intricate in its design and function, yet ever so complex in disease. Neurodegenerative disorders constitute a group of diseases that share in common neuronal loss in characteristic brain regions, often with pathological hallmarks, which are present in both genetic and sporadic forms, the latter frequently demonstrating genetic susceptibility traits. Importantly, as a group, they are age-associated such that their incidence rises with advancing age. As such, the increasing prevalence of this group of diseases is an unwanted outcome of the rise in human longevity. Tremendous advances have occurred over the past two decades in our understanding of the pathophysiology of many human diseases, including neurodegenerative diseases. From these advances, it has become evident that misfolding and aggregation of proteins—such as amyloid β (Aβ) protein, tau, TAR DNA-binding protein 43 (TDP-43), superoxide dismutase-1 (SOD-1), α-synuclein, or huntingtin—underlie most of the neurodegenerative disorders. Furthermore, aggregated proteins most likely play pivotal roles in initiating the disease processes. Yet, as with many other diseases, effective treatments often lag many years behind breakthroughs in basic research. Thus, with the possible exception of multiple sclerosis, whose pathophysiology is quite different, effective treatments for neurodegenerative diseases have been frustratingly elusive. Against the backdrop of several negative pivotal clinical trials over the past several years, the community has not surprisingly begun to question not only our current approaches but also the hypotheses underlying disease pathophysiology.

This Ernst Strüngmann Forum was convened to review the present state of knowledge of neurodegenerative diseases and to generate new ideas about research directions in basic pathophysiology and new paths for preclinical as well as clinical research. This chapter discusses some of the major gaps in our understanding, as well as needs and capabilities in preclinical and clinical research, and posits where we should be looking. Illustrative examples, where used, will be taken liberally from the Alzheimer disease (AD) field, as research in this area has exploded over the past decade in comparison to other neurodegenerative diseases.

Gaps in Preclinical/Clinical Research

Are the Proposed Cellular Pathways the Correct Targets to Test in Humans?

What Are the Right Clinical Targets?

Historically, studies on the pathophysiology of neurodegenerative diseases focused on characteristic morphological changes which helped define the disorders. For example, senile or amyloid plaques and neurofibrillary tangles are diagnostic for AD, as are Lewy bodies in the case of Parkinson disease (PD). Does this mean that these structures are diagnostic and that their very presence also points to the cause of the disease? As a corollary, would approaches to reduce the formation of these pathologic changes lead correspondingly to improvements in clinical symptomatology? Early efforts in isolating the proteins that make up these structures—Aβ in senile plaques, tau in neurofibrillary tangles, and α-synuclein in Lewy bodies—did not establish whether these proteins play a causative role, not until mutations were subsequently discovered in the genes that encode the respective proteins. Yet, discovering the mutant gene and protein product does not automatically point to potential therapeutic targets. For example, knowledge of the expanded polyglutamine repeats in trinucleotide diseases, the most notable representative being Huntington disease, does not lend itself readily to an obvious "druggable" target. On the other hand, in AD, identification of mutations in the amyloid precursor protein (APP) gene strongly implicated the key role played by the proteolytic product of APP, namely Aβ, in AD pathogenesis. The location of the APP gene on chromosome 21 also dovetailed nicely to prior knowledge that AD neuropathology invariably develops in individuals with trisomy 21, and thus provided further support to the proposed seminal role of Aβ in AD. The race to inhibit Aβ production began in earnest when β- and γ-secreteases (the two proteases involved in cleaving APP to liberate Aβ from its parent molecule) were discovered, thus providing clear targets for pharmaceutical companies to pursue. Inhibitors to β-secretase turned out to be a difficult challenge due to the structure of the enzyme, but promising compounds have now entered Phase III testing. γ-secretase inhibitors (GSI) turned out to be considerably easier to develop. The uncomfortable surprise, however, is that nonspecific inhibition of γ-secretase activity by GSIs led to unacceptable adverse effects, likely due to inhibition of the constitutive cleavage of other γ-secretase substrates, which now number more than 50, notable examples being Notch or ErbB4. This explanation likely accounts for the negative outcomes of the late phase trials of two GSIs: not only were the primary end points unreachable but adverse side effects, including worsening cognition, were noted in subjects given semagacestat chronically (Doody et al. 2013). It can be argued that these adverse outcomes could have been predicted. One lesson from the semagacestat trial is that in our attempts to fulfill the unmet needs for effective treatments, enthusiasm must be tempered by appropriate assessment of risks and proper consideration of the biological pathways (Blennow et al. 2013). Otherwise, additional negative outcomes in pivotal trials will do more harm to the field, especially if wrong conclusions are drawn from the negative results.

When Should the Suspected Targets Be Treated?

A theme common to virtually all neurodegenerative diseases is that mutations (often missense mutations), increased gene dosage (from gene duplication or triplication), or ineffective removal in sporadic diseases magnified during aging trigger a cascade of events to bring about full pathologic manifestation of the diseases. This cascade hypothesis only predicts that if accumulation of these aggregated and misfolded proteins—such as Aβ in AD, tau in various tauopathies, α-synuclein in PD, or TDP-43 in amyotrophic lateral sclerosis/frontotemporal dementia (ALS-FTD) spectrum—can be aborted or prevented, then the subsequent disease development will be attenuated or blocked. Sadly, this plausible prediction has yet to be adequately tested in any one neurodegenerative disease. In the meantime, it is unclear whether any of the pathological processes can be halted or slowed once a trigger is initiated, leading to clinical improvement. For example, it is now well established that the full spectrum of pathologic changes are evident once an individual is symptomatic with AD (Bateman et al. 2012). We also now know that these changes occur gradually over one to two decades before clinical symptoms develop, reflecting perhaps the insensitivity of purely neuropsychological and cognitive measures. Thus, where these changes are protracted but the cellular trigger(s) occurred long ago, it is unlikely that targeting the trigger of the disease (e.g., Aβ deposition or α-synuclein aggregation) will demonstrate substantial benefit by altering disease course because significant pathology has already developed at the time of diagnosis. Similarly, in PD, it is estimated that up to 70–80% of neurons are lost in the substantia nigra before an individual is symptomatic. Thus, reducing α-synuclein aggregation in symptomatic individuals will likely have only marginal efficacy in reversing preexisting neuronal damage.

This discussion presupposes that we have a solid concept of what the precise disease triggers are. Is this supposition really correct? Knowledge of a genetic mutation does not imply an understanding of disease causation, nor do we know whether nongenetic, sporadic forms of the same disease share identical pathophysiology. Further, have the secondary or downstream changes been recognized? If the latter were better defined to prevent downstream effects and mediators could attenuate disease progression, then more targets would be available for drug discovery efforts. Some of these secondary changes, such as tau accumulation or oxidative damage in AD, may follow different rates of development, thus providing additional windows of opportunities for therapeutic interventions. Consequently, major conversations have appropriately arisen to argue for treatments being given as early as possible to be effective, especially in the setting where the drugs target the putative disease triggers or very proximal mechanisms to the triggers of the disease cascade. Addressing this major gap in our understanding will fundamentally impact testing and delivery of effective treatments in the near future.

Finally, the idea that effective treatments fall neatly into symptomatic or disease-modifying categories, while important or critical from the pharmaceutical industry perspective, is likely too simplistic in practice. An early example was the "DATATOP" study designed to determine whether long-term treatment with deprenyl, a MAO-B inhibitor, together with vitamin E could delay the need for levodopa treatment due to progressive motor disability in PD; in other words, delay disease progression. Results from the trial initially appeared to demonstrate precisely this desired effect; that is, subjects on treatment took longer to require levodopa therapy (Parkinson Study Group 1989). However, the symptomatic benefit of deprenyl was not taken into account such that in the subsequent full analysis of the trial results, it was concluded that deprenyl did provide clinical improvements but that the evidence did not support a neuroprotective effect (Parkinson Study Group 1993). Another example can be seen in a tetracycline-regulatable model of mutant tau transgene overexpression in mice to drive neurofibrillary tangles formation in neurons. Surprisingly, in these mice, cessation of tau expression by administration of doxycycline improved cognition and halted further neuronal degeneration, but accumulation of tangle pathology persisted nonetheless (SantaCruz et al. 2005). Thus, there was symptomatic benefit even in the presence of unremitting pathology. In this context, in AD, could toxicity of oligomeric Aβ be attenuated and produce symptomatic benefit if Aβ production could be pharmacologically inhibited in spite of possibly continued disease progression? Conversely, cholinesterase inhibitors such as donepezil have shown some evidence of mild slowing in the rate of progression in individuals with mild cognitive impairment, even when the major benefit is primarily symptomatic (Petersen et al. 2005). In short, it is not always possible to draw a clear distinction between symptomatic and disease-modifying treatments. From a patient's standpoint, this may be an academic debate, especially if a particular treatment is truly efficacious.

Why Are There Not More Prevention Trials?

Given the preceding discussion that early treatments in preclinical individuals may be necessary to see efficacious outcomes, an obvious question is: Why are there not more prevention trials? A simplified answer might include the following: lack of reliable early diagnosis, lack of well-defined patient populations, lack of good drug candidates, safety, cost and duration, appropriate end points, trial design, and subject selection (Golde et al. 2011).

First, how does one choose a treatment to test? What evidence of target engagement is necessary to embark on a prevention study? How does one obtain the necessary evidence: is demonstration of efficacy in symptomatic individuals required or, at the minimum, target engagement in the intended treatment group, such as Aβ removal from brain with anti-amyloid strategies in AD subjects?

Second, safety is an obvious issue when a number of individuals given the treatment may not develop the disease, so harm must not come to those individuals. Whether individuals with incurable diseases are willing to tolerate higher risks is unclear, especially when the disease course is measured in years to decades.

Third, given the necessary duration, trials are expensive. In this regard, there is an urgent need for surrogate biomarkers that track disease progression, preferably ones that reflect improvement in actual pathophysiological measures (see below). Further, is it enough to delay onset or will prevention be required?

Lastly, trial designs are not simple. An end point that demonstrates an absolute reduction in the number of new cases will be long and require a large cohort, even with enrichment strategies. Thus, methods to shorten the duration by assessing biomarkers in hereditary cases, as in two recently started AD trials (the Dominantly Inherited Alzheimer Network Trial[1] and the Alzheimer's Prevention Initiative[2]), may achieve the desired goals quicker than conventional clinical outcomes in a normal elderly cohort. An alternative approach was seen recently in PD treatment, where a novel delayed start design was used to test whether rasagiline could slow disease progression (Olanow et al. 2009). Conflicting outcomes between two different drug doses, however, led to justifiable skepticism about the possible effectiveness of this treatment. Perhaps the therapeutic intervention would have been successful with a longer delay (36 weeks) or a longer follow-up period (36 weeks). Regardless, even if there was slowing of clinical progression, it does not necessarily address whether the underlying natural history of the disease has been altered, as there was no way to evaluate pathologic status definitively.

Have We "Thrown the Baby Out with the Bath Water"?

If early treatment will indeed be more effective in neurodegenerative diseases given its protracted course, one might speculate whether the negative results from many Phase III trials imply that these approaches will be equally non-efficacious if given during the prodromal stages of the diseases, when only the barest changes are present in brain. The Alzheimer Disease Cooperative Study,[3] the largest academic clinical research organization in the United States,

---

1 http://www.dian-info.org/ (accessed April 14, 2014)

2 http://banneralz.org/research-plus-discovery/alzheimers-prevention-initiative.aspx (accessed April 14, 2014)

3 http://www.adcs.org (accessed April 14, 2014)

was formed in 1991. Since then it has conducted 23 drug trials but only the first one, tacrine, was unequivocally positive. The others (such as estrogen, NSAIDs, statin, B vitamins) were ineffective as AD treatment in individuals who suffered mild to moderate AD. Would any of these compounds be effective if they were given in the presymptomatic setting or, better yet, as a preventive measure prior to occurrence of the disease triggers? At a time when the development of a successful new drug is in the range of one billion dollars, might a careful reevaluation of previously failed drugs, given much earlier, be a fruitful exercise to undertake? For AD, active vaccination of AD subjects with Aβ as an immunogen led to detectable removal of amyloid from brain but no overt clinical improvement (Holmes et al. 2008). The trial was aborted due to development of encephalitis in ~5% of the research subjects. However, if a safer vaccine were to be developed, should the testing be conducted right away as a secondary prevention (i.e., following onset of first symptoms) or as primary prevention? The U.S. National Institutes of Health, in part with industry support, is now studying whether the repurposing of drugs or drug-like compounds might yield additional uses beyond their primary indications. The last decade witnessed considerable success in drug repurposing, such as metformin and thalidomide (especially newer analogs), in the treatment of various cancers. Even in neurological diseases, there is the example of amantadine, which is now used in PD treatment but was initially developed as an antiviral drug. Would a concerted repurposing strategy in neurodegenerative diseases bear fruit? Will old drugs continue to find new uses in brain diseases?

Good and Bad News from Animal Models

Limitations of Animal Models

At present, no single animal model is able to replicate all of the key phenotypes of any one neurodegenerative disease. While this group of diseases is often age-associated, aged animals (including large animals or nonhuman primates) unfortunately do not phenocopy the human disorders, although some animal species (e.g., bears, dogs, elephants, monkeys) naturally develop amyloid deposits with age. However, the amyloid deposits seen spontaneously in aged animals are not accompanied by the full spectrum of brain changes that occur in disease. There is much recent excitement over the development of transgenic nonhuman primates (Sasaki et al. 2009; Niu et al. 2014), but it is too early to tell how useful these models will be, given the expense, resources, and time required to develop and age these monkeys. Rodent models remain our mainstay, as advances in the manipulation of rodent genome has led to many mouse and increasingly rat transgenic lines that demonstrate one or more pathological features of the desired disease. Many of these lines, however, suffer from several common deficits, whose impact is difficult to assess.

First, most lines exhibit only a subset of the changes in the brain that are actually seen in human pathology. In AD, transgenic mice did not develop neurofibrillary pathology when only APP (with or without Presenilin encoding disease-associated mutations) was expressed in brain, and they generally showed a relative paucity of neurodegeneration, a necessary key feature of the disease. Therefore, it is felt that virtually all APP transgenic mouse lines represent early preclinical stages of AD pathophysiology (Zahs and Ashe 2010). In the increasingly complicated ALS-FTD spectrum of diseases, many aspects of motor neuron degeneration have been recapitulated in SOD-1 mutant mice, but the same has not been true in mutations identified in the TDP-43 or FUS/TLS (fused in sarcoma/translocated in sarcoma) genes (Ling et al. 2013; McGoldrick et al. 2013). In the latter cases, there is a conspicuous paucity of typical TDP-43 or RNA-binding protein FUS/TLS aggregates in transgenic animals, unlike in humans.

Second, most rodent lines expressed mutant gene constructs rather than the normal gene. This may be a nonissue in diseases such as trinucleotide repeat disorders, in which the cases are always genetic in origin. But there is always a question whether sporadic or nongenetic presentation of a disease follows similar pathophysiological pathways as the genetic form.

Third, most lines depend on overexpression to demonstrate the anticipated pathology. This is due partly to our own impatience to see the desired phenotype as early as possible, which has often been addressed by driving transgene expression to abnormally high levels, but also because expressing transgenes at endogenous levels often produces little to no phenotypes. Potentially, this can lead to artifacts related to overexpression, although there are now examples of gene duplication in AD and PD, as well as in neurodevelopmental disorders. Accordingly, in this context, it is sometimes difficult to ascertain which of the many abnormally elevated proteins represent the key triggers of brain pathology. For example, in spite of more than a dozen mouse and rat transgenic lines that overexpress various APP mutations using different promoters, it remains unclear whether overexpression of APP itself, rather than elevated Aβ levels, was responsible for some of the brain abnormalities. Unanticipated insights have been gained, such as the findings of epileptiform activity in some APP-overexpressing mouse lines which have also been seen in human cases (Vossel et al. 2013). Again, the situation is quite different in trinucleotide diseases because "knocking in" expanded polyglutamine repeats into the rodent endogenous huntingtin gene leads to many of the expected abnormalities and, further, expression is controlled by the endogenous murine promoter (Pouladi et al. 2013).

Fourth, virtually all rodent models bring with them considerable limitations, which investigators tend to minimize if not ignore. For example, the genetic background of the rodent strains can be a significant confound. Most investigators prefer inbred mouse strains, which reduces variance but can simultaneously also amplify unwanted or unrelated phenotypes. Whether using outbred lines is necessary or desired is unclear and may depend on a number of factors which cannot be controlled *a priori*, such as insertion site of the transgene, the particular mutant gene being expressed, toxic interactions between the transgene with certain mouse genetic loci, etc. (Krezowski et al. 2004). Even gender can be important. Many lines of transgenic mice have shown substantial gender variability, much more so than seen in human diseases when the mutant gene is not X-linked. Furthermore, the age span of rodents is vastly different from humans, and the assumption that reduced life span in these animal models reflects compressed human life span is not necessarily correct and may even be erroneous. Some cellular changes may result from disease triggers that require time to manifest themselves, and these stochastic processes simply cannot be compressed into the lifespan of rodents, canines, or even nonhuman primates. One example may be the infrequency of neurofibrillary tangles outside of aged human brains or in AD. Unlike amyloid deposits found in many aged mammals and even aged salmons (Maldonado et al. 2002), presence of tangle pathology is so infrequent that each time this is detected in a different aged animal species, it is noteworthy enough to warrant a publication (Rosen et al. 2008).

Fifth, are investigators biased in choosing which pathological change(s) should be present? Given that the rodent models rarely, if ever, phenocopy the human disease, one often "picks" one or more phenotype to select for when evaluating the animal models during their development phase. For example, are intracellular aggregates of α-synuclein, tau aggregates, or amyloid plaques the best measure of whether a particular transgenic mouse line is a faithful model of the disease? Or, for example, are abnormal animal behaviors reminiscent of a particular neurodegenerative disease an accurate representation of the human phenotype and derived from a similar cellular basis? Put another way, does an animal model that demonstrates some but not all of the expected brain pathology represent a reliable model to study disease pathogenesis and treatment response in support of the discovery and development of adequate therapies?

Lastly, while the situation is improving, many of the preclinical studies reported to date are not designed with sufficient statistical power. The standard practice taken by biostatisticians and clinical trialists to perform power calculations before commencing a study is often not undertaken or ignored in preclinical studies. One consequence of this is that results cannot be replicated by other labs because the initial result was spurious due to inadequate sample sizes or open-label in design, and this was not discovered until much later. Indeed, although a systematic study has not been done, there is evidence from other disciplines that many preclinical studies cannot be replicated, and this hinders the ability of translating candidate drugs into the clinic if the initial assumption of validity of the cellular target is erroneous (Ioannidis 2005; Prinz et al. 2011a). Even if these errors did not exist, there is always the need to be cautious when extrapolating or "jumping" from rodents to humans, given the significant evolutionary differences that have evolved between the two species, a key example of which is in innate and adaptive immunity (Shay et al. 2013).

Misalignment of Preclinical to Clinical Studies

Another unintended bias that has been introduced in preclinical studies is the initiation of treatment before or at the onset of pathology. This, in many ways, is representative of a primary or secondary prevention study, and not a treatment study, as typically carried out in humans where the pathology is well developed at the time of diagnosis. On the other hand, treatments given to mice after the onset of pathology are generally less effective. In this setting, extrapolating results from animal studies to humans is potentially flawed because trials in humans follow a different testing paradigm and would, at first blush, be predicted to be less efficacious. Furthermore, even when the preclinical and clinical studies are perfectly aligned, caution is always necessary given the suboptimal nature of the preclinical models, as discussed above. In this way, one major gap in our knowledge is the problem in translating misaligned preclinical studies in imperfect animal models to the clinical setting. Despite the technological prowess in rodent gene targeting technologies, the leap from rodents to humans is still fraught with dangers, and there should be renewed search for intermediate model systems to span this gap. In the end, until one "success" is realized from a treatment that is positive in both rodents (or another animal model) and humans under comparable testing conditions, the predictability of any of the myriad of animal models to neurodegenerative diseases remains completely unknown.

Biomarkers

Biomarkers for Accurate Diagnosis

Neurodegenerative diseases with a genetic basis are reliably diagnosed by genetic testing where available. In cases without a genetic cause, the situation is considerably more difficult as the usual "gold standard" is postmortem examination of the brain. Unlike in cancer, brain biopsies are not routinely conducted, and clinical diagnoses are prone to errors. Sensitive measures of brain atrophy by volumetric magnetic resonance imaging (MRI) are not specific for any one degenerative disease, as there are many causes of loss in brain volume. Fortunately, advances in imaging have been highly informative, especially from PET scanning, such as ^{18}F-Florbetapir to detect Aβ deposits or DaTSCAN (^{123}I-Ioflupane) to detect loss of dopaminergic innervation in PD. This was highlighted in the recent Phase III bapineuzumab trial where 36% of apoE4 noncarriers (compared to 6% in apoE4 carriers) diagnosed with mild to moderate AD were not positive on amyloid imaging, arguing against AD as the underlying cause of dementia (Salloway et al. 2014b). Whether routine use of amyloid imaging is warranted remains controversial in light of the fact that approximately 30% of cognitively normal individuals over 70 years of age are amyloid positive (Johnson et al. 2013). Does this mean that these are preclinical AD subjects, or does this represent incidental findings in a subset of cases whose clinical outcome is benign? Until the time when longitudinal follow-up reveals whether the cognitive status in these individuals will deteriorate in a fashion consistent with AD, a reliable noninvasive diagnostic marker is still wanting.

The desire to begin treatment at the earliest possible disease stage puts further challenges on the requirement for accurate biomarkers. If all we need is to treat individuals before the onset of clinical symptoms (i.e., in individuals who may have already developed changes indicative of disease, such as loss of dopaminergic markers suggestive of preclinical PD or presence of tau aggregates or amyloid deposits), then current biomarkers may well suffice in this setting of secondary prevention. This is because the full manifestations of neurodegeneration may only be present at the later symptomatic stages of disease. On the other hand, if we need to begin treatment prior to the onset of characteristic pathology or initiate treatment at a time when the disease triggers are just beginning, then this sets a much higher bar for success. Currently, only primary prevention studies will address this scenario as we do not know when the disease cascade, for example, begins but the accepted pathologic hallmarks have yet to develop (Sperling et al. 2011b). On the other hand, if secondary prevention is successful, then this speculation is moot.

Biomarkers for Comorbidities

Age is one of the biggest risk factors for neurodegenerative diseases, as are other chronic systemic illnesses, many of which have manifestations in brain. These illnesses lead to symptoms suggestive of neurodegenerative diseases, to complicate matters, and they occur concomitant with the underlying neurodegenerative disorders. For example, diabetes and cardiovascular diseases are highly prevalent in the elderly, perhaps even more so than AD, and they result in microvasculature changes which compromise neural function, including dementia. These comorbidities are important because they can aggravate the underlying neurodegenerative disease, complicate diagnoses, or confound response to treatment. In the latter case, agents targeting Aβ would not be predicted to improve microvascular damage due to chronic diabetes. This may mask some, and perhaps all, of the beneficial effects of drugs that have achieved appropriate target engagement. Therefore, an additional current gap is the need to diagnose these comorbid conditions accurately when they affect brain function and to take them into account in trial design and in measuring treatment responses.

Biomarkers as Readouts for Treatment Response

The need for sensitive and specific outcome measures in clinical trials is obvious. As mentioned, both the AD and PD research communities have recognized the importance of starting treatment as early as possible to maximize the potential beneficial effects of therapy. However, trials in presymptomatic individuals will need large numbers of subjects and take many years to complete without strategies to enrich subject selection and the use of validated surrogate markers. In this context, multiple sclerosis (MS) offers many instructive lessons. MS is a disease with heterogeneous presentations and unpredictable clinical course such that disease activity was incompletely or even inaccurately assessed by using only clinical measurements. Prior to MRI, results from treatment trials can be difficult to interpret in the face of unpredictable clinical course. Routine use of MRI in MS has had a major impact on diagnosis and in our understanding of underlying pathophysiology, and its incorporation as one outcome measure of disease activity has significantly altered MS treatment in a positive way (McFarland et al. 2002). Furthermore, the advent of disease-modifying therapies for MS has led to the acceptance that institution of treatment early in the disease course will lower long-term disability to the extent that treatment is now recommended even after only one single demyelinating event ("clinically isolated syndrome"), as this may delay or postpone development of subsequent multifocal disease (Goodin and Bates 2009).

Similarly, the AD community has recently incorporated imaging modalities into diagnostic criteria, at least in the research setting (Dubois et al. 2010; McKhann et al. 2011). The ongoing AD neuroimaging initiative, established in 2005, was designed to evaluate brain and other biomarkers that can accurately follow disease progression and monitor treatment response. The hope was that these imaging biomarkers would qualify as outcome measures by regulatory agencies. Significant data have been generated toward the former goal, one example of which is the demonstration that use of imaging end points can, in theory, significantly lower the sample size required to detect response in drug treatment trials (Holland et al. 2012). Nonetheless, whether any surrogate measurement provides the necessary sensitivity and accuracy awaits the arrival of the first successful disease-modifying trial.

Additional Considerations in Clinical Trials

From the above discussion, it is apparent that there is growing consensus that potentially disease-modifying treatments for neurodegenerative diseases have to be tested as early as possible to exact the best outcomes, preferably in the primary or secondary prevention stages (Holtzman et al. 2011). Beyond costs and safety, this shift to earlier testing places an increasing burden on discovery and validation of surrogate markers that reflect and guide treatment response and also reduce sample sizes. However, prevention trials raise a number of ethical issues that must be considered *a priori*. For example, an individual may not be aware or may not wish to be aware of his/her high-risk genetic status. The latter would happen if the subject is enrolled in a trial that specifically tests an individual with such a risk, e.g., apoE4 homozygotes or presenilin (*PSEN*) mutation. The Alzheimer's Prevention Initiative trial in Colombia tests an anti-Aβ antibody in individuals with a *PSEN* E280A mutation, implying that subjects enrolled in the trial must carry this familial AD mutation. To counter this assumption, a third arm testing placebo in control (nonmutation carrier) subjects was appropriately added so that one cannot assume that an enrollee carries the *PSEN* mutation (Reiman et al. 2011).

We must also consider whether giving placebo for long periods of time to at-risk individuals is acceptable. Early AD treatment trials were as short as six weeks. Now, it is not uncommon to see trials lasting up to two years or longer, especially in prevention trials. Since the 1950s, it has been the accepted standard practice to study response to drug treatment in randomized placebo-controlled clinical trials. Is there a place for a placebo alternative, both to reduce size and cost, but also to address the ethics of withholding potentially effective treatments that may delay disease onset in a prevention study? Can computer modeling, based on data obtained through careful monitoring of at-risk or early stage patients, provide sufficient statistical power to predict progression accurately in a way that will reduce the size of the placebo arm or even the heretical thought of eliminating placebo control in highly selected circumstances (Spiegel et al. 2011)? Can other improvements in clinical trial designs improve our current way of testing new therapies, both to maximize success and to address emerging ethical issues? Ultimately, there is perhaps little disagreement that we need to discover "new ways of doing business" as, to date, the standard practice has not been successful in developing effective treatments for neurodegenerative diseases.

Conclusions

This review has highlighted some of the challenges, gaps, and needs in current preclinical and clinical research toward finding effective treatments for various neurodegenerative diseases. It is by no means comprehensive, focusing rather on clinical targets, issues in preclinical research, and the need to find sensitive and accurate biomarkers to accompany clinical trials as they move into preclinical stages of the respective diseases. It is also admittedly biased toward AD research. But in many ways, current research has been hampered by the lack of a single truly successful treatment trial, symptomatic or disease-modifying, for any neurodegenerative disease, with the exception of motor complications of PD. The seminal success, when it comes, will hopefully provide new and critical insights as to both the right and the wrong directions taken in the past and to take into the future. Although the discussion has emphasized disease-modifying treatments, there is an important need for symptomatic treatments, not to mention the difficulties in drawing a clear distinction between these two therapeutic approaches.

Regardless of the initial presentation, virtually all neurodegenerative diseases have cognitive and/or behavioral symptoms in the later disease stages. These can be both burdensome for caregivers and costly for the healthcare system. For example, treatment of the motor complications of PD have improved immeasurably over the past decade and, as individuals survive longer with this disease, there is increasing recognition of the nonmotor complications of PD (e.g., behavioral and cognitive problems) that need to be addressed.

Further, we should rigorously ask whether there is a role for nonpharmacological approaches. In terms of preclinical studies, renewed effort should be placed on developing more sophisticated large animal models, such as transgenic minipigs or nonhuman primates. Can induced pluripotent stem cell technology pay handsome dividends in helping to understand pathophysiology, develop alternative disease models, or aid in drug discovery?

Continuing to look at other diseases will be important as they may offer practical lessons and roadmaps to impact research in neurodegenerative disease pathophysiology and treatment. Greater attention should be paid to neuroprotection, as these strategies may be applied to more than one neurodegenerative disease. In addition, are there ways to "de-risk" the drug discovery process, including clinical trial designs, to lessen cost, minimize unrealistic expectations, and maximize the value of pivotal but expensive Phase III trials? Finally, how should we approach polypharmacy? Given the complex nature of neurodegenerative disorders, it seems naïve to assume that targeting a single cellular pathway will achieve comprehensive therapeutic success.

If we are to obtain a clear understanding of the mechanisms, sequence, and disease progression of neurodegenerative diseases, supporting science must be better grounded. Genetic mutations have shed light on some of the contributory factors to disease, but not enough progress has been made. A concerted effort is needed between genetics, genomics, proteomics, circuit understanding, connectomics, etc., if we are to gain a clearer understanding of what can and does go wrong. These are daunting challenges, but they must be faced in view of the inevitable onslaught of age-associated disorders, such as neurodegenerative diseases, that is already at our doorsteps.

Acknowledgments

This work was supported, in part, by National Institutes of Health Grants AG20206, AG32179 and NS084324. EHK is co-inventor on a patent relating to AD therapeutics and has served as a consultant for Pfizer Inc. (including Wyeth Research), GlaxoSmithKline, and Roche.

First column (top to bottom): Mene Pangalos, Karl Broich, Lennart Mucke, Nancy Ip, Eddie Koo, Mene Pangalos, Pierluigi Nicotera
Second column: David Holtzman, Maria Grazia Spillantini, Eddie Koo, David Holtzman, Walter Koroshetz, Pierluigi Nicotera, Mareike Schnaars
Third column: Karoly Nikolich, Eliezer Masliah, Walter Koroshetz, Karl Broich, Mareike Schnaars, Maria Grazia Spillantini, Lennart Mucke

7

Neurodegenerative Diseases

What Is to Be Done?

David M. Holtzman, Karoly Nikolich,
Menelas N. Pangalos, Karl Broich, Nancy Y. Ip,
Edward H. Koo, Walter J. Koroshetz,
Eliezer Masliah, Lennart Mucke, Pierluigi Nicotera,
Mareike Schnaars, and Maria Grazia Spillantini

Abstract

Neurodegenerative diseases, including Alzheimer disease, Parkinson disease, and many others, lead to significant morbidity and mortality. As medical care for other disorders (e.g., cardiovascular disease and cancer) has improved and people are living longer, neurodegenerative diseases have become more prevalent because age is a major risk factor for most of them. There have been tremendous advances in our understanding of the scientific underpinnings of neurodegenerative diseases over the last thirty years. Nonetheless, with a few exceptions, very few effective treatments are available to delay the onset or affect the course of these diseases. This Forum brought together leaders in the field of neurodegeneration from different disciplines and tasked them with defining areas in need of more attention—areas where focused work is needed to reveal a better understanding of these disorders. A time frame of 5–20 years was the goal within which to develop more effective diagnostics and treatments. This chapter identifies eight areas to address:

1. Major specific pathologies and circuit dysfunction in neurodegenerative diseases need to be pinpointed over the life span, and dysfunctional circuits require identification.
2. Utilization of genetically well-defined patient populations will likely offer a better chance for therapeutic success.
3. Therapies affecting neurotransmitter systems and signaling pathways should be further explored utilizing defined patient populations and disease-affected nodes.

4. Better ways need to be developed to understand protein aggregation processes, from the formation of misfolded proteins to the critical clearance pathways that regulate their levels and toxicity, including understanding the mechanisms which underlie protein aggregate spreading in the brain as this could lead to novel therapeutics.
5. Better understanding is needed on the role of human apolipoprotein E (apoE), lipoproteins, and lipid biology under normal conditions and in neurodegenerative diseases.
6. To increase understanding of disease and facilitate drug/biological delivery, more information on the blood-brain barrier, the neurovascular unit, and other barriers separating CNS from non-CNS compartments is required.
7. The role of the innate immune system and other immune mechanisms that contribute to progression of neurodegeneration must be better understood.
8. Regardless of the upstream processes, it may be possible to activate neuroprotective mechanisms using defined factors, signaling pathways, or via cell-based methods.

With significant progress in each of these eight areas, substantial changes in the diagnosis and treatment of neurodegenerative diseases should be possible over the next twenty years. Given the current cost of these diseases to society and the increase in their prevalence with no additional progress, a major worldwide effort should be made to address these issues immediately, with the highest of priorities.

Introduction

Neurodegenerative diseases, including Alzheimer disease (AD), Parkinson disease (PD), dementia with Lewy bodies, frontotemporal dementia (FTD), amyotrophic lateral sclerosis (Xia et al. 2013), and Huntington disease (HD), have devastating effects on the nervous system that lead to progressive cognitive, behavioral, and motor dysfunction. With improvement in overall medical care, the importance of these age-related diseases to society is rising. In the United States and Europe, it is estimated that the cost of dementia care alone is already equal to or greater than that of cardiovascular disease and cancer (Hurd et al. 2013). The number of affected individuals and the cost of caring for them are expected to triple over the next forty years in the absence of effective disease-modifying treatments. Thus, developing a better understanding and treatment for these disorders has become paramount. While several useful treatments exist to treat PD and multiple sclerosis, no treatments have been shown to delay the onset of these diseases, and few have significantly affected disease progression. In our discussions, several major areas emerged which we believe are important to make progress over the next 5–20 years, if we are to begin to decrease the incidence of these diseases and develop better treatments for those affected. Eight specific topics are discussed in this chapter, which we believe are critical in making progress.

Need to Pinpoint Major Specific Pathologies and Circuit Dysfunction in Neurodegenerative Diseases over the Life Span and to Identify Dysfunctional Circuits

There is clear data that aggregation of specific proteins is a signature feature of most neurodegenerative diseases (Caughey and Lansbury 2003): amyloid-β (Aβ) and tau in AD; α-synuclein and Aβ in dementia with Lewy bodies; α-synuclein in PD; tau in various forms of FTD, progressive supranuclear palsy, corticobasal degeneration, and chronic traumatic encephalopathy; TDP-43 in amyotrophic lateral sclerosis (Xia et al. 2013); and huntingtin in HD. For most of these disorders, however, there is strong data that the proteins found in aggregates play major roles in disease pathogenesis. Importantly, in AD, PD, and HD (and probably in the other diseases), these protein aggregates begin to accumulate many years prior to the onset of symptoms—at least 15 years, for example, in the case of Aβ and AD (Jack and Holtzman 2013). By the time symptoms appear, there is already significant neuronal cell loss in certain regions of the CNS as well as synaptic, axonal, and dendritic loss and dysfunction. These findings argue that if effective treatments are going to be developed, identification of pathology and regional brain dysfunction *in vivo* prior to the onset of symptoms and signs of these diseases is critical to the initiation of primary or secondary prevention. In addition, specific neurological syndromes can result from different underlying pathologies. Thus, in individuals with different symptomatic stages of a disease process, identifying the actual underlying pathology is critical to an accurate diagnosis, enrollment in clinical trials, and ultimately treatment. To establish the exact pathological process that is taking place, as well as staging of the disease, an assessment of what is occurring in the CNS (e.g., by imaging, biofluid, and electrophysiological analysis) needs to begin during the asymptomatic period (e.g., at middle age), to identify preclinical disease, as well as once symptoms and signs emerge for both prognosis and to monitor disease progression (Holtzman et al. 2011; Sperling et al. 2011a). In addition to genetic factors, aging plays a big role in disease risk and thus needs to be accounted for in the assessment and interpretation of the biomarkers for neurodegenerative diseases. As testing of asymptomatic individuals for the presence of disease becomes more widespread, in both research and for clinical use, appropriate education and informed consent will be required.

In regard to the detection of protein aggregates, use of molecular imaging has been very useful. There are now good ways to detect the presence of fibrillar Aβ by both imaging and by measuring cerebrospinal fluid Aβ42 (Klunk et al. 2004; Fagan et al. 2006). In recent studies, tau aggregates are now detectable with molecular imaging (Maruyama et al. 2013; Xia et al. 2013). For the other major aggregates that occur in these diseases, including α-synuclein, huntingtin, and TDP-43, molecular imaging is not yet possible and should be developed. Cerebrospinal fluid or blood tests for the presence of any of these aggregates in the CNS are also not yet possible and would be very helpful. In addition, for both molecular imaging and biofluid assessment, it would be extremely helpful to be able to detect oligomeric forms of these proteins, which may be the forms that mediate toxicity (Benilova et al. 2012). There is strong evidence that many of these proteins that aggregate form different types of aggregates in different diseases. For example, aggregated tau has both different ratios of 3 or 4 repeat isoforms as well as different structure in AD versus progressive supranuclear palsy (Mandelkow and Mandelkow 2012). For each of the proteins that aggregate, development of methods to detect the specific type of aggregate as well as the cell type within which they are present (glia versus neurons) would help classify the specific molecular features that are present for each proteinopathy. Other types of biomarkers that would be potentially very helpful to assess and develop, to add to the information about protein aggregates, involve the status of microRNA in biofluids, given recent data showing their role in disease pathogenesis (Mendell and Olson 2012). It is also worth determining if fluid biomarkers or the earliest cellular triggers precede protein aggregation.

Important issues to consider in biomarker development include the necessity to develop reproducible, reliable, and cost-effective tests. Before widespread use, biomarkers need to be well validated in large populations in longitudinal studies. As effective treatments emerge, the risk and cost-benefit ratio of the test versus the treatment will emerge. The more invasive the test, the more demand there will be for having effective treatments. In non-Mendelian neurodegenerative disorders, it will be necessary to develop sensitive measures for population screening which can then be paired with more specific secondary tests to identify persons for treatment. A measure that is not as costly (e.g., a blood test) is certainly preferable, but there is already a precedent for more invasive measures (e.g., colonoscopy for colon cancer screening); dementia is certainly in a similar category in terms of burden of illness. Of course, a less invasive test will increase acceptance and utilization. Along those lines, assessments of plasma, serum, peripheral blood cells, and skin cells should be further assessed with the more advanced "–omics" and other technologies. Finally, as the cost of complete genome sequencing has reduced dramatically, the assessment of each person's genome in combination with the other imaging and biofluid tests could prove very useful: the integration of genetic and biochemical information from an individual offers greater predictive value.

In all the neurodegenerative diseases under discussion, specific regions and circuits in the brain are selectively vulnerable. It is thus critical to both utilize and expand upon current methods to detect synaptic and circuit dysfunction, as well as to pick up changes that are occurring prior to, during, and after the development of protein aggregation in the CNS, but still prior to the onset of clinically detectable symptoms and signs of disease. Both structural and functional connectivity magnetic resonance imaging combined with assessments of brain metabolism with FDG-PET are showing changes in some of these diseases in asymptomatic individuals (Jack and Holtzman 2013). It would thus be very useful to optimize both of these methods, especially their quantitative aspects. Data coming from the Human Connectome Project may provide new methods for quantitative tract tracing, which would prove useful (Van Essen et al. 2013). Using these and other imaging methods, it will be important to determine if any changes occur in brain regions that precede protein aggregation. This will permit a better understanding of pathogenesis as well as early detection of disease-related cause or dysfunction (e.g., precuneus/posterior cingulate cortex in AD; Greicius et al. 2004).

In addition to current neuroimaging methodologies for assessing brain function, assessment of synaptic and circuit function via other methods might provide important new ways to get at pathogenesis and diagnosis. For example, neuroimaging ligands that would permit detection of specific synaptic markers would be very useful. This has been developed for the dopaminergic system (Varrone and Halldin 2012) but not to similar degrees for other major neurotransmitter systems or for proteins present in excitatory versus inhibitory synapses. Other methods that would detect activity within specific neuronal populations would also be very useful. Knowledge from methods currently utilized in animals that allow for expression profiling of specific neuronal populations (e.g., BAC-TRAP; Heiman et al. 2008) may provide information to allow assessment of specific neuronal populations *in vivo*. Electrophysiological testing with qEEG during alert and sleep phases as well as other methods should be further studied to determine how early circuit and synaptic changes might be detectable. In many cases, current application of neuropsychological testing in humans may not be able to detect some of the earliest functional changes. Longitudinal cognitive assessments and cognitive tests in areas such as spatial memory and attention in presymptomatic populations might permit earlier changes to be detected (Sperling et al. 2011a).

To achieve the aspirational goal of disease prevention, a number of steps should be taken in preparation for the testing of promising therapies. To demonstrate prevention benefit, it is necessary to be able to identify clear, quantifiable end points, often of conversion from presymptomatic to symptomatic stages. It is also critical to be able to identify a cohort for study in whom there is high probability of conversion within the time period of a study, generally 1–5 years. Natural history studies are therefore required ahead of a trial to construct and validate end points and formulae that accurately predict expected time to reach these end points in a specific population. The PREDICT and PHAROS studies in Huntington disease provide a model for this exercise: imaging and neuropsychological changes have been followed over time in individuals with known cytosine-adenine-guanine expansion who are close to expected time of conversion to symptomatic HD (Biglan et al. 2013). Given the time frames needed to follow individuals in a prevention trial, the current patent lifetimes are a major disincentive and these need to be revisited by policy makers.

Utilization of Genetically Well-Defined Patient Populations May Increase Chance for Therapeutic Success

Once we are able to diagnose and assess neurodegenerative diseases better, which populations should clinical trials focus on over the next 5–20 years? The negative results obtained through the testing of drugs for the treatment of neurodegenerative disorders from the general non-enriched population, in which genetics or biomarkers have not been utilized for enrollment criteria, has been both disappointing and discouraging. This continuing approach reflects the standard conventional practice of testing of neurological drugs and market-driven decisions: to seek as broad an indication as possible, if successful, and to position the drug in ways that differentiate it from competing drugs. The recent testing of the anti-Aβ monoclonal bapineuzumab in apoE4 carriers and apoE4 noncarriers provided very useful information (Salloway et al. 2014a). It was one of the first times that a potential Alzheimer treatment was tested in genetically defined groups (by apoE status). If the treatment had been successful in one but not both subgroups, this would have set the precedent that a treatment is not intended for the affected population at large. While still novel in the Alzheimer field, this practice follows the lead from oncology, where the genetic makeup of a particular tumor can show predictable outcomes to targeted therapies. The recent start of several secondary prevention trials in individuals with various rare hereditary mutations in the *APP* and *Presenilin* genes (Moulder et al. 2013; Ayutyanont et al. 2014) as well as in apoE4 homozygotes continues this new trend to examine genetically defined populations to test mechanism-based therapies in subjects who may either benefit most from the treatment or offer insights not available from unselected populations.

Genetically well-defined populations can offer a clearer route to demonstrate pathophysiological mechanisms or provide proof-of-concept for the mechanism of drug action. For example, the recent positive response to either tafimidis (Coelho et al. 2012) or diflunisal (Berk et al. 2013) in the rare familial amyloid polyneuropathy, due to transthyretin accumulation in peripheral tissues, was a seminal success in a previously untreatable neurodegenerative disease. This trial offers a number of instructive lessons:

1. Use of objective measurements (in this case, tests of nerve and muscle function) can provide meaningful assessment of disease progression.
2. Treatment can be initiated after onset of disease and still demonstrate a positive effect.
3. Early treatment demonstrated greater benefit than when treatment was instituted later in disease course.
4. Results represent a dramatic proof-of-concept of the proposed underlying disease mechanism (inhibition of aggregation of transthyretin).

This benefit may, in the future, extend to similar success in the more common systemic amyloidosis, which does not have a genetic basis. Therefore, this example suggests that targeting a "purer" population can provide more clear-cut answers, albeit in a very select group of individuals. This idea could be expanded to a much larger nongenetic disease-affected group of individuals. Following the genetic enrichment lead, one can envision similar approaches applied to genetic forms of other neurodegenerative diseases, such as with the *LRRK2* and *GBA* mutations associated with PD and mutations in various genes in FTDs. In the latter, progranulin mutations are particularly interesting as disease pathophysiology appears to be related to haploinsufficiency (Cenik et al. 2012). Thus, one can readily envision boosting progranulin levels as a therapeutic approach.

Genetically well-defined groups offer quick validation of surrogate markers that track disease stages or progression. They also provide confirmation of validity of various end points used to assess treatment effects. Further, new biomarkers could be discovered that represent changes in brain circuits. The latter are particularly informative as they may represent new clinical measures of disease symptomatology useful for quantification of treatment outcomes.

One shortcoming of this approach is the rarity of individuals carrying various mutations. Mounting a successful biomarker or treatment study will require a consortium of national or international sites, a precedent established by the Dominantly Inherited Alzheimer Network (DIAN) study (Bateman et al. 2012; Moulder et al. 2013). In the DIAN study, a number of *PSEN* and *APP* mutations were included, since any one mutation is exceptionally rare. Fortunately, the different mutations all appear to induce perturbations in Aβ metabolism, aggregation, or production as the underlying disease pathophysiology. This may not be the case with other neurodegenerative diseases, such as in the ALS-FTD spectrum, where mutations have been identified in different genes that seemingly appear to affect very different cellular pathways (Perry and Miller 2013).

In AD, in addition to apoE4 homozygotes, trisomy 21 (Down syndrome) offers another genetically well-defined population to study Alzheimer pathophysiology that arguably is a resource that has not been taken advantage of fully. As these individuals invariably develop Alzheimer brain changes by the fourth decade of life and dementia in later years, they offer the possibility to correlate markers of disease initiation and progression as well as to provide another genetically defined group for treatment trials. Further, drugs affecting other neurotransmitter pathways, such as pentylenetetrazol acting as a GABA-A antagonist, may also provide cognitive benefit in these individuals, perhaps distinct from an effect on dementia (Fernandez et al. 2007). Of course, use of different instruments to assess cognition will likely be required, given that all individuals suffer mental retardation to varying degrees at baseline.

Finally, many neurodegenerative disorders have both genetic and nongenetic causes and are likely to share multiple disease mechanisms. Complicating this is the tendency for the diseases to be age related. As such, comorbidities are frequently present in the elderly, and these concurrent systemic disorders complicate drug testing. These comorbid conditions can contribute to disease risk, confound treatment outcomes (as these conditions can be nonresponsive to the treatments) and, in so doing, contaminate the treatment pool or decrease diagnostic accuracy. The use of genetically well-defined populations can overcome some of these complicating factors. On the other hand, some of these comorbid diseases (e.g., hypertension, cardiovascular disease, or diabetes) can be effectively treated, and treatment may well reduce risk of, for example, dementia or AD (Schrijvers et al. 2012). Thus, for multiple reasons, comorbidities should not be ignored.

Explore Therapies That Affect Neurotransmitter Systems and Signaling Pathways through Defined Patient Populations and Disease-Affected Nodes

Over the past few years, efforts have been directed toward the development of strategies to address the pathogenic mechanisms underlying diseases such as AD and PD. Strategies to prevent or remove accumulation of Aβ, tau, and α-synuclein are in the early stages of testing. Utilizing these approaches, a clear-cut success in humans has not yet been obtained, but newer trials of agents directed against these targets have identified potentially more appropriate populations of patients earlier in the course of disease, optimizing chances for success. The reason that previous trials of potential disease-modifying therapies targeting Aβ in individuals with mild to moderate dementia have not yet been successful may be due to a fundamental problem with targeting Aβ accumulation in patients with overt manifestations of disease—the intervention may be already too late. However, there is reason to believe that delaying disease is possible in individuals at early clinical stages of diseases such as AD (mild cognitive impairment, very mild dementia) or in asymptomatic subjects with preclinical AD. Progress has been made in targeting tau pathology in animal models (Boutajangout et al. 2011; Yanamandra et al. 2013), which may be applicable to a variety of neurodegenerative disorders. Similar strategies are thought possible for PD, and common approaches include antibodies to prevent the spreading of protein aggregates in brain.

In addition to potential therapies that target molecules involved directly in disease pathogenesis, several other possibilities address symptoms of disease in AD, PD, and other disorders, and have the potential to delay disease progression and improve quality of life, even in patients where disease is already advanced. Arguably, these can be considered "disease-modifying" therapies. Delaying or ameliorating disease progression can produce substantial benefits. Historically, symptomatic treatments that have been developed and widely utilized include cholinergic/dopaminergic agents. While useful in AD, cholinergic agents have only been moderately effective (Birks 2006). New formulations and combinations with other therapeutics might increase effectiveness. Testing early symptomatic treatments in patients without overt disease manifestations might offer better chances for success. In particular, delaying the onset of disease by targeting early synaptic dysfunction might significantly delay other major disease symptoms. Some of the major neurotransmitter systems in the CNS that have not been fully explored therapeutically in neurodegenerative diseases include agents to modify the glutaminergic, GABAergic, serotonergic, and noradrenergic systems. In addition, development of better cholinergic and dopaminergic agents is likely still possible. One potential example of a novel type of neurotransmitter/excitability modulator comes from the use of levetiracetam, a drug used for treatment of epilepsy. At lower doses, it has shown cognitive benefits in mouse models of Aβ-amyloidosis that have epileptiform features and brain network abnormalities (Sanchez et al. 2012). Here, the difference between symptomatic and disease-modifying therapies, if similar effects are seen in humans, would only become apparent when neurodegenerative processes progress beyond the manifestations of the initial synaptic dysfunction observed. Symptomatic treatments might also be directed to early symptoms which are dependent or independent of disease, such as sleep, which could optimize memory performance (Ju et al. 2014). Critical to a research roadmap is careful attention to the problems that are most important to quality of life in affected individuals and their caregivers. More effective means of ameliorating symptoms is a critical area for research. Symptomatic trials could be performed more quickly and potentially at lower cost. For instance in PD, nonmotor symptoms have been identified as most troublesome in the modern age of effective treatments for bradykinesia (Chaudhuri et al. 2006). Cognitive impairment, gait disturbance, and dyskinesias rise to prominence. In AD and HD, disorders of circadian rhythm and behavioral disorders frequently exhaust a family's ability to care for an affected individual at home. Pharmacologic or nonpharmacologic means of addressing these concrete problems should be a focus of research and could lead to success in the short term. It will also be important to determine whether responses to treatments are seen only in patients with specific characteristics, so as not to miss effects.

Another major area for investigation that can offer a strong therapeutic prospective is the modification of *APOE*. This area is largely underfunded and understudied, despite it being recognized as the major genetic risk factor for AD and other diseases (see below). Inflammation and the innate immune system offer another symptomatic/disease-modifying treatment target in both AD and PD. Inflammation appears to be involved in disease progression in AD, PD, ALS, FTD, and HD at different disease stages. Identifying the window of opportunity to intervene with drugs that reduce or prevent activation of infection-related inflammation or innate immunity could also prove fruitful. Finally, treatments that would enhance or restore plasticity including transcranial magnetic stimulation, deep brain stimulation, or nonpharmacological interventions (exercise, diet, sleep) have the potential to produce significant benefits and should be further explored.

Some limitations to the development of new symptomatic or disease-modifying therapies include:

- Current design of clinical trials (which have so far been considering individuals with overt disease over presymptomatic individuals with high risk).
- Poor outcome measures (e.g., better tools to assess cognition).
- Lack of multiple biomarkers: Better and more molecular, pathogenesis-linked biomarkers need to be identified and followed over time during, before, and after disease onset to detect dynamic changes. To this end, new useful biomarkers need to correlate with existing biomarkers related to disease progression.
- Funding agencies need to support areas of research which, although not widely represented within the scientific community, offer strong potential for therapies (e.g., *APOE*). Regulatory agencies need to be more open to the acceptance of preventive trials and new functional biomarkers.

Understand Protein Aggregation Processes and Spreading Mechanisms to Spur Novel Therapeutics

In terms of ultimately developing disease-modifying therapies for neurodegenerative diseases, it is critical to develop a better understanding of the normal metabolism as well as the pathophysiology of key proteins that form aggregates and accumulate in the brain. Most proteins implicated in the pathogenesis of neurodegenerative disorders can exist in a dazzling array of different assemblies and conformational states, and for most, if not all, it remains to be determined which conformation has the greatest impact on neuronal function and survival. To fill this important knowledge gap, there is an urgent need to develop methods that can reliably quantitate distinct assemblies of pathogenic proteins in brain tissues of humans and experimental animal models, ideally through noninvasive means. Relevant protein species include soluble oligomers, which have emerged as potentially critical but rather elusive culprits in a number of diseases (Benilova et al. 2012). With such methods, it should be possible to define not only how many of these forms of proteins are present but also how much the most pathogenic subspecies have to be reduced to improve significantly brain function at the synaptic, network, and clinical levels. This information could have an important impact on the design and interpretation of clinical trials. For example, none of the anti-Aβ trials completed to date have shown whether treatment significantly lowered Aβ oligomer levels in relevant brain regions, nor is it clear whether, and by how much, the levels of these assemblies would have to be lowered to improve synaptic function or slow down neuronal loss.

Based on a large body of published data, it is likely that pathogenic protein assemblies or conformations can trigger diverse molecular cascades (Hayden and Teplow 2013). The most proximal downstream mechanisms may involve both receptor-dependent and receptor-independent processes. This diversity of protein species and downstream pathways makes it unlikely that targeting any one of them by itself will have a major impact on the disease overall. Blocking the production of the abnormal proteins, inhibiting their aggregation, or enhancing their clearance may be more effective strategies. Far greater efforts appear to have been invested to date in inhibiting production of specific proteins, such as Aβ, as opposed to enhancing their clearance. Indeed, modulation of clearance mechanisms clearly deserves greater attention as this may be one of the main overall mechanisms by which aging leads to elevated risk of protein aggregation. Some of the abnormalities in clearance pathways which may give rise to disease and be targetable include degrading enzymes, chaperone-mediated clearance, autophagy, uptake into glial cells, better utilization of interstitial fluid or cerebrospinal fluid flow pathways, and transfer across the blood-brain barrier into the circulation. Determining which of these mechanisms are most effective, most druggable, and most affected by disease could result in the identification of novel entry points for therapeutic interventions.

Closely related to this quest is the issue of whether the spread of pathogenic proteins, such as tau and α-synuclein, from one cell to another, might involve unique processes that are amenable to specific therapeutic interventions. In cellular and animal models, cell-to-cell spread appears to involve converting the normal, non-aggregated form of a protein into an aggregated form with high β-sheet content by interacting with an already seeded/aggregated form of the same protein (Frost and Diamond 2010; Lee et al. 2014). In the case of tau and α-synuclein, this process appears to move from cell to cell in what some scientists call a prion-like manner. However, to what extent this type of spread via template seeding contributes to the progression of neurodegenerative disorders in both animals and humans remains uncertain and is currently under intense investigation. One condition that may be a particularly good example to test more extensively is chronic traumatic encephalopathy, where there appears to be spreading of tau aggregates over many years between different brain regions (Stein et al. 2014). There also is a need to understand the mechanisms by which pathogenic proteins impair the function of brain cells and the networks they form. In particular, in AD and dementia with Lewy bodies, there is strong evidence that Aβ accumulation leads in some way to further accumulation and spread of tauopathy in AD (Musiek and Holtzman 2012) and synuclein in dementia with Lewy bodies (Masliah et al. 2001). The mechanisms (e.g., cellular signaling pathways) that underlie these effects are unknown and if sorted out, could lead to new ideas for treatment. From a treatment standpoint, it would also be valuable to complement ongoing efforts to detect the accumulation of abnormal proteins in the brain with intensified efforts to develop and incorporate novel methods into clinical trials to detect the emergence, progression, and potential therapeutic reversal of synaptic and network dysfunction.

Understand the Role of ApoE, Lipoproteins, and Lipid Biology under Normal Conditions and in Neurodegenerative Diseases

Of the common diseases in humans, one of the strongest genetic risk factors is the link between *APOE* and AD. ApoE4 dose dependently increases risk for AD and cerebral amyloid angiopathy whereas apoE2 protects against AD (Strittmatter and Roses 1996; Kim et al. 2009). *APOE* genotype may also modulate risk for other CNS disorders, such as recovery after head injury, stroke, and other forms of neurodegeneration (Mahley and Huang 2012; Wolf et al. 2013). Despite the strong effects of *APOE* genotype on AD, our understanding of exactly how apoE has these effects is incomplete, and this has limited development of ways to target apoE-related pathways. Despite the importance of apoE in AD risk, there is a paucity of academic and industry-related efforts and investment in this important topic. Significantly more effort and resources need to be prioritized here, given the strong impact of this gene which appears to be mediated by the protein product.

There are two general categories by which apoE proteins appear to modulate AD and neurodegeneration: Aβ-dependent and Aβ-independent. For Aβ-dependent mechanisms, there is overwhelming evidence that apoE isoforms influence whether and when Aβ aggregates and accumulates in the brain and contributes to its toxicity (E4 promotes Aβ accumulation, E2 retards it relative to E3) (Holtzman et al. 2012). The general mechanisms appear to be due to the ability of apoE to interact directly with Aβ to influence its aggregation as well the ability of apoE to influence soluble Aβ clearance. In terms of which receptors modulate apoE-related Aβ clearance, this has not been entirely resolved. However, there is strong genetic and biochemical evidence that increasing apoE lipidation (Koldamova et al. 2010) as well as decreasing levels of apoE3 and apoE4 can decrease Aβ accumulation (Kim et al. 2011; Bien-Ly et al. 2012). ApoE lipidation may be targetable by liver X receptor and retinoid X receptor agonists (Cramer et al. 2012), although side effects from this approach would need to be overcome. Decreasing apoE levels can be accomplished by increasing activity of apoE receptors, such as low-density lipoprotein receptor (LDLR) and LDLR-related protein (LRP1) (Holtzman et al. 2012). There may be ways to increase function of these receptors using a biological or small molecule approach. Knocking down apoE3 and apoE4 via an antisense oligonucleotide approach should also be considered. Experiments with anti-apoE antibodies show promise in decreasing Aβ deposition (Kim et al. 2012) and should be further explored and developed. Inhibiting the interaction between apoE and Aβ is another target that has been attempted successfully in animal models, using peptides that might be approachable with small molecules. In addition to apoE's effect on Aβ, it may also influence tauopathy via direct interactions with tau or via an indirect mechanism. This needs further exploration as a biological mechanism as it offers new ways to target this interaction.

In regard to other non-Aβ-related mechanisms, apoE has been shown in various *in vitro* and animal models to have the following effects, which may be relevant to neurodegeneration (Mahley and Huang 2012; Wolf et al. 2013):

- effects on neurite outgrowth and synaptic plasticity,
- formation of apoE fragments in neurons that can be toxic,
- effects on mitochondrial function,
- effects on neuroinflammation, and
- effects on reverse cholesterol transport and lipid scavenging.

Further exploration of these effects *in vivo* in different models of neurodegeneration and in different laboratories will be critical to determine how important each mechanism is. Investigators at the Gladstone Institute (Chen et al. 2012) suggest that small molecules act as "structure correctors" to convert an apoE4 structure close to that of apoE3 or apoE2. This should be further explored in regard to the influence on both Aβ- and non-Aβ-dependent effects of apoE.

Two areas of apoE biology should be fruitful: further characterization of apoE structure and neuroprotective effects of apoE2. Although the structure of apoE has been determined, it was done in the nonlipidated state. Since almost all apoE *in vivo* is lipidated, understanding apoE structure on high-density lipoprotein-like lipoproteins, as it occurs in the brain, should assist in modeling its various interactions. ApoE2 is strongly protective against AD, yet our understanding of the mechanism underlying this effect is poor. ApoE2 results in lower Aβ deposition; increased expression of apoE2 in the brain (via a viral vector approach) decreases Aβ deposition (Hudry et al. 2013; Dodart et al. 2005). Further experiments utilizing such an approach should be pursued in animal studies and potentially taken into humans if successful.

To Facilitate Disease and Drug/Biological Delivery, More Information Is Needed on the Blood-Brain Barrier, Neurovascular Unit, and Other Barriers Separating CNS from Non-CNS Compartments

In AD, PD, NeuroAIDS, and in other neurodegenerative disorders, there is evidence of dysfunction of the neurovascular unit (NVU), which is composed of neurons, astrocytes, pericytes, and endothelial cells that surround blood vessels in the brain (Iadecola 2004). This dysfunction may contribute to CNS dysfunction as well as accumulation of neurotoxic molecules. It has been suggested that NVU abnormalities might contribute or even trigger the neuropathology of AD and other diseases (Zlokovic 2008). The NVU plays a key role in regulating the permeability of the blood-brain barrier, cerebral blood flow, and the activity of neuronal circuits. Better understanding of how the NVU and blood-brain barrier normally function as well as how alterations in the NVU are involved in clearance of misfolded proteins, neuroinflammation, and other aspects of degeneration will provide important new insights. In addition, the brain's intrinsic drainage pathway, the recently named gliolymphatic system (Iliff et al. 2012), needs to be better understood both under normal and disease conditions, in relation to all the issues just mentioned. How the neurons are interacting with endothelial cells and astroglia at the NVU is very important and a key area for further research. Clarifying the nature of the alterations in the NVU in neurodegeneration is likely to be important in the elucidation of treatments designed to target the NVU as well as in understanding how this alters the traffic of drugs and biologics into and out of the CNS. The concept that the blood-brain barrier is damaged in CNS disorders has been around for a while, but whether this contributes to disease pathophysiology remains controversial. Still, recent evidence has shown that AD changes affect pericytes and the blood-brain barrier (Zlokovic 2008), because vascular amyloid is a known association that leads to hemorrhage and potentially to impaired autoregulation. In this context, there is a need for better tools to quantify blood-brain barrier leakage in both animals and humans.

The traditional drug release systems that deliver drugs systemically fail to transport biologics and drugs effectively into the brain. As most therapeutic agents for neurodegenerative disorders need to likely reach the CNS to be effective, approaches that will selectively target the CNS more efficiently need urgent development. Growth factors, neuroprotective and anti-inflammatory peptides, antibodies directed against neurotoxic protein aggregates, and enzymes are some examples of biologics with potential application in the treatment of neurodegenerative disorders. Major challenges when considering the delivery of such biologics into the CNS are how to target them specifically to the brain. This has the potential to reduce off-target effects by lowering the drug dose as well as by other mechanisms. Trafficking of proteins and peptides into the CNS involves the interactions of the macromolecules of interest with receptors located on the luminal and/or abluminal surfaces of the brain endothelial cells. Both endocytosis and transcytosis play key roles in the trafficking of such macromolecules across the blood-brain barrier.

Multiple strategies deliver proteins of therapeutic interest into the CNS, including the use of peptides or antibodies that bind to endothelial cell receptors as well as nanoparticles, liposomes, and other container-type of carriers and viral vectors (e.g., adeno-associated virus, rhabdovirus). All have advantages and disadvantages. In the case of charged nanoparticles, ensuring specificity has been a challenge. With viral vectors, questions have been raised as to side effects and long-term toxicity. Overall, each strategy needs to be categorized according to specificity, capacity, mechanism of action, and application. The so-called "one-fit-for-all" approach is unlikely to work; thus we recommend that each strategy be considered independently and specific applications developed. We also recommend that combinations of strategies to carry biologics into the CNS be considered. For example, nanoparticles can be combined with selected peptides to improve specificity. Given the potential complications with rabdoviruses and some nanoparticles, other strategies have been considered, in particular those involving receptor-mediated delivery.

The main receptors that mediate endocytosis and transcytosis of antibodies, peptides, and proteins across the blood-brain barrier include LDLR, LRP1, transferrin receptor (TfR), Fc receptor (FcRn), as well as the insulin receptor (IR) and insulin growth factor-1 (IGF-1) receptor. Compared to other strategies, the use of targeted small peptides or antibodies to transfer macromolecules into the CNS has the advantage of being minimally invasive. One of the most advanced strategies involves the use of monoclonal antibodies against TfR or against IR. An interesting application of this strategy for the potential treatment of AD has been the development of a divalent antibody in which one arm of the antibody targets TfR (to facilitate blood-brain barrier permeability) while the other arm targets β-secretase-1 (BACE1), a protease involved in the processing of amyloid precursor protein (APP) into Aβ (Atwal et al. 2011). These types of antibodies have been shown to facilitate the blood-brain barrier with great efficacy and to reduce Aβ levels in both mice and nonhuman primates. Antibodies against TfR have also been used to enhance the trafficking of antibodies in which one arm targets Aβ. This approach has been tested in APP/PS1 transgenic models that develop Aβ deposition and was more effective than using a standard anti-Aβ antibody in reducing Aβ levels (Niewoehner et al. 2014). More recently, utilizing phage display, two single domain antibodies, FC5 and FC44, were identified that display high permeability across the blood-brain barrier (Haqqani et al. 2013). Both have been proposed for development as vectors for brain delivery. Overall, utilizing this or analogous approaches offers great hope that one can increase the entry of proteins such as antibodies in the brain that will be able to bind to and target specific proteins and cells in the CNS.

Another approach is to use selective targeted small peptides that bind lipoprotein receptors present on the surface of endothelial cells in the blood-brain barrier. For example, in recent years small peptides derived from apolipoproteins (B and E) that bind LDLR have been shown to have the capacity of increasing peptides and antibody entry into the CNS. Through the fusion of the LDLR-binding domain of proteins, such as apoB (38 amino acids) and apoE (19 amino acids), to cargo proteins, therapeutic promise has been shown in various neurodegeneration models for AD, PD, Gaucher, and Sly disease. For example, by coupling the LDLR-binding domain of apoB to neprilysin (a metalloprotease that degrades Aβ), it has been shown that the systemically administered fusion protein crosses the blood-brain barrier and accumulates in the neocortex and hippocampus, thus reducing Aβ levels in the brain as well as behavioral deficits and neurodegeneration (Spencer et al. 2011).

Other molecular Trojan horses, which use a lipoprotein receptor, are peptides derived from the apoE sequence, Angiopep-2, and receptor-associated proteins that target LRP1 (Demeule et al. 2008). This lipoprotein receptor has been shown to mediate the endocytosis of Aβ peptides across the blood-brain barrier (Zlokovic 2008). Aprotinin, a pancreatic trypsin inhibitor that contains the Kunitz protease inhibitor (KPI) sequence, is a ligand for LRP1 and can cross the blood-brain barrier. Aligning the sequence of aprotinin with the KPI domain, a family of peptides (named Angiopeps) was developed. Angiopep-2, in combination with micelles, has been tested to increase the CNS penetration of drugs such as the antifungal amphotericin B (Shao et al. 2010). The amphoterocin B-incorporated, angiopep-2 modified micelles showed highest efficiency in penetrating into the CNS. Another peptide has been tested as a receptor-specific carrier, namely cross-reacting material 197 or CRM197, which is a nontoxin mutant of diphtheria toxin. CRM197 increased pinocytotic vesicles and vacuoles in brain microvascular endothelial cells and enhanced caveolin-1 protein expression in brain microvessels (Wang et al. 2010b). Regarding the apoE strategy, peptides of this molecule have been fused to α-L-iduronidase (IDUA), a lysosomal enzyme being evaluated for efficacy in a mouse model of mucopolysaccharidosis type I (Miller et al. 2013b). After 5 months of treatment, apoE-IDUA was found to have normalized brain glycosaminoglycan and β-hexosaminidase in an mucopolysaccharidosis type I model (Wang et al. 2013a). These strategies offer great promise as they are further developed. Targeting receptor-mediated transport via receptors such as TfR has potential side effects. For example, long-term use can theoretically trigger deficits in iron transport. Thus, due to the potential side effects for chronic use, other receptors should be considered. For example, LRP1 may have a higher capacity, better trafficking abilities, and be more efficient than TfR. Some controversy has emerged over whether LRP1 is present in endothelial cells in the blood-brain barrier; further investigation is thus needed.

Nonreceptor-mediated approaches are also being considered for targeting the CNS. The use of the HIV protein Tat as a carrier has been tested alone as well as in combination with liposomes to deliver antibodies, peptides, and growth factors. In addition, approaches that involve nanoparticles and liposomes are being tested to deliver small molecules into the CNS. In terms of delivering biologics to the brain, it is important to consider how the biologic will be delivered to the correct compartment, how specific brain regions can be targeted, and how the trafficking of the biological can be measured.

It will be crucial to develop imaging tools that allow measurement of the trafficking of the therapeutically administered bioactive proteins more effectively. Some work has shown that selectively tagged (coded) methods using RNA sequences can be used to target neuronal populations selectively. Once a therapeutic bioactive protein is targeted to the correct brain region and cellular population, the next problem is for the protein to find the correct cellular compartments. Using molecules that target a receptor, such as LRP1, has a unique advantage: once the protein is transcytosed, it is targeted to the lysosome, which may be useful for specific metabolic diseases. This approach has also been used with antisense oligonucleotides tagged for delivery into the CNS to treat neurodegenerative disorders.

Other questions to consider concern the advantages and disadvantages of invasive versus noninvasive approaches. Is it preferable to deliver a biological or medicinal therapy into one specific brain region site or diffusely into the brain? For example, for some years, intranasal administration has been considered a potential approach for getting certain compounds into the CNS effectively. An intranasal trial of insulin in AD is currently in progress: the advantage here is that it appears to influence solely the CNS and does not change blood glucose levels. Is intranasal administration still a valid approach for drug delivery? Alternatives include considering whether compounds can be delivered into the brain more effectively in areas lacking a blood-brain barrier (e.g., hypothalamus). It may also be possible to utilize cell-mediated delivery of material into the brain via cells such as monocytes to deliver drugs and biologics in the CNS. Finally, direct CNS delivery via the lumbar or ventricular cerebrospinal fluid as well as into the brain parenchyma via convection-enhanced delivery under appropriate circumstances may be very effective and, while more invasive, may still provide a good risk-benefit ratio. This approach is being explored for gene delivery via adeno-associated virus (AAV) as well as administration of antisense oligonucleotides (Miller et al. 2013b). Overall the NVU, blood-brain barrier, and gliolymphatic systems are being actively investigated under normal conditions and in the setting of neurodegeneration. Clearly, more work needs to be done in this area.

Developing a Better Understanding of the Role of the Innate Immune System and Other Immune Mechanisms That Contribute to Progression of Neurodegeneration

The occurrence of microglial activation in AD and all of the neurodegenerative diseases has been observed for many years. Over the course of these diseases, microglia change their phenotype, retract their ramified processes, and adopt an amoeboid cell shape. These morphological changes are accompanied by the accumulation of microglia in areas of injury as well as in and around AD amyloid-β plaques. Microglial activation is accompanied by secretion of pro-inflammatory cytokines (Meyer-Luehmann et al. 2008; Prinz et al. 2011b). However, questions remain as to whether and how the main innate immune effector cells, microglia, contribute to disease progression. In addition to activated microglia and their potential role in neurodegenerative diseases, there is a striking increase in certain complement proteins. This is interesting given the emerging role that complement proteins such as C3 appear to play in synaptic pruning during development (Schafer et al. 2012), suggesting that increased complement in disease may contribute to synaptic damage. This area needs further exploration as complement may prove to be an attractive target for intervention.

Further highlighting the role of microglia, specifically in AD, are recent genetic data. Genome-wide association studies as well as whole genome sequencing have led to the identification of additional confirmed genetic risk factors for AD: among them are CD33 (Bertram et al. 2008; Hollingworth et al. 2011; Naj et al. 2011) and TREM-2 (Guerreiro et al. 2013; Jonsson et al. 2013). Both receptors are specifically expressed by microglia in brain. These may offer future specific insights into both mechanism and treatment opportunities.

There are a number of additional questions that will be important to ask regarding the role of the innate immune system and microglia in neurodegeneration:

- We need to understand the inefficient role of microglia in amyloid clearing in AD. While antibodies to Aβ and other stimuli can enhance microglial clearance of Aβ *in vivo*, can other methods be utilized to activate this pathway?
- How do molecules released by microglia, such as cytokine and chemokines, influence CNS function? These may be good therapeutic targets.
- What is the time course of inflammation? When is the right time to intervene? Magnitude of the anti-inflammatory intervention as well as the specific inflammatory target will be important to resolve.
- Recent evidence suggests the importance of microRNAs in modulating microglial function. Can this be better understood to develop therapies?

In the healthy brain, microglia continuously scan their microenvironment for pathogens and inflammatory stimuli by sending out membrane processes. This surveillance activity is thought to support tissue homeostasis. Moreover, microglia play a role in monitoring synaptic activity and assist in synaptic pruning, perhaps via complement receptors which they express (Prinz et al. 2011b; Tremblay and Majewska 2011). Recent animal data show that the chronic accumulation of Aβ within the brain drives inflammation. This effect appears to be mediated by the activation of the microglial NALP3 inflammasome and the subsequent release of interleukin-1β that leads to the shift of microglia toward a pro-inflammatory M1-type phenotype (Heneka et al. 2013). In turn, Nalp knockout animals in the APP/PS1 model background show reduced interleukin-1β activation as well as enhanced Aβ clearance and a shift of the microglial phenotype toward the anti-inflammatory M2-like phenotype (Heneka et al. 2013). Moreover, it was shown that inflammation not only results in a gain of microglial function, the protective function that microglia exert in tissue is impaired in mouse models of amyloidosis (Krabbe et al. 2013). These findings raise the following questions:

- Is it possible to shift microglia back toward a phagocytic, surveillance M2-like phenotype that would improve neurodegeneration and postpone disease onset?
- If we understand the transition between phenotypes, what are further trigger events that could be targeted?
- Is NALP3 a good target for intervention?
- How do we target microglia? Can nanoparticles be used to deliver therapeutics?

Other important questions to be addressed in the future in this area include:

- In addition to microglial activation, astrocytes are activated in neurodegenerative diseases. What is the role of this activation? Do astrocytes contribute to the pathology? Can this activation be harnessed in a protective fashion?
- Is there any role for an adaptive immune system in neurodegenerative disease outside of multiple sclerosis? If so, how does this occur?

Since most of the mechanistic data on the effects of the immune system in neurodegeneration has been generated through mouse models, it will be important to understand the differences in innate and adaptive immunity of mice and humans.

Neuroprotective Mechanisms through Defined Factors, Signaling Pathways, or Cell-Based Methods

Extensive dendritic atrophy and synapse loss correlates with the severity of cognitive and memory impairment in neurodegenerative diseases. Neuronal damage (e.g., the loss of synapses) is believed to be the pathophysiological consequence of neurotoxic agents, such as oligomeric and other forms of Aβ, tau, and α-synuclein in AD, tauopathies, and synucleinopathies. Since these early deficits are likely reversible, strategies targeting the stabilization and potential restoration of dendritic arbors and spines are expected to modify disease progression. Based on current understanding of the pathways and regulatory mechanisms that are disrupted by such neurotoxic agents, a variety of therapeutic approaches may activate neuroprotective mechanisms in the brain.

Neurotrophic factors (e.g., BDNF, NGF, GDNF, Neurturin, IGF1, and BMP9) have been shown to be supportive of dying and injured/stressed neurons in animals. Some of them are also effective in reducing Aβ and plaque burden in animal models of AD. However, direct delivery of neurotrophic factors into the human brain has proved to be very challenging. To date, two main approaches have been used: (a) the factors were infused into brain as recombinant proteins and (b) viral vectors carrying the genes encoding the trophic factors were directly injected into specific brain areas. Mixed results have been obtained in human studies, and there is much to learn from these past experiences if we are to improve future clinical trial design. For example, GDNF was found to be beneficial for PD in a small, open-label study (5 patients). However, Amgen conducted a systematic study (34 patients) where no clear activity was found (Lang et al. 2006). It is important to improve the technology for delivery of GDNF and other growth factors. Importantly, in these and other trials, the patients treated had advanced clinical disease. Conducting studies at earlier ages, when there is less cell death, would seem more likely to yield benefits. More recently, gene therapy has been used to treat PD by introducing the neurturin gene via a viral vector approach. A Phase II trial by Ceregene did not show a statistically significant benefit; however, further work with this type of approach seems warranted (Bartus et al. 2014). AAV-NGF is also being tested in AD patients and appears to show some promise (Rafii et al. 2014). In addition to direct brain delivery, intranasal trophic factor administration has been tested for decades using NGF, BDNF, FGF-2, IGF-1, and insulin. Outcome measures using nasal delivery of trophic factors has been modest and acceptance limited and controversial. Overall, optimization of protein delivery, better understanding of pharmacokinetics/pharmacodynamics, and beginning trials early in disease course seem to be critical factors for future trials. Further, implementation of larger trials with such approaches need to be justified by careful Phase I results that show strong target engagement.

Attempts have also been made to design or identify small molecules from screening that activate neurotrophic factor receptors, such as TrkB receptor, or their signaling pathways (Obianyo and Ye 2013). For some small molecules, efficacy data were not reproducible, calling for a need to use unified protocols. While the ability of these small molecules to mimic the neurotrophic factors remains to be validated, some have been orally effective and show promise as a basis to develop novel therapeutics for neurodegeneration. Rigorous pharmacological profiling of these small molecules is warranted. In addition to classical growth factor signaling, other intracellular signaling pathways have the potential to be exploited to prevent axonal, dendritic, and synaptic degeneration. These include pathways involving Nmnat and the sirtuin pathways (Araki et al. 2004). Further exploration of their role in preventing degeneration in the CNS seems warranted.

Cellular transplantation offers an alternative to provide a source of growth or other trophic factors as well as a potential cell replacement strategy. Further experiments to sort out how iPS cells differentiated into neurons or glia might best serve in this role should be explored. Use of neural precursor cells that can differentiate into specific neural cell types, such as interneurons, may prove useful in that they can migrate to the appropriate layers within regions such as cortex as well as modulate regional neuronal activity.

The relevance of nonpharmacological manipulations (e.g., physical exercise, sleep, cognitive activity, and specific lipids) to neuroprotection needs further exploration and may be better understood as neuroprotective strategies.

Identification of genes/factors, such as apoE4 homozygotes, which allow high-risk individuals to escape disease, might provide insight into new neuroprotective mechanisms (Jonsson et al. 2012). In addition, the use of iPS cells coupled with genomic analysis of such subjects could provide a way to explore this area.

New Tools to Drive Future Therapies

8

How Might We Get from Genes to Circuits to Disease?

Tobias Kaiser, Yuan Mei, and Guoping Feng

Abstract

Recent advances in the identification of risk genes for psychiatric disorders have set the stage for functional interrogation of disease related-circuits and underlying mechanisms of pathophysiology. Still, investigators face significant challenges: (a) hundreds of genes may contribute to pathogenesis of a given disorder (polygenicity and genetic heterogeneity), (b) risk alleles may only cause the disease in combination with other factors (reduced penetrance), and (c) commonly used rodent models may have significant limitations in studying psychiatric disorders, due to differences in brain structure and function between rodents and humans.

To address these challenges, high-throughput functional assays should be developed in combination with induced pluripotent stem cells (iPSC) technology and novel genome-engineering technologies, as these will help identify common pathways and mechanisms onto which multiple risk genes may converge. To overcome limitations of current animal models in psychiatric research, novel genome-editing technologies provide an opportunity to generate better animal models (e.g., the common marmoset) to dissect disease-relevant circuit dysfunction. Finally, powerful new tools (e.g., CLARITY, dense neural circuit reconstruction, and optogenetics) may help identify and test the relationship between distinct circuit defects and abnormal behaviors observed in these animal models. Such approaches are needed to span the gap between emerging genetic information and the symptomatic description of neuropsychiatric disorders.

Introduction

Large-scale genetic studies of psychiatric disorders are identifying an increasing number of risk genes for psychiatric disorders, such as autism and schizophrenia (Craddock and Sklar 2013; Giusti-Rodríguez and Sullivan 2013; Kendler 2013; McCarroll and Hyman 2013; McCarroll et al. 2014; Owen 2012a,b; Rosti et al. 2014). For the first time in history, it is now possible to study, with high confidence, a large number of genes associated with psychiatric disorders. Although genetics is not the only factor that contributes to the development of psychiatric disorders, it is one of the most important factors, given the high heritability of these diseases (Cardno et al. 1999). Thus, the recent discovery of high-confidence risk alleles that are being mapped to disease-associated genes provides an unprecedented opportunity for neuroscientists to dissect neurobiological mechanisms of psychiatric disorders.

The genetic data, however, also present several challenges. The first concerns polygenicity and genetic heterogeneity. Current data suggest that hundreds of genes may cause or contribute to the pathogenesis of each common psychiatric disorder. An important step is therefore to identify potential converging pathways and mechanisms that are shared by the multiple risk alleles. The second challenge is the limited penetrance of each risk allele. Each risk allele has a small effect and acts in combination with other genetic and nongenetic risk factors to produce disease symptoms. Thus, identifying the multiple genetic variants and nongenetic risk factors that render a particular risk allele pathogenic will be critical to the successful modeling and subsequent understanding of the disorder. Another challenge is that genetic data do not directly provide cell type-specific information. Systematic mapping of cell type-specific effects of risk alleles on gene expression and function will be essential for dissecting their pathogenic roles. Finally, although protein-coding DNA sequences are fairly conserved through evolution, regulatory sequences are far less conserved. Moreover, it is difficult to model deficits of higher brain function in animals such as rodents, due to the inherent differences in brain structure, function, and physiology between species. The recent development of new genome-editing technologies enables genetic manipulation in many species (Boch et al. 2009; Cong et al. 2013; Mali et al. 2013b; Zhang et al. 2011) and thus may lead to the generation of better animal models for psychiatric research. In addition, other new technologies for mapping and probing circuit connectivity and function will greatly facilitate the study of neural circuit mechanisms of psychiatric disorders.

From Risk Genes to Converging Pathways and Mechanisms

Although psychiatric disorders are genetically heterogeneous, each disorder has certain defining behavioral abnormalities, such as social communication deficits in autism and psychosis and cognitive impairment in schizophrenia. This suggests that heterogeneous genetic factors converge on certain common mechanisms controlling these behaviors. Bioinformatic analyses of genetic data are beginning to reveal clustering of risk genes to certain signaling pathways or cellular domains (e.g., postsynaptic density). This convergence of action could happen at multiple levels, including molecular and cellular pathways, circuit connectivity, and network dynamics.

Mapping a large number of risk genes onto functional pathways requires medium- to high-throughput assays to measure neuronal functions at multiple levels, including protein-protein interaction, neuronal morphology, synaptic function, and neuronal connectivity. While gene knockdown approaches have traditionally employed RNA interference or morpholinos in cultured neurons and model organisms (e.g., *Caenorhabditis elegans*, *Drosophila*, and zebrafish) to investigate gene function, new genome-editing technologies (e.g., TALEN and CRISPR) provide highly efficient ways to make precise genetic manipulations in a variety of systems (Fontes and Lakshmipathy 2013; Mali et al. 2013a; Hsu et al. 2014). By introducing the exact variants from human patients, these new technologies allow the interrogation of functional consequences and directionality (loss or gain of function) for each risk allele. Importantly, genome editing by these new technologies is not limited to coding sequences; this allows risk alleles to be investigated with sequence variants in noncoding regions, which may affect epigenetic modification, transcription, and splicing.

Multiple assays, such as high-density imaging (Sharma et al. 2013), multielectrode arrays (McConnell et al. 2012), and calcium imaging (Chen et al. 2013), can be used to detect changes in signaling pathways, neuron morphology, synapse number, receptor trafficking, synaptic transmission, and neuronal activity. However, we do not have scalable assays that would allow us to detect circuit-specific defects, which might be a key converging mechanism. One possible solution is to use embryonic stem cell-derived organoid culture (Eiraku et al. 2008; Lancaster et al. 2013) to develop local circuits for functional analysis. For some evolutionary conserved circuits, it is also possible to use simpler model organisms (e.g., zebrafish) for high-throughput disease-relevant circuit assays (Stewart et al. 2014).

The lack of brain tissues from patients has been a major obstacle in psychiatric research. With the rapid advance in iPSC technology (Takahashi and Yamanaka 2006; Takahashi et al. 2007; Zhang et al. 2013), it might soon be possible to develop functional assays using human embryonic stem cells or iPSC-derived neurons (Imaizumi and Okano 2014). When combined with the CRISPR genome-engineering technology, these assays will likely permit functional studies of diverse risk alleles, both individually and in combination. Although promising, further development is needed to resolve several key issues, including the lack of robust methodologies to produce distinct subtypes of neurons and glia efficiently. Another main issue to be addressed is the immature nature of induced pluripotent stem cells, which is evident through the paucity of spines and synapses in the derived neurons.

How to Dissect Polygenic Mechanisms

Recent genome-wide association studies have identified a large number of risk alleles. Each risk allele likely has a small effect and acts in combination with other genetic and/or nongenetic factors. Thus, identifying other genetic and nongenetic factors that render a particular risk allele pathogenic is key to understanding the disorder.

One possible approach is to build protein interactomes involving the risk genes. This can be achieved by combining proteomic analysis of protein complexes and bioinformatic analysis of existing protein-interaction databases. Once the interactome is established, risk alleles (individually or in combination) could then be tested for their effects on the assembly, and possibly the function of the protein complexes. Such data would provide a framework to identify other genes that function in the same pathway as the risk gene, and thus are more likely to work in concert with the risk gene. Ultimately, these proteomic studies may reveal novel molecular and cellular mechanisms of gene function and yield potential drug targets.

The human brain contains a very large number of neuronal and glial subtypes. Each subtype displays unique structure, connectivity, and function that arise from, or are mediated by, distinct patterns of gene expression, signaling pathways, electric properties, and neurotransmitter utilization. Having new genomics and related technologies available that offer precise profiling of gene expression and cellular function means that hosts of diverse cell types are likely to be identified in the coming years. Combining this approach with interactome studies will be an exciting but challenging step toward the understanding of cell-specific interactomes.

Another promising approach is to use genome-wide screens to identify functional modifiers of risk genes. This will require the development and validation of medium- to high-throughput assays for risk allele-associated cellular phenotypes. One could then use whole-genome RNAi knockdown or CRISPR knockout approaches to perform the genome-wide screen to identify modifier genes of cellular phenotypes of risk alleles. The outstanding strength of this approach is that it not only aims at the interactions of one gene product with another, but also may identify parallel pathways that converge on regulating the same cellular function, and thus help to build a "functional interactome," which may be the molecular correlate of polygenic diseases.

Genome-wide mutagenesis screens for genetic modifiers of gene functions have been successfully used in model organisms such as *C. elegans*, *Drosophila*, and zebrafish (St. Johnston 2002; Kettleborough et al. 2013). For risk genes that are evolutionarily conserved, this approach can be very powerful. In the light of high costs, however, mutagenesis screens of phenotypic enhancers in mice, even for a handful of risk genes, are not realistic. A new approach is to take advantage of the large resource of new mutation- or variant-carrying inbred mouse strains created at the Collaborative Cross (Churchill et al. 2004). By systematically crossing mutant mice harboring risk alleles with various inbred mouse strains bidirectionally (i.e., either to ameliorate or exacerbate), genetic modifiers of particular phenotypes can be rapidly mapped and identified.

Together, combined large-scale data from these approaches may be able to define the biological pathways that contribute to the polygenic nature of psychiatric disorders. These proteomic and functional gene-gene interaction data would also be extremely helpful in interpreting rare variant data from exome and whole genome sequencing. Once the genetics and pathways of polygenic traits are identified, it will be possible to truly model these disorders in cells and animals, and then to take advantage of these valid models in drug screening.

Better Animal Models for Circuit Analysis

Due to the limited access to patient brain tissue, animal models will continue to play key roles in understanding how disease-associated molecular and cellular changes lead to neural circuit dysfunction underlying clinically relevant cognitive and behavioral abnormality. Currently, genetically engineered mice are the most commonly used model organism; they have been very informative for studying neural circuits and brain regions that are structurally and functionally conserved in evolution, such as amygdala function in fear memory (LeDoux 2014) and the basal ganglia circuit in repetitive behavior (Welch et al. 2007; Graybiel 2008). However, modeling deficits of higher brain function, such as mood and cognition, has been difficult in rodents due to significant differences in the structure and physiology of the brains between humans and rodents. One such example is the significantly smaller size of the prefrontal cortex in rodents compared to that of humans. Crucially, structural and functional defects in dorsolateral prefrontal cortex of the brain of schizophrenia patients are considered an important cause of working memory deficits (Volk and Lewis 2010). This anatomical structure is unique to primates (Preuss 1995), making rodent studies difficult to interpret. Indeed, the lack of suitable preclinical models for brain disorders has been considered a major obstacle to the development of new drugs for CNS diseases such as autism, depression, Alzheimer disease, and many others. Thus, there is an urgent need to develop better animal models that more closely capture human pathophysiology.

The development of highly efficient new genome-editing technologies, such as TALEN and CRISPR, has enabled genetic manipulation in primates (Liu et al. 2014; Niu et al. 2014), raising the possibility of generating primate models for psychiatric disorders. Decades of nonhuman primate research have been central to our understanding of physiological brain function. Most such work is done with rhesus macaques, but because of their size, cost, and long generation time, it is not ideal to use macaques as a routine genetic model. An attractive alternative is the common marmoset *Callithrix jacchus* (Carrion and Patterson 2012). Marmosets are New World monkeys; they are highly social with a strong family structure, and show complex vocal behaviors that are seen as an expression of and model for social communication and social cognition (Schiel and Huber 2006; Dell'Mour et al. 2009; Miller et al. 2010; Takahashi et al. 2013). From a practical and genetic engineering perspective, the common marmoset has considerable advantages: small body size and low weight (~350 g), comparatively early sexual maturity at 12–18 months, and biannual births which produce 2–3 offspring from each pregnancy.

The neuroanatomy of the common marmoset is well described. Like macaques, but unlike rodents, marmosets have a well-developed prefrontal cortex, a region that is critical for many cognitive functions that are impaired in human psychiatric disorders. The small size and smooth surface of the marmoset brain are also experimentally advantageous for neuroscience research. Although less studied than macaques, the marmoset has been a popular model organism for studies of auditory physiology. With its fast reproductive cycle, the common marmoset has the potential to anchor the next generation of genetically engineered model organisms for brain disorder research (Okano et al. 2012).

New Technologies for Circuit Analysis

In addition to advances in the genetic analysis of psychiatric disorders, ongoing development of new technologies for probing circuit connectivity and functionality will also greatly facilitate the dissection of neurobiological substrates of psychiatric disorders. These technologies include long-range circuit imaging techniques, cellular resolution connectomics, and optogenetic tools for precise spatiotemporal manipulation of neural activity.

CLARITY

Functional interrogations of complex biological systems are particularly challenging since primary defects, which occur on the molecular and cellular level, may produce symptoms and impairments at the level of neural circuits. Understanding the full nature of the pathophysiology thus requires the integration of high-resolution molecular and cellular observations with functional imaging of neural circuits in their native three-dimensional structure.

A recently developed technique, termed CLARITY, meets this need by enabling the preservation of the three-dimensional framework of the mouse brain while rendering the entire tissue optically transparent (Chung et al. 2013). The method removes the light-scattering lipid bilayers from the brain and replaces them with a new physical framework composed of a tissue-hydrogel hybrid, which facilitates the penetration of both light and macromolecules such as antibodies and *in situ* hybridization probes (Chung et al. 2013). Using CLARITY, biochemical information can be obtained in the context of cellular structure and long-range connectivity with single cell resolution. These unique qualities of CLARITY could provide new invaluable insights into differences in specific circuit structure and wiring between normal and diseased brains. For example, preliminary results from CLARITY revealed abnormal ladder-like dendritic arborizations in parvalbumin-positive interneurons in an autism postmortem brain (Chung et al. 2013). In addition, new insights may originate from functional integration of *in vivo* activity recording with calcium imaging or from optogenetic manipulation of specific neuronal populations with *post hoc* analysis of their respective long-range connectivity.

Dense Neural Circuit Reconstruction Using Electron Microscopy

To gain insight on a level of neuronal ultrastructure in a network context, high-resolution two-dimensional imaging using either serial block face scanning electron microscopy (SBF-SEM; Briggman et al. 2011) or automated serial-section tape-collection scanning electron microscopy (ATUM-SEM; Hayworth et al. 2006) followed by three-dimensional reconstruction can be performed. These labor-intensive approaches allow reliable detection of cell bodies, neurites, and synapses. They thus yield cellular resolution connectomics data, which may eventually shed light on how dysfunction of distinct neural circuits arises from genetic variants linked to psychiatric disease.

Specifically, genetically engineered rodent models for these disease conditions could be subject to sequential two-dimensional imaging and three-dimensional reconstruction to reveal altered connectivity on microcircuits. These analyses could potentially lead to the identification of new distinct cell types that have been previously missed by lower-resolution approaches, such as sparse labeling and fluorescence microscopy. Identification of these subclasses may be of great importance to unravel subtle differences in neural architecture in disease conditions. In addition, while classical labeling methods cannot provide the three-dimensional information needed to track synapses back to one neuron involved in the connectivity with another, dense neural circuit reconstruction allows a count of the number of synapses on a specific neuron as an indicator of the strength of interaction. Furthermore, since electron microscopy permits the number of vesicles at a given synapse to be determined, not only the number of synapses but also the strength of interaction can be evaluated. Given that altered information processing in psychiatric disorders may arise from dysfunctional wiring and skewed synaptic interaction strength, high-resolution connectomics information, when generated with high throughput, would greatly improve our understanding of these conditions.

Optogenetics

After risk-associated genes are identified and valid animal models generated, brain areas that underlie the pathophysiology can be identified using correlative approaches, such as brain activity marker labeling, calcium imaging, and electrophysiology. Just as the observation of the genetic association has to be causally supported by the generation of an animal model, so too must the observation that an underlying circuit appears to be abnormally active, and therefore potentially driving abnormal behavior, be tested through experimental manipulation. Optogenetics, a technology based on light-activated ion channels, enables researchers to conduct these functional interrogations by controlling defined circuits in a bidirectional temporally and spatially precise manner (Boyden et al. 2005). Importantly, transgenic mouse lines with cell type-specific opsin expression or specific Cre-driver mouse lines that can be transduced with conditional opsin-encoding adeno-associated viruses allow precise manipulation of select neural subsets (Zhao et al. 2011). In addition to cell type-specific control, high spatial precision allows not only the stimulation of distinct anatomical regions but also projection-defined activity control. Thus, observational resolution is improved, since a given nucleus in the brain may send long-range projections to multiple downstream regions with different, potentially even oppositional, downstream effects. For example, while stimulation of region-defined cell populations in the basolateral amygdala elicits anxiogenic effects in accordance with previous findings, high-resolution manipulation using projection-specific activation of lateral basolateral amygdala to central amygdala connections revealed an anxiolytic effect of this circuit (Tye et al. 2011). In an experimental setting, one can employ retrograde rabies- or herpes-virus based approaches to transduce nerve terminals at a projection site and then focally stimulate the respective upstream region. Alternatively, transduction methods based on adeno-associated viruses can be used in a defined region to study projection-specific effects after opsin transport into axon terminals and focal activation of the terminals in a given downstream region (Tye and Deisseroth 2012).

Another exciting optogenetic approach takes advantage of bistable kinetics, a novel class of engineered opsin displays. These bistable channels conduct cations upon illumination with blue light and remain open for extended periods of time, on the order of tens of minutes, until conductance is terminated with a flash of yellow light (Yizhar et al. 2011). The key advantage is that activating bistable opsins does not readily trigger spiking, but rather causes subthreshold depolarization. Given that specific circuits in psychiatric diseases may neither be fully shut off, nor in a highly hyperactive state, and that we lack knowledge regarding spiking representations of behavioral state features, bistable opsins are particularly valuable in studying the facilitation of endogenous activity in defined circuits in a minimally artificial manner. In addition, more complex behavioral tasks can be performed upon detachment of the optical fiber or even minimally invasive transcranial opsin activation, which is possible in rodents due to enhanced light sensitivity of bistable opsins (Yizhar et al. 2011).

Further interesting optogenetic tools available to study disease-related circuits are the recently developed opsins Chronos und Chrimson, which allow independent stimulation of two distinct neural populations at a time (Klapoetke et al. 2014). While both Chrimson and Chronos extend the rich toolbox of activating optogenetic tools, there is a strong demand for more refined silencing tools than halorhodopsin and ArchT. Intriguingly, comprehensive remodeling of a cation-conducting channelrhodopsin variant resulted in chloride-conducting opsins for neural silencing with high temporal resolution (Wietek et al. 2014; Berndt et al. 2014). Building on these findings, further molecular engineering extending the optogenetic toolbox with opsins that display desired properties, such as high light sensitivity or bistable kinetics, can be anticipated. Applying these new tools to circuit studies in models for neuropsychiatric diseases, ideally nonhuman primate models which are being developed (Liu et al. 2014), will substantially inform our understanding of these debilitating conditions.

9

How Do We Define and Modulate Circuits in Animals That Are Relevant to Pathophysiology in Humans?

Joel S. Perlmutter

Abstract

How do investigators define and modulate circuits in animals that are relevant to the pathophysiology of human neurodegenerative diseases? Animal models provide a variety of advantages for translational investigations related to neurodegenerative diseases, with the best models revealing new understanding of the mechanisms of disease and stimulating new therapeutic interventions. However, animal models have also led to many missteps with investigators following false paths. This review addresses the purpose of these animals models, describes how to identify relevant circuit abnormalities, provides examples of how such models relate to human neurodegenerative diseases, warns of critical limitations of such models, and finally suggests that better tools be developed for these translational investigations.

Introduction

How do investigators define and modulate circuits in animals that are relevant to the pathophysiology of human neurodegenerative diseases? To address the series of relevant issues about such animal models and circuits, it is helpful to begin by reviewing the purpose of animal models and what we mean by circuits. Different model systems may be optimal or totally inadequate to address selected research questions, and an investigator must understand the various intricacies that determine the appropriate selection of the right model system. Animal models provide a platform for translational studies that cannot be done safely or as efficiently in humans. Such models provide a variety of advantages for investigations related to neurodegenerative diseases, with the best models revealing new understanding of the underlying pathophysiology of disease and stimulating development and testing of new therapeutic interventions. In the past, most clinical-anatomic correlation studies focused on specific regional deficits and the corresponding changes in behavior. Recent studies now focus on how such regional deficits affect brain circuits. In this context, a brain circuit could refer to any linked parts of the brain or perhaps, more appropriately, to a pathway with a start and finish that coincide. Of course, the designated "start" or "finish" can just be any arbitrary nodes along a pathway; they do not necessarily imply initiation or completion. However, the notion of circuits has extended to networks with recent development of techniques that directly identify or assess the integrity of networks. Thus, studies now commonly investigate changes in brain networks or circuits which underlie normal or abnormal behaviors without obvious isolated focal defects. Animal models have revealed insights into dysfunction of such circuits, but they have also led investigators astray. In this chapter, I review the purpose of defining these animals models and describe how relevant circuit abnormalities can be identified. Examples are provided on how such models relate to human neurodegenerative diseases. Critical limitations of such models are discussed and the suggestion is made to develop better tools to permit translational investigations.

Purposes of Animal Models

The selection of an animal model for an investigation requires matching the model system with the research goals. Early animal models focused on identifying biochemical or pharmacologic effects produced by known defects found in human neuropathology. For example, the loss of nigral neurons and subsequent striatal dopamine deficiency found in brain tissues from people with Parkinson disease (Hornykiewicz and Birkmayer 1961) eventually led to toxin-induced models of nigrostriatal dopaminergic neuron injury that mimicked that which occurred in Parkinson disease. Defects in nigrostriatal dopaminergic neurons can be replicated in animals through direct injection, into the relevant brain region, of a neurotoxin (e.g., 6-OHDA) or through direct intracerebral or systemic administration of a protoxin like MPTP or the complex I inhibitor rotenone. Such toxin-induced models have been used to investigate the function of the nigrostriatal pathway and effects of loss of nigral input on basal ganglia circuits, as revealed through changes in glucose metabolism (Wooten and Collins 1981). The changes in circuits found in such models eventually led to new models of the functional anatomy of basal ganglia circuits (Albin et al. 1989; Mink 1996).

Similarly, genetic mutations found in humans with neurodegenerative disorders have been used to develop genetic animal models that permit identification of the biochemical consequences of these gene defects. Dramatic advances in genetics have led to the identification of genetic mutations that cause a variety of neurodegenerative disorders, and knowledge of these mutations provides the basis for a host of different types of genetic animal models. This seemingly simple approach is actually far more complex. Genetic models can include knocking out the gene that is defective in humans, knocking in a mutant gene, inducing chromosomal rearrangements, and inserting conditional gene modifications (that limit expression of the mutant gene to specific brain regions and at selected times). Genetic background also influences the effects of specific genetic manipulations (Heiman-Patterson et al. 2011). Knockouts may demonstrate behaviors unrelated to the specific genetic-related behavior in humans with development of other phenotypes, like greater aggressiveness or blindness. Conditional expression of genes can be controlled by the presence or absence of a drug, such as the antibiotic tetracycline through the use of tet/on and tet/off mutations. Tet/on would express the mutation in presence of tetracycline whereas the tet/off would express the mutation in the absence of tetracycline. Knockin models can vary by the degree of expression, which can be even more complicated for modeling triplet repeat disorders (e.g., Huntington disease), in which the number of triplet repeats of CAG can vary from person to person. Thus, different genetic models of Huntington disease with varying numbers of triplet repeats may produce different biochemical consequences, an important issue to keep in mind when interpreting such studies (Pouladi et al. 2012). These models may, however, permit related changes in the physiology of pertinent circuits to be identified (Cepeda et al. 2013; Kravitz et al. 2010). In fact, optogenetic animal models may be useful in teasing out the specific function of relevant circuits whose dysfunction contributes to neurodegenerative disorders (Tye et al. 2013). Many genetic models do not exhibit behaviors that clearly correspond to the human condition, thereby challenging investigators to identify appropriate readouts of pathologic processes. Nevertheless exploration of relevant pathways, identification of cofactors that may contribute to development of pathology, and potential testing of interventions which could interfere with the etiopathology of disease demonstrate some of the incredible translational utilities of these animal models.

Animal models also permit us to test hypotheses about the relationship of known pathologic defects and brain networks. For example, optical imaging has been used to measure cortical network activity in wild-type mice and transgenic mice that exhibit abnormal amyloid beta (Aβ) deposition in brain (histologically visualized as plaques with no measure of soluble Aβ). These studies revealed a correlation between the amount of regional Aβ amyloid deposition (which accompanies Alzheimer disease in humans), changes in functional connectivity, and aging (Bero et al. 2012). Similarly, people with Huntington disease have loss of medium spiny striatal projection neurons. The consequences of loss of two different types of medium spiny neurons—those inhibited by D2 dopamine receptors or excited by D1 dopamine receptors—have been investigated using animal models with optogenetic approaches that selectively activate one or another of these neuronal populations (Cepeda et al. 2013). This type of strategy can reveal changes in networks related to selected genetic defects with subsequent pathological defects relevant not only to Huntington disease but also to normal function of these basal ganglia circuits.

Once identified in animal model systems, pathologic circuits can be used to identify targets for therapeutic interventions or to determine target engagement by selected interventions. A rather straightforward example is the use of the multiple toxin models of nigrostriatal injury which produce behavioral abnormalities that vary in closeness to human Parkinsonism. The conceptual simplicity of such models, however, gives way to varying complexities, depending on the precise model and how it is used. Rodent models have motor behavioral deficits that have some similarity to human Parkinsonism, whereas nonhuman primate models recapitulate much more closely many of the motor and nonmotor features (Irwin et al. 1990; Tabbal et al. 2012; Brown et al. 2012, 2013; Schneider et al. 1988, 2013). Levodopa, commonly used to treat Parkinson disease in humans, may ameliorate some of the motor manifestations, but it can also exacerbate nonmotor features in nonhuman primate models, as it may do in human Parkinson disease, further validating this model system (Schneider et al. 2013). Multiple drugs to treat Parkinsonism were first tested in these animal models based on behavioral responses. In addition, these models can be used to screen other potential therapeutic agents by demonstrating pharmacologic target engagement or change in a measure of severity of brain pathway dysfunction. The key for these studies is to develop and validate appropriate measures (e.g., occupancy of the relevant receptor or changes in activity in a brain circuit). To do this, neuroimaging methods have been used with varying degrees of success. Although specific targets can be identified with diverse imaging methods in these models systems, care must be taken to ensure that the measures truly reflect the underlying pathologic processes; otherwise results obtained can be misleading (Ravina et al. 2005; Karimi et al. 2013; Brown et al. 2013). The circuit consequences of specific defects can be explored with a variety of methods, including MR-based resting state functional connectivity (Helmich et al. 2009; Hacker et al. 2012), PET-based covariance networks (Ma et al. 2012), and optogenetic methods (Ozden et al. 2013). Such changes in networks may provide a means to infer target engagement by a therapeutic agent, as has been suggested for changes in functional connectivity induced by memantine in rodents (Sekar et al. 2013), as a prelude to potential application to neuropsychiatric conditions. Changes in networks also may be used to investigate the mechanisms of a nonpharmacologic therapeutic modality like deep brain stimulation. The physiologic responses to deep brain stimulation in Parkinson disease have been investigated with multiple technologies, such as changes in ^{18}F-fluorodeoxyglucose (FDG) PET-based covariance identified networks (Asanuma et al. 2006) or optogenetic methods (Gradinaru et al. 2009). Premature application of imaging methods can, however, lead to substantial errors in interpreting of the effects of interventions. For example, several studies to test new therapies to slow progression of Parkinson disease employed molecular imaging of presynaptic dopaminergic striatal neurons as a measure of disease severity, since these nigrostriatal dopaminergic neurons degenerate in this disease. In several studies, changes in the striatal uptake of these radioligands seemed to indicate slower progression of the disorder in one group of subjects, whereas clinical measures of progression provided evidence for a more favorable response in the other group (Ravina et al. 2005). These molecular imaging biomarkers had not been properly validated prior to implementation; that is, no one had determined "standard curves" to demonstrate that these measures truly reflect changes in severity of loss of nigrostriatal pathways until just recently. We now know that in a model system of toxin-induced damage to the nigrostriatal pathway, all of these striatal measures do not consistently reflect the severity of injury. Instead, these striatal measures approach zero as nigral cells decrease to 50%, yet motor Parkinsonism continues to progress as more nigral cell bodies die. Thus, previous studies using striatal measures of uptake likely found discordant results with clinical measures since striatal measures of more severely affected individuals were mostly noise (Karimi et al. 2013). Thus, validation in animal models prior to application in clinical studies seems prudent.

Animal models can also be used to develop and validate biomarkers that can reflect an underlying pathophysiologic process. The potential value of this type of biomarker is that it can help with diagnosis, if it:

- is found to be sufficiently specific and sensitive,
- measures disease severity,
- determines target engagement (which in this case refers to appropriate pathway modification) of a therapeutic agent (as discussed above),
- provides insight into mechanism of a therapy, or
- can be used, with the agreement of regulatory agencies (e.g., the U.S. Food and Drug Administration or the European Medicines Agency) as a surrogate endpoint for a clinical trial to potentially limit the number of participants needed for such a trial or to shorten the needed length of a trial.

These types of biomarkers frequently must be validated in appropriate animal models that permit careful control of the relevant variables and allow *in vitro* measures that may not be possible in humans to compare with the biomarker-based measures. Validation of biomarkers can be tricky and must include well-described reproducibility, accuracy, sensitivity, and specificity for the underlying biologic process. Without these critical steps, biomarker candidates may yield confounding results (Ravina et al. 2005; Karimi et al. 2013).

Identification of Circuit Dysfunction Due to Different Defects

A variety of approaches can be employed to define and modulate circuits in animal models that are relevant to the pathophysiology of human neurodegenerative disorders. Circuits or networks of coordinated brain regions can be identified with neuroimaging or electrophysiologic methods. Neuroimaging tools for this include structural imaging that can identify selected patterns of atrophy for specific dementing conditions (Kim et al. 2007), which may stimulate exploration in animal models of spread mechanisms for pathologic proteins (Walker et al. 2013). Functional imaging, with either resting state or activation-induced changes in regional metabolism or blood flow (Ma et al. 2012), provides an alternative strategy. These functional measures include molecular imaging-based calculations of resting state covariance patterns of regional activity (typically FDG or blood flow PET measures) or determination of patterns of regional activation induced by specific tasks or pharmacologic manipulations. MR-based methods that focus on blood oxygen level dependent (BOLD) include resting state and task/pharmacologic activation. Resting state data, particularly BOLD-based measures, can be analyzed using a variety of approaches, and each has its advantages and disadvantages. One common approach is seed-based analysis: seed regions in the brain are selected and all other brain regions which have resting state BOLD signals, correlating with the activity in the seed regions, are identified. Seeds can be based on a canonical standard set of brain regions (there are many such standard region sets): regions from the classic resting state networks, regions known to be affected by a specific pathologic condition (Hacker et al. 2012), or regions found in multimodal studies such as molecular imaging (Park et al. 2013) or MR-based diffusion imaging. For example, seed regions placed in the posterior putamen (i.e., the part of the striatum preferentially denervated in people with Parkinson disease) led to the identification of dysfunctional networks extending into thalamus, upper brainstem, and cerebellum (Hacker et al. 2012). Correlating the strength of these networks with severity of Parkinsonism, as measured by clinical rating scales, demonstrates face validity for the relevance of this circuit. Thus, this type of analysis could be used to investigate a variety of specific manifestations of Parkinson disease. Further studies in animal models may provide greater insights into underlying biochemical or physiologic changes that lead to these circuit changes. Resting state data may also be analyzed using dynamic functional connectivity approaches (Hutchison et al. 2013), graph-based methods (Wang et al. 2010a), or independent component analysis. Each approach has its advantages and limitations. In general, MR neuroimaging methods have high spatial resolution but poor temporal resolution. Molecular imaging has about a thousandfold greater sensitivity than MR for detecting changes in concentrations of chemical moieties (like a labeled ligand).

Electrophysiologic methods have much better temporal resolution than neuroimaging methods and can be used to investigate circuits in animal model systems. These types of translational studies can be particularly relevant as electroencephalography (EEG), event-related potentials, magnetoencephalography (MEG), and local field potentials have been used to investigate neurodegenerative disorders in humans (Rossini et al. 2007; Oswal et al. 2013), and the animal model studies can help identify the mechanisms and pathophysiology underlying observed electrophysiologic changes. For example, local field potentials were used to identify changes in basal ganglia beta oscillations in various rodent models of Parkinson disease (Lobb et al. 2013); such animal studies provide important clues to the functional effects of deep brain stimulation in the treatment of Parkinson and related disorders. Another interesting example from a nonhuman primate model of Parkinson disease is the change in EEG synchronization of beta and gamma band frequencies after injury to nigrostriatal neurons, found with simultaneous scalp EEG and neuronal activity in pallidum and subthalamic nucleus (Gatev and Wichmann 2009). This approach revealed a marked disruption of basal ganglia-cortical connections.

Relevance to Neurodegenerative Conditions

These neuroimaging methods have been applied to humans with neurodegenerative conditions and then animal models of these disorders. Surprisingly, some of the standard resting state networks have been found in nonhuman primates or even rodents, despite the fact that the animals were anesthetized at the time of the imaging (Vincent et al. 2007; Mantini et al. 2011). The effects of anesthesia may not be trivial and remains a major point of investigation; some suggest that light sedation is better than deep anesthesia (Guilfoyle et al. 2013; Kalthoff et al. 2013). Alternatively, studies can sometimes be performed in awake animals. While this adds a substantial level of complexity and difficulty, it may prove critical for some types of studies. For example, levodopa-mediated circuits in nonhuman primates are completely altered by anesthesia (Hershey et al. 2000).

Each neuroimaging method has relatively limited temporal resolution compared to electrophysiologic methods such as MEG, which has poorer spatial resolution but much better temporal resolution (Hall et al. 2013). In fact, combining different imaging modalities may yield a more complete view of brain networks than any one individual method. For example, it may be possible to combine multiple imaging methods like molecular imaging (with thousandfold greater sensitivity compared to MR methods) with MR-based resting state connectivity, tractography, or both. Similarly, these methods could be combined with electrophysiology, which offers greater temporal resolution. Multimodal studies have been gaining increasing traction and will clearly play a greater role in future studies.

Once seemingly relevant and statistically significant networks have been found in animal models, there are still several criteria that must be met to determine relevance to the human condition. Networks found in animal models that resemble those found in patients with specific neurodegenerative disorders suggest that the model may prove to be valid. Next, it would be helpful to know whether changes in the strength of network connectivity relates to a change in the relevant animal behavior. In this way, one could consider using network strength as a biomarker for testing the therapeutic efficacy of a new proposed intervention. The value of such a biomarker could include greater sensitivity compared to behavioral measures, potentially greater objectivity of the measure, and insight into the underlying mechanism of the intervention. In particular, a biomarker or change in network function that became abnormal prior to behavioral or symptomatic manifestations of a disorder could be particularly helpful in the development of an intervention that could forestall or prevent disease progression, or even the development of any manifestations. It is important to note that an intervention could have substantial behavioral benefit, even though its mechanism of action could bypass the function of the brain network proposed as a biomarker. A simple example of this would be the application of a molecular imaging of midbrain uptake of radiotracers for nigrostriatal dopaminergic neurons (reflecting severity of nigrostriatal injury that occurs in Parkinson disease and correlates with severity of motor Parkinsonism) to measure the efficacy of deep brain stimulation of the subthalamic nucleus (Tabbal et al. 2012; Brown et al. 2013). The stimulation effects are downstream of the component of the circuit measured with that particular molecular imaging biomarker (Hershey et al. 2000). In this case, the network biomarker would not be an appropriate measure of the efficacy of the intervention, even though it might be used to assess other interventions with different mechanisms of action. Thus, one must approach this type of investigation with caution and have clear goals before selecting the right biomarker.

The key steps in development and validation of a relevant biomarker for use in neurodegenerative diseases require an animal model system. The readout or biomarker applied to such an animal model must be robust, validated, and relevant to either the underlying mechanisms of disease or the behavioral manifestations of the disorder. Validation includes demonstration of a standard curve, just like any assay applied in a lab, especially if the biomarker is to be used to assess therapeutic interventions. To make a standard curve, one needs to be able to produce a graded deficit in the animal model system and then ensure that the measurement of the biomarker (or in this case, network function) reflects graded deficits in a clear manner. Alternatively, a biomarker or change in network function could be intact or not be intact—a categorical designation, potentially important for group classifications for research studies. This more limited descriptor still can have value if it is a more sensitive indicator of relevant dysfunction than overt behavior.

Once validated in an animal model system, a biomarker might serve several functions in human studies. If more sensitive than behavioral measures, it could provide an endophenotype for selecting patients for clinical trials (stratifying or enhancing the homogeneity of patient groups) or genetic studies (distinguishing affected from unaffected subjects) (Racette et al. 2006). Another strategy is that analysis of various circuit abnormalities can provide sites for target engagement. For example, Bergman et al. (1994) used an MPTP-induced nigrostriatal deficit in nonhuman primates to identify increased single unit recordings with increased bursting in the internal segment of the globus pallidus. These direct electrophysiologic recordings provided rationale to further pursue pallidotomy and deep brain stimulation of either the subthalamic nucleus (which provides a direct glutamatergic input to globus pallidus) or direct stimulation of globus pallidus for treatment of Parkinson disease (Wichmann et al. 1994, 1999). In fact, much of the work on mechanisms of deep brain stimulation has come from studying changes in circuits in animal model systems (Miocinovic et al. 2013).

Investigations using multiple modalities may provide greater insights into the etiopathologic mechanisms of disease than studies which use a single metric of brain function. Examples include investigations of what causes regional vulnerability of selected brain regions to Aβ amyloid pathology, as identified by *in vivo* amyloid PET imaging in mouse models (Bero et al. 2012), or the relationship of various pathophysiologic changes that precede dementia onset in people with dominantly inherited mutation destined to develop the disease (Bateman et al. 2012; Benzinger et al. 2013). An important finding may be that these brain regions are key components of the default mode network and have high intrinsic brain activity (Sheline et al. 2010). Loss of intra- and internetwork resting state functional connectivity may accompany the progression of Alzheimer disease (Brier et al. 2012). Multimodality studies in animal models of abnormal amyloidosis may permit identification of the key pathologic dysfunctions that lead to these circuit abnormalities; however, these types of studies necessitate an analysis of the underlying changes in brain function, which requires careful *in vitro* biochemical analyses (Bero et al. 2011, 2012). Thus, the greatest insights into mechanisms of disease may arise from studies that integrate findings from multiple approaches to reveal what biochemical or physiologic processes underlie changes in abnormal brain circuits.

Caveats on Relevance of Animal Model to Human Neurodegenerative Disease

Several caveats must be considered when animal models are implemented to investigate human neurodegenerative disease. As noted, animal behaviors and neuroanatomical complexity frequently require extrapolation to corresponding human behavior or brain circuits. This is less true for nonhuman primates, particularly in neurodegenerative disorders, but clearly applies to rodents, flies or zebrafish. Imaging studies in animal models frequently employ anesthesia or sedation, which may or may not confound the interpretation of results. Some resting state networks seem relatively impervious to this, but caution is clearly warranted as pharmacologic activation may vary substantially in nonhuman primates, depending on anesthesia (Hershey et al. 2000). Perhaps, most importantly, the selection of an animal model system must match the goals of an investigation. Some models offer the means to test potential symptomatic treatments (like toxin models), whereas others may be better suited to investigate underlying changes in biochemical pathways (like genetic models). These are not hard and fast rules but rather important guidelines to consider. The development and validation of model systems require an iterative process. Insights from human pathology, physiologic dysfunction, and genetics may provide the initial impetus for development of an animal model and investigation of changes in brain pathways and circuits. However, refinement of the model almost always yields a better tool. Finally, careful validation of these tools is critical to the interpretation of studies of either the model systems or biomarkers developed using these models. Each investigator that applies such tools needs to be responsible for understanding their advantages and limitations.

What Tools Do We Need?

Genetic models provide useful model systems to investigate underlying brain mechanisms. Advances in genetic engineering based on TALENs and CRISPR (Wang et al. 2013b) have made development of such models much easier. These studies can be done in mice, faster in zebrafish, and even faster in flies. However, behavior of animals such as zebrafish and flies may be difficult to align with human behavior, depending upon the complexity of the behavior. Development of sophisticated nonhuman primate genetic models could have a substantial impact on applicability of these model systems. Use of nonhuman primates is likely to be slower and more expensive, but the value and utility of such models may justify the additional effort for some applications (Hutchison and Everling 2012).

Additional work on functional connectivity methods remains to be done. Issues regarding preprocessing these types of data, including potential confounding factors such as the effects of global regression or tolerance for movement during data collection, need further evaluation. Clearer roles for different methods of data analysis will help. Substantial efforts to refine functional connectivity with resting state BOLD and anatomical connectivity with tractography through connectome projects offer important opportunities to compare connectivities (Blumensath et al. 2013; Ugurbil et al. 2013) in animal models; some have already started doing direct comparisons between human and nonhuman primates. Application of additional multimodality approaches will help integrate various measures and provide a more complete view of underlying pathology. This represents a particularly critical need, as multimodal approaches can take advantage of the greater sensitivity of molecular imaging, the high spatial resolution of MR-based methods, and the faster temporal resolution of electrophysiology. In addition, further genetic studies in humans with neurodegenerative diseases will help provide genetic mutations upon which to base additional animal models.

The development of biomarkers that measure circuit function or other brain dysfunctions impervious to typical symptomatic treatments would be a major advance for development of interventions that could slow or halt disease progression for conditions like Parkinson disease. This would allow us to test such therapies in patients taking symptomatic therapy and offers two advantages: accuracy of diagnosis frequently increases with longer follow-up and recruitment of treated patients is far easier than recruiting those not yet treated. These advantages translate into substantially lower clinical trial costs. Of course, one must consider whether pathologic progression has exceeded the potential to reverse or slow disease processes at that point. This will depend on the degree of injury that occurs at the time of symptom onset and may vary greatly for different neurodegenerative diseases. Finally, enhanced educational opportunities, improved collaboration between basic investigators and clinician scientists, as well as greater funding for this type of research are all critical if progress is to be exacted. This poses a particular challenge in times of financial constraint.

Acknowledgments

The author's work was supported by NIH (NS050425, NS058714, NS41509, and NS075321); the American Parkinson Disease Association (APDA) Center for Advanced PD Research at Washington University; the Greater St. Louis Chapter of the APDA; the McDonnell Center for Higher Brain Function; and the Barnes-Jewish Hospital Foundation (Elliot Stein Family Fund for PD Research and the Parkinson Disease Research Fund).

First column (top to bottom): Gül Dölen, Guoping Feng, Katja Kroker, Tobias Boeckers, Rob Malenka, Katja Kroker, Nils Brose
Second column: Rob Malenka, Isabelle Mansuy, Richard Frackowiak, Ilka Diester, Bruce Cuthbert, Tobias Boeckers, Joel Perlmutter
Third column: Joel Perlmutter, Alvaro Pascual-Leone, Nils Brose, Guoping Feng, Isabelle Mansuy, Alvaro Pascual-Leone, Gül Dölen

10

Pathophysiological Toolkit

Genes to Circuits

Gül Dölen, Robert C. Malenka, Joel S. Perlmutter,
Nils Brose, Richard Frackowiak, Bruce N. Cuthbert,
Ilka Diester, Isabelle Mansuy, Katja S. Kroker,
Tobias M. Boeckers, Alvaro Pascual-Leone, Guoping Feng

Abstract

Understanding the etiology and pathophysiology of neuropsychiatric disease requires the development of new tools (ranging from evolving diagnostic strategies to biomarkers) that can address the unique challenges of neuropsychiatric disease, including the current lack of tractable interfaces between what we can learn in the clinic and the tools available using model systems in the laboratory. This chapter outlines some of these tools, addressing pitfalls and opportunities, while acknowledging the iterative nature of bridging the gaps between different levels of inquiry.

How Do We Diagnose Human Brain Diseases?

The diagnosis of human brain diseases has historically relied on relating anatomical pathology to clinical presentation. In cases where gross anatomical changes are not apparent (e.g., psychiatric disorders), diagnosis is made exclusively based on signs and symptoms. Mismatches between these measures, as well as the more recent appreciation of extensive pleiotropy across brain disorders, underscore concerns about the lack of the etiological and pathophysiological validity of current diagnostic classification systems. Not surprisingly, then, pharmaceutical companies have recently withdrawn *en masse* from drug development for brain disease, citing the lack of good targets with genetic and pathophysiological validation that can be related to the symptom-based diagnostic categories and the heterogeneity of patients within a given disease category. If we are to proceed successfully with research on the etiology and pathogenesis of brain diseases, improve individual patient diagnosis, and develop novel mechanism-based therapeutics, we need new approaches to disease classification that take these concerns into account. In recognition of this need, two wide-scale projects have recently been initiated: the Research Domain Criteria (RDoC) project, based in the United States, and the Human Brain Project (HBP), based in Europe. Here, we will briefly outline the basic approaches of these projects and highlight advantages and limitations of each.

The Research Domain Criteria Project

In early 2009, the National Institutes of Mental Health (NIMH) initiated the RDoC project[1] with the intent of creating a research system to support studies that will ultimately improve diagnosis and treatment. Mental disorders are increasingly being viewed as neurodevelopmental disorders that become manifest within and among neural circuits as a result of a fundamental biological risk and its interaction with various environmental contexts across development. RDoC reflects this perspective and is built around neural systems posited within five major domains of function:

1. negative valence (i.e., systems which respond to threats and other aversive situations),
2. positive valence (related to various aspects of reward and appetitive behavior),
3. cognitive systems,
4. social processes, and
5. arousal/regulatory systems.

These functional systems have been termed "constructs" to denote their status as noncomputable concepts, the nature and perceived function of which change over time with advancing research. Because the constructs are inherently dimensional, RDoC research designs emphasize using a wide range of subjects for research studies, to include both those with levels of psychopathology of comparable severity to disorders defined by current nosology (i.e., the Diagnostic and Statistical Manual of Psychiatric Disorders, 5th edition, DSM-5) as well as subjects representative of a broad range along the distribution of function within the population. Investigators are expected to specify their sampling frame clearly and to measure each construct with an array of measures. This might include genetics, molecular/cellular processes, neural circuit measures, other physiological measures, behavior, and self- or interviewer-reports, which, at the extreme nonfunctional end of a distribution, would comprise symptoms.

[1] http://www.nimh.nih.gov/research-priorities/rdoc/index.shtml (accessed March 2, 2015)

The withdrawal of pharmaceutical companies from psychiatric drug development occurred coincidentally at about the same time as the RDoC project was initiated. As a result, the RDoC framework has somewhat unexpectedly become an important component of the new experimental medicine approach to early clinical trials at NIMH. This paradigm emphasizes an RDoC-like specification of the clinical sample and a strong hypothesis-based approach to a particular CNS mechanism. Under this rubric, an investigator must:

1. demonstrate adequate engagement of the drug under investigation with the hypothesized molecular target, e.g., with a positron emission tomography (PET) ligand (Wietek et al. 2014), but with other CNS measures when PET ligands are not available;
2. determine appropriate dosing;
3. ideally demonstrate that adequate levels of target engagement result in a relevant change in a brain or behavioral measure (e.g., change in fMRI signal, event-related potential, or behavioral task); and
4. relate these measures to a preliminary clinical efficacy signal.

A "fail" at any stage of this "fast-fail" paradigm results in the discontinuation of the study series, thus saving time and funds for another target. More importantly, this allows the hypothesis concerning the proposed modification (e.g., agonist, antagonist) of the chosen target to be tested. Too often, in the recent history of CNS pharmacology, there has been no convincing test of such hypotheses (e.g., due to lack of evidence of target engagement or pathway modification), and this has led companies to return repeatedly to the same target, even in the face of multiple negative studies. Critically, emphasis is on targeting specified neural mechanisms and their behavioral outputs, rather than heterogeneous DSM disorders. This feature is highly consonant with the RDoC framework. A critical aspect of the early trials is to enroll subjects on the basis of their measurable deficits or impairment in the systems being tested, not simply because they are members of a particular diagnostic category. Clinical outcome is measured by assessing the mechanism of interest in addition to clinical symptom reduction or improvements in functioning. In this manner, NIMH hopes to provide more homogeneous patient groups to improve the chances of detecting a successful outcome during the early stages of treatment development.

An important caveat to this approach is that the RDoC is a research framework, not a completely specified system. Thus, the constructs and measurement units should be considered as a guide to research, rather than a replacement for extant diagnostic paradigms. Although the RDoC project is still at an early stage, the number of funded grants using RDoC constructs, instead of DSM categories, has increased steadily and is expected to accelerate in the future.

The Human Brain Project

Launched in 2013, and funded by the European Union, the HBP[2] is an initiative to build a working computer model of the human brain by 2023. Drawing on technical concepts pioneered in other areas of science and technology (e.g., simulation modeling, distributed query engines, real-time data visualization, statistics, and data mining algorithms), this approach aims to federate and integrate existing clinical and basic neuroscience data, with the proximate goal of redefining disease diagnosis—moving away from a symptom-based classification to one based on patterns of biological abnormality. Ultimately, the HBP aims to use information about how the brain is structured and organized to facilitate design and production of neuromorphic computing systems.

One initial focus of the HBP is to analyze and cluster available patient data to develop diagnostics and personalized treatments by identifying unique biological signatures of disease. Such signatures, it is hoped, will permit the "cleanup" of populations recruited for clinical trials. This "big-data" approach may also help in redefining nosology, by pointing out symptom clusters that have not been considered in the current framework. A recent study by Denny et al. (2013) has demonstrated the feasibility of a large-scale application of the phenome-wide association study paradigm within electronic medical records. Especially when linked to computational approaches, such as those implemented in the HBP, and incorporating neuroimaging methods, this could highlight new mechanisms and syndromal clusters that are relevant for treatment.

Future iterations of the HBP approach will attempt to use simulations based on such disease signatures to enable development of new drug discovery paradigms. The HBP will make its tools and data freely available to participating scientists, clinical researchers, and industries, who are prepared to share their data within a dedicated network through six technology platforms that provide a new model for very large-scale international scientific collaboration in the life and health sciences. Despite enthusiasm for this approach, some have voiced concern that available data sets, particularly for neuropsychiatry, are information-poor, and that the results of this model-free analysis will be insufficient to provide a mechanistic understanding of human brain disorders.

Other Considerations

Current diagnoses are based on the assumption that a criterion set is being matched to the clinical picture observed in an interview with the patient. However, there is good evidence that the clinical picture is highly variable over time and that there may be much interesting information, not so much in the cross-sectional picture at any given point in time, but rather in the variability of psycho(patho)logy in relationship to the ongoing environmental experience.

[2] https://www.humanbrainproject.eu (accessed March 2, 2015)

Advances in environmental assessment technology, including wearable computing devices, intelligent textiles, and the technology referred to as the "internet of things," should be harnessed to give clinicians a richer understanding of what their patients actually experience in the real world, and how their behavior changes in response to treatment.

Conclusions

The long-term goal of diagnosis is treatment, prognosis, and continuing the dialogue between clinical neuropsychiatry and basic neuroscience. Historical divisions between psychiatry and neurology should eventually be abandoned. Data collection should be informed by recent advances in genetics and neurobiology and prioritized over re-sorting extant data. Techniques that incorporate the molecular basis of disease into traditional diagnostic tools should continue to be developed.

How Do We Identify and Utilize Biomarkers?

Broadly defined, a biomarker is a measured characteristic that serves as an indicator of disease, a pathophysiologic process, or response to therapy. Biomarkers can be used in many different settings which may not be mutually exclusive. For example, a biomarker may report the status of the underlying pathophysiology, serve to delineate subcategories of disease, act as a reporter of a therapeutic intervention, or, in some cases, even act as a surrogate endpoint. This last category will receive special consideration here, since the definition of endpoints for clinical trials has been a major limitation for development of new therapies for both neurodegenerative and neurodevelopmental disorders.

Selective biomarkers may provide surrogate endpoints, but this requires that the biomarker meets specific criteria. To qualify as a surrogate, the biomarker must reflect the underlying pathophysiologic process of the disease. In addition, the tested intervention must affect the biomarker, and the intervention must affect the biomarker without altering the underlying intervention. The degree of the effect of the biomarker must be sufficient to measure clinical outcome and not just reach statistical significance. The timing of the effect on the biomarker must be appropriate for the length of the clinical trial. Finally, a surrogate must reflect toxicity of an intervention. This last criterion frequently prevents a biomarker from acting as a surrogate endpoint for a clinical study. Nevertheless, even if this criterion is not met, a biomarker could provide an important measure of target engagement or efficacy of the intervention.

In comparison with psychiatric diseases, neurological diseases have been more amenable to biomarker identification and utilization, although few are used in routine clinical practice. To provide a roadmap and to highlight pitfalls for discovery, a number of examples from the neurology literature are examined below to illustrate the problem and possible solutions.

Parkinson Disease

Molecular imaging of the nigrostriatal pathway has used several tracers, including FDOPA, DAT (dopamine transporter) radioligands, or VMAT2 (vesicular monoamine transporter type 2) radioligands to measure presynaptic terminals of nigrostriatal neurons in the dorsal striatum. These have been used in clinical trials to provide insights into the possible efficacy of drugs that may reduce progression of Parkinson disease (PD). However, almost all of these clinical trials found discordant results between changes in the striatal uptake of these radiotracers and a clinical endpoint, as measured by a change in a PD motor rating scale. Several problems may have contributed to this. Of course, the rating scales are far from perfect. In addition, a convincing validation of any of these radiotracers, which would demonstrate that they were sensitive to changes in numbers of nigrostriatal neurons, has been lacking. In fact, this has only been completed recently in a nonhuman primate model of nigrostriatal neuronal loss induced by variable amounts of the selective dopaminergic neuronal toxin MPTP (Karimi et al. 2013). These studies demonstrated that the PET measures of the striatal uptake of each of these tracers correlates well with striatal dopamine as measured by high-performance liquid chromatography, with *in vitro* quantitative autoradiography of DAT or VMAT2 sites in striatum and with each other. However, all of these striatal terminal field measures reached near zero when the number of nigral cell bodies decreased by 50%. Most importantly, the Parkinsonian behavior of these animals correlated fully with the nigral cell counts, whereas the striatal PET measures hit the flooring effect with relatively mild Parkinsonism. This was surprising; however, Kordower et al. (2013) ascertained similar findings in postmortem brains of people with PD who died at various stages of disease severity. Once Parkinsonism was mild to moderate, they found that striatal measures of dopaminergic neurons were down to zero. Fortunately, aiming the PET at the midbrain provides measures of uptake of either the DAT or VMAT2 radioligand, which correlates well with nigral cell bodies and with Parkinsonism motor severity (Brown et al. 2013). The key point of these studies is that biomarkers used to measure disease severity must be adequately validated prior to the implementation of clinical trials and that animal model systems can play an important part in this validation.

Alzheimer Disease

Although pathological and neurodegenerative biomarkers can be identified before symptom onset, they do not share the same time course. For example, pathologic amyloid in the brain can be detected 15 years before symptom onset, at which time the accumulation of this biomarker has peaked. Increases in tau protein can be detected five years before symptom onset, and tau continues to accumulate as symptoms progress. Although each of these biomarkers can independently predict future outcomes in symptomatic patients, in presymptomatic patients they are only predictive in combination. Currently, it is unknown whether these biomarkers will predict treatment outcome, which may depend on the mechanism of therapy.

Multiple Sclerosis

Gadolinium-enhanced lesions, reflecting inflammatory pathologic processes in brain, occur more frequently than individual clinical relapses of symptoms in patients with multiple sclerosis. The appearance and number of these lesions have thus been used as a biomarker of disease activity in therapeutic trials in remitting-relapsing multiple sclerosis. Biomarker use has the potential to reduce the length of time needed for clinical trials, if this meets the criteria for a surrogate endpoint (as discussed below). Nevertheless, European regulatory agencies still require a clinical endpoint for therapeutic trials. A critical limitation of this approach is that once a patient with multiple sclerosis develops a chronic progressive course, magnetic resonance-based measures may not adequately reflect the progressive neurodegenerative process.

From these examples we can infer that although basic mechanisms are informative, useful biomarkers can be developed even in the absence of a complete understanding of the pathogenesis of disease. Furthermore, even when the biomarker does not act as an endpoint measurement, it can serve to inform an understanding of pathogenesis. For example, PET measures of striatal uptake of radioligands may provide valuable information about the pathophysiology of Parkinsonian conditions (Criswell et al. 2011). Similarly, PET measures of the amyloid-β radioligand PiB may reveal coexisting abnormal deposition of amyloid-β in people with PD and cognitive impairment that is different from those with cognitive impairment due to Alzheimer disease (Campbell et al. 2013).

Future Directions: Psychiatric Disease

Neuropsychiatric diseases are susceptible to placebo effects, since the desired outcome is almost invariably a change in the patients' subjective experience of the world. For this reason, development of surrogate endpoints and biomarkers is of critical importance to translational research in this area. Previous attempts to identify biomarkers using cerebral spinal fluid and blood markers, microRNA changes, or REM latency measures have failed. These failures may reflect the etiologic complexity and phenotypic heterogeneity across cell types and indicate that a better understanding of pathogenic mechanisms will be required for biomarker development.

Despite these hurdles, studies with resting state glucose metabolism, measured with ^{18}F-fluorodeoxyglucose PET, reveal changes in limbic frontal circuits that reflect responses to pharmacotherapy or cognitive therapy in people with major depression (Seminowicz et al. 2004). Similarly, MRI-based measures of BOLD (blood oxygenation level dependent) signal activation in prefrontal or anterior cingulate cortex can predict efficacy for both antidepressant (Frodl et al. 2011) and cognitive behavioral therapies (Klumpp et al. 2013). Resting state network measures with BOLD MRI show great promise in identifying new biomarkers that reflect the various domains affected in neuropsychiatric disorders (Greicius 2008). The number of studies in this area has exploded over the last several years. However, careful attention to technical details of the imaging and detailed characterization of patient symptoms is necessary to determine how helpful these potential biomarkers will be for clinical trials.

Electrophysiological measures have also been explored as potential biomarkers in psychiatric disorders. Event-related potentials (ERPs), mismatched negativity (MMN), gamma oscillations, and electroretinograms are promising candidates. Electrophysiological methods provide a tool that reflects functional brain dynamic changes within milliseconds and may also be used as an ensemble of biomarkers. Furthermore, these methods have the advantage of being low-cost, fast, and well-tolerated by patients.

ERPs are characterized by a positive voltage deflection that occurs approximately 50 ms after the stimulus (P50), a negative voltage deflection at 100 ms after the stimulus (N100), and two more positive voltage deflections at approximately 200 ms (P200) and 300 ms (P300) poststimulus, respectively. In the paired-click paradigm, component amplitudes elicited by the first stimulus are normally greater than equivalent potentials elicited by a second stimulus. Schizophrenic individuals exhibit similar response amplitudes to both stimuli, yielding a lower ratio between them (Adler et al. 1982; Boutros et al. 1991). This gating deficit may result from a decreased response to the first stimulus and/or a failure to inhibit the second stimulus. ERP components are affected by certain pharmacological treatments and stimulus manipulations (Maxwell et al. 2004). Deficits in ERP components are reliable and robust findings in schizophrenia, indicating that ERP may be a potential biomarker for this disease.

MMN is a component of ERP and a promising biomarker candidate for psychotic disorders such as schizophrenia (Michie 2001; Michie et al. 2000; Näätänen and Kähkönen 2009; Rissling et al. 2010; Turetsky et al. 2007a; Umbricht and Krljes 2005). MMN, an exaggerated negative voltage deflection following the N100, is elicited when the qualitative features of a novel, or deviant, tone fail to match the pattern of a previous series of repetitive stimuli (Javitt et al. 2000; Light and Braff 2005). Reduced MMN amplitude is one of the most robust findings in schizophrenia (Lawrie et al. 2011), and the mean effect size is approximately 0.99 (Umbricht and Krljes 2005). Furthermore, MMN is regarded as a reliable, sensitive index of central auditory system plasticity, with important relationships to cognition and psychosocial functioning

(Rissling et al. 2013). For example, it has been reported that MMN amplitude is associated with social function (Light and Braff 2005), social cognition (Turetsky et al. 2007b; Wynn et al. 2010), and executive function (Toyomaki et al. 2008), whereas phonetic MMN amplitude is associated with verbal memory (Kawakubo et al. 2006) and social skills acquisition (Kawakubo and Kasai 2006; Kawakubo et al. 2007). Previous studies have shown that antipsychotic medication has little effect on MMN. However a recent report (Zhou et al. 2013) showed that antipsychotics such as aripiprazole improve MMN amplitude reduction in schizophrenia. MMN might offer promise for contributing to the continued development of pharmacologic and nonpharmacologic therapeutics (Light and Näätänen 2013; Näätänen 2008; Nagai et al. 2013).

Frequency oscillations have recently emerged as an important measure of brain activity in the study of schizophrenia. Oscillations represent the coordinated firing of clusters of neurons and are thought to help synchronize activity within and between brain regions. The theta and gamma bands have both been shown to be abnormal in schizophrenia (Ozerdema et al. 2013), and these deficits have been linked to the disease via heritability studies (Hall et al. 2011; Hong et al. 2008). Theta power is reduced in patients with schizophrenia and is associated with memory deficits seen in the illness (Brockhaus-Dumke et al. 2008; Schmiedt et al. 2005). Interestingly, symptoms of schizophrenia have been reported to correlate with increased synchronization of gamma oscillation, although mean gamma synchronicity is lower in patients with schizophrenia than in control subjects (Lee et al. 2003; Uhlhaas and Singer 2006). As oscillation is a biologically and clinically significant index of schizophrenia; it has the potential to serve as a biomarker for the disease. However, it remains to be seen whether this measure covaries with treatment.

Electroretinography (ERG) measures the electrical responses of various cells types of the retina to visual stimuli, including photoreceptors as well as bipolar, amacrine and ganglion cells. Excitatory synapses onto these cells, like most other neurons, contain postsynaptic density proteins, anchoring proteins such as Shank and Homer, and cell adhesion molecules such as neuroligin (Hoon et al. 2011; Stella et al. 2012), many of which have been implicated in autism (Dölen and Bear 2009; Grabrucker et al. 2011). Owing to its location outside of the skull, the unique access afforded by the retina may allow measurement of synaptic phenotypes with ERG recording in both model organisms and human patients. Furthermore, because ERGs are already in clinical use for the diagnosis of various retinal diseases, phenotypes captured with this technique might readily serve as biomarkers. Indeed, this opportunity is currently being explored by the Innovative Medicines Initiative through the European autism interventions project.

Neurodevelopmental diseases present unique challenges to the design of outcome measures. For example, improvements in cognition can be reflected in the "products" of learning (e.g., number of words a patient can utter) or in the "processes" that underlie learning (e.g., an accelerated rate of learning new words) (Berry-Kravis et al. 2013). This distinction is critical because it seems likely that an intervention, which completely reverses impairments in learning capacity assayed by process measures, might require years of learning under conditions of restored capacity to accumulate improvements assayed by product measures, such as intelligence quotient. The evolving KiTAP test (Test of Attentional Performance for Children; Zimmermann et al. 2004) attempts to overcome this challenge by utilizing computer-based cognitive games designed to assay learning across multiple cognitive domains over time. Recently, KiTAP has been validated in fragile X patients across a wide range of cognitive ability, thus raising the hope that this computer-based cognitive assay will become a practicable behavioral outcome measure in future clinical trials (Knox et al. 2012).

Conclusions

The types of imaging, electrophysiological, and behavioral measures described above may provide biomarkers to help stratify or classify patients for clinical studies or symptomatic treatment. Additional work is required to determine whether these biomarkers can be useful for determining efficacy of a therapeutic intervention or target engagement, either as categorical or continuous variables. It will also be important to implement protocols for standardization and harmonization. In the next section we outline opportunities for developing novel biomarkers by focusing on tractable interfaces between basic science and clinical investigations.

How Do We Design "Tractable Interfaces" between Studies in Animal Model Systems and Patients?

Significant advances in understanding the pathogenesis of neuropsychiatric diseases have been aided by new key discoveries in genetics, which is beginning to provide important molecular entry points for the modeling of the respective diseases in genetically modified animals. However, translation of insights into pathogenic mechanisms from such model organisms back to the clinic has had only limited success. One main reason for this, hopefully transient, failure is that the broad spectrum of methods and readouts that can be employed in animal studies cannot usually be recapitulated or emulated in patients, where noninvasive methods are required. What is needed are "tractable interfaces": methods and readouts that can be employed in model organisms and patients alike, or readouts that can at least be emulated or approximated in the respective other realm. Furthermore, a deeper understanding of the human nervous system—leveraging experimental neuroscience approaches—promises to enable more targeted, controlled interventions.

As outlined above, ERPs, MMN, network oscillations measured by EEG, and ERGs belong to the category of tractable interfaces. They can reflect alterations in the function of defined circuits and neuronal networks that underlie the pathophysiology of major neuropsychiatric disorders. In addition, brain stimulation techniques combined with various functional readouts hold particular promise as tractable interfaces, because they can be applied in animal models and patients alike.

Brain stimulation techniques offer a fascinating opportunity to modulate specific neural networks, identify neural substrates, affect behavior, and address symptoms of disease. At the same time, brain stimulation represents a controllable input that can be clearly characterized so as to gain novel insights into the integrity and dynamics of neural networks. It has the potential to help patients while advancing scientific insights as it can bridge the gap between human and model systems. In addition, noninvasive brain stimulation approaches—compared to invasive ones—are safe, relatively easy to apply, and cost effective. Because they lack surgical invasiveness, application is possible in normal subjects as well as in patients across the entire age span.

To date, methods based in electromagnetic and electric stimulation, notably transcranial magnetic stimulation (TMS) and transcranial current stimulation (tCS), are the two most popular and best-studied noninvasive brain stimulation tools. There is, however, a fast-growing array of other noninvasive techniques to query and modulate brain activity, including methods that utilize ultrasound or light combined with transgene expression via viruses or other vectors. Noninvasive brain stimulation techniques can effectively modify, suppress, augment, or disrupt brain function, depending on stimulation, individual, context, and environmental characteristics. Applications include investigations into fundamental principles of brain function, efforts to translate insights from animal models to humans, research on brain-behavior relations, and novel therapeutic approaches for neurological and psychiatric illnesses.

Opportunities for Induction of Synaptic Plasticity-Like Changes in Humans

As mounting evidence from studies in model organisms suggest, an important pathogenic mechanism in neuropsychiatric disease is disruption of synaptic plasticity (Dölen and Bear 2009; Grabrucker et al. 2011; Lüscher and Malenka 2011; Paoletti et al. 2013). In animals, mechanistic studies of synaptic plasticity are typically carried out *ex vivo*, although whole cell recording techniques have more recently been adapted to enable *in vivo* synaptic plasticity studies (Chu et al. 2014). Synaptic plasticity has also been demonstrated in human hippocampal tissue excised from patients undergoing temporal lobe surgery for intractable epilepsy, and seems to share essential mechanisms described in animal models (Beck et al. 2000; Chen et al. 1996; Testa-Silva et al. 2010). Direct assays of *in vivo* synaptic plasticity are not currently feasible in humans;

however, measurements of circuit-level changes, which are thought to reflect these synaptic changes, can be obtained using several emerging noninvasive techniques (Bliss and Cooke 2011; Nitsche et al. 2012).

Noninvasive techniques used in human studies typically mimic electrical stimulation patterns used to induce long-term potentiation or depression (LTP and LTD, respectively), or spike timing-dependent plasticity (STDP), whereas other approaches attempt to modulate the mechanisms known to be required for plasticity. Below, we briefly review some of these including repetitive transcranial magnetic stimulation (rTMS), transcutaneous electrical nerve stimulation (TENS), sensory tetanization, interventional paired-associative stimulation (IPAS), direct current stimulation (DCS), and priming. (For further details, see Bliss and Cooke 2011; Nitsche et al. 2012.)

Repetitive Transcranial Magnetic Stimulation

rTMS delivers relatively small electrical currents generated by fluctuating magnetic fields administered over the skull using a magnetic coil. Single TMS pulses evoke ERPs, measured by EMG, EEG, or fMRI, and can be used to quantify cortical reactivity before and after a given intervention. Potentiation or depression of ERP amplitude can be induced by delivering either high- or low-frequency trains of rTMS pulses, mimicking stimulation patterns for LTP and LTD induction in animals (Goldsworthy et al. 2012; Rothkegel et al. 2010; Ziemann et al. 2004, 2008). Significantly, changes in excitability induced by rTMS last for minutes to hours (Rossi et al. 2009; Siebner et al. 2009; Thut and Pascual-Leone 2010; Ziemann et al. 2008). Currently, this technique is limited to excitation of neural circuitry in structures relatively close to the brain surface and cannot yet replace invasive techniques such as deep brain stimulation. Noninvasive stimulation of deep targets (e.g., the thalamus, basal ganglia, and brainstem nuclei) may soon be enabled by emerging "deep" rTMS technologies (Bersani et al. 2013) and neuro-navigated TMS protocols (Julkunen et al. 2009).

With rTMS, it is also possible to deliver two pulses—a test pulse and a conditioning pulse—applied in rapid succession to the same cortical region. Although this method has been termed "paired-pulse TMS," it should not be confused with paired pulses recorded with local field potential and whole cell electrodes. Changes in paired-pulse ratios after induction of synaptic plasticity have been used to localize the expression to the pre- or postsynapse (Granger and Nicoll 2014; Manabe et al. 1993); however this inference can only be made under the condition that what is being recorded is a monosynaptic response (Voronin 1994). In contrast, paired-pulse TMS is polysynaptic, thought to be mediated by GABAergic transmission, and is used to measure intracortical inhibition and facilitation. Three variants of this protocol exist: short-interval intracortical inhibition, long-interval intracortical inhibition, and intracortical facilitation. At this time, it is unclear whether ERP changes induced by these protocols share mechanisms in common with synaptic plasticity. (For a more detailed discussion, see Ziemann et al. 2008.)

Transcutaneous Electrical Nerve Stimulation

Although interpreting synaptic changes that occur across multiple nodes of a circuit is complicated, previous studies have shown that high-frequency TENS delivered directly to peripheral nerves through the skin in the forearm induces plasticity of cortical ERPs (van den Broeke et al. 2010). Similarly, in animal models, removal of a digit induces nociceptive changes that are transmitted across at least three synapses, ultimately causing induction of LTP in the anterior cingulate cortex (Chen et al. 2014a; Wei and Zhuo 2001). These parallels raise the possibility that TENS-evoked changes in cortical ERP response properties may have the potential to serve as readout of polysynaptic plasticity in the human brain.

Sensory Tetanization

Studies of *in vivo* experience-dependent modification of sensory-evoked responses have demonstrated that plasticity can be induced in primary sensory cortices of humans without electrical stimulation. For example, the amplitude of the visually-evoked potential can be potentiated in the visual cortex using a photic tetanus that consists of flashing or phase-reversing visual stimulus presented on a computer screen (Ross et al. 2008; Teyler et al. 2005). Similarly, other studies have shown that tetanic auditory stimulation can evoke potentiation of the auditory-evoked potential (Clapp et al. 2005; Zaehle et al. 2007). Importantly, many of these circuit-level plastic changes have been shown to be stimulus specific, localized to a particular synapse, and to share induction and expression mechanisms with synaptic plasticity (Bliss and Cooke 2011). Although only a handful of studies have used similar measures to identify pathogenic changes in animal models of neuropsychiatric disease (Dölen et al. 2007; Tropea et al. 2009), the identification of sensory tetanization-induced plasticity in humans, in conjunction with the development of stimulus-specific response potentiation biomarker assays in mice, raises the very real hope that these measures could serve as a tractable interface between models and disease (Cooke and Bear 2012).

Interventional Paired-Associative Stimulation

IPAS is a refinement of the rTMS method, which combines sensory- and electrically-evoked stimulation. Induction protocols aim to mimic STDP by timing the delivery of electrical stimulation (rTMS) to either before (LTD-like) or after (LTP-like) the height of the sensory-evoked ERP (Sowman et al. 2014; Wolters et al. 2003, 2005). Although the parallels to STDP are necessarily imperfect (since IPAS activates polysynaptic pathways), this approach offers the advantage of regional specificity (Bliss and Cooke 2011), likely due to the fact it requires lower frequency stimulation than rTMS alone and that the activated pathways represent the subset recruited by relevant sensory circuits.

Direct Current Stimulation

An older approach, DCS applies continuous weak current (rather than frequency modulated pulses, as in rTMS) using two electrodes fitted on the scalp surface, and has been shown to induce plastic changes in the ERP (Cheeran et al. 2008). Although the exact mechanisms of this plasticity are unclear (Bliss and Cooke 2011), low-frequency stimulation, which would ordinarily induce LTD, when combined with DCS, has been shown in rodents to instead induce LTP (Fritsch et al. 2010). Furthermore, DCS causes the release of brain-derived neurotrophic factor (BDNF) (Figurov et al. 1996), and BDNF knockout mice show impaired motor learning and absence of DCS plasticity (Fritsch et al. 2010). Interestingly, these phenotypes are recapitulated in patients who carry a polymorphism associated with reduced BDNF concentrations and show diminished motor cortical circuit plasticity induced by DCS/rTMS (Antal et al. 2010; Cheeran et al. 2008; Kleim et al. 2006).

Priming

Synaptic metaplasticity (i.e., the plasticity of plasticity; Hulme et al. 2013) has also been reproduced in humans by brief application of DCS or a train of low-frequency rTMS to the region of interest prior to the induction of LTP-like plasticity with rTMS or motor training (Bütefisch et al. 2004; Carey et al. 2006; Cosentino et al. 2012; Lang et al. 2004; Nitsche et al. 2007). Significantly, synaptic metaplasticity has been shown to be disrupted in a variety of animal models of neuropsychiatric disease (Hulme et al. 2013), suggesting that development of this assay as a biomarker would be of value to translational neuroscience.

As exciting as these parallels between noninvasive-evoked modifications in human brain and synaptic plasticity are, caution should be exercised in equating these phenomena. For example, *ex vivo* field potentials are recorded at the site of excitatory postsynaptic potential (EPSP) generation, and are only used when the anatomical arrangement of the brain area in question produces synchronous synaptic currents that can be attributed to a single input pathway. In contrast, ERPs in humans are recorded at a distance, and necessarily reflect the summation of polysynaptic EPSPs and action potentials; as a result, a change in the ERP amplitude may reflect some combination of synaptic LTP or LTD, increased or decreased inhibitory tone, as well as alterations of the intrinsic excitability of the underlying cell population (Bliss and Cooke 2011). This mismatch may have important consequences for extrapolating findings in animal models to the clinical setting. For instance, exaggerated mGluR5 LTD has been observed in hippocampal field potential recordings in the knockout mouse model of fragile X syndrome. However, because this change is modest, region and induction protocol specific (Sidorov et al. 2013), and the hippocampus is relatively distal to the surface of the human brain, this phenotype would likely be obscured in an ERP recording. As a result, it is likely that testing the translatability of therapeutic targets identified in mouse models using these assays must await development of technologies to address these limitations.

Noninvasive Imaging

Noninvasive imaging provides another tractable interface between top-down studies on symptomatic and syndromic features of human disease and molecular, cellular, or circuit-based analyses of disease-related pathophysiological mechanisms in animal models. PET, fMRI, MRS, electrophysiological techniques, and, as discussed above, TMS can all be used in animals as well as in humans. However, given the differences in spatial and temporal scales, further technical developments are necessary and caveats regarding the interpretation of comparisons between model organisms and patients will have to be overcome. Specific disruptive techniques (e.g., TMS, deep brain stimulation, ultrasound and gamma rays) can also be used to lesion reversibly or irreversibly the brain with high spatial fidelity in humans and animals. These techniques can be combined and used for investigative purposes and treatment, separately or in combination. The applicability of PET depends on finding specific molecular or metabolic targets as well as on the question of whether pharmacokinetics and drug distributions or mechanisms are being explored. MRI in functional and structural modes is developing rapidly. The BOLD technique remains very useful in structural imaging, and the quantitative measurement of tissue characteristics through the use of a series of specific MR sequences provides an exciting opportunity for future work. Electrophysiological studies—both invasive, in the form of electrode arrays in patients or animal models, and noninvasive, in the form of EEG or event potential recordings—are further techniques with high temporal resolution that can, in principle, be used in humans and animals alike.

Optogenetics

As each of the new technologies described above is further developed in humans, an important iterative process is needed to confirm the observed mechanisms in model organisms. Here, optogenetic tools (Boyden et al. 2005) can help to identify relevant pathways in animal models and to understand in more detail which parts of the brain need to be electrically stimulated to achieve a desired effect. Contrary to electrical stimulation, where one cannot be certain that only a local area or a particular pathway is affected, optogenetic manipulations allow for very precise stimulation. By stimulating or inhibiting axons while simultaneously measuring the impact on neural activity and on behavior, it is now possible to measure the impact of manipulating activity in a particular pathway very specifically. This will help identify disease-related projection pathways in the brain and determine where to place ideally a stimulating electrode for a particular purpose. Further, it can also help explain effects seen with electrical stimulation (e.g., by cell type-specific optogenetic stimulation) and reduce side effects of electrical stimulation by refining the target site.

Conclusions

An important appeal of each of the methods described above lies in their potential to help patients while gaining critical insights into fundamental questions of brain function and pathophysiology of nervous system diseases. It is likely that scientific and clinical interest in noninvasive brain stimulation, imaging, and optogenetics will continue to expand as the spectrum of their applications is nearly inexhaustible. The most transformative approach in this context would be to truly integrate technological, basic, and clinical neuroscience with therapeutic efforts. This could lead, for example, to an exploration of combined behavioral and brain stimulation interventions that are individually targeted, perhaps based on more sophisticated understanding of individual circuit dysfunctions as well as individual genetic and epigenetic factors.

How Do We Model Neuropsychiatric Disease?

Understanding the etiology and pathophysiology of complex human neuropsychiatric disorders will require modeling disease in systems ranging from humans to stem cells. While this will enable insights appropriate to each model system, the development of novel therapeutics ultimately requires that these models converge at tractable interfaces. In the previous discussion, we laid out a roadmap for developing technologies to achieve this convergence. Here we describe these model systems in greater detail, in an effort to highlight the appropriateness of each to answering a particular question. Rather than stratifying models into "top-down" and "bottom-up" heuristics, we highlight opportunities to interface between systems, as well as the iterative nature of this process.

Role of Human Genetics for Developing Models

Genetic technology has exploded over the last several years, and soon we will have the ability to obtain full genetic information for each person. Gene manipulation in model systems can provide a platform for investigations of the consequences of genetic mutations on neurochemistry, plasticity, circuits, and behavior in a manner that cannot be done in humans. Advances in human genetics have facilitated tremendous progress toward identifying pathogenic or risk genes for several neuropsychiatric disorders, and this information has been applied to model systems. The genetic variations in autism spectrum disorders (ASD), for example, are quite diverse, encompassing rare and common alleles, sequence and structural variants, and *de novo* as well as transmitted mutations. Moreover, in addition to gene discovery in what is often referred to as "common idiopathic" forms of these disorders, there are also multiple examples of established "syndromes" caused by monogenic mutations, including fragile X and Rett syndromes, highly penetrant copy number variants, including 22q11 deletion syndrome (also known as velocardiofacial syndrome), and rare recessive syndromes. It is important to note that the clinical presentation of these genetic disorders can vary widely and that distinctions between syndromic and idiopathic categories are largely historical. It also remains to be determined whether the underlying pathophysiology of the relevant psychiatric manifestations differs as a result of divergent transmission modes.

The characterization of differences in penetrance, effect size, and likelihood that a given risk gene may be implicated in more than one disease state will inform both the selection of genes for modeling in nonhuman systems and the strategies that will need to be applied. For example, mutations carrying relatively large effects and mapping to coding segments of the genome may be suitable to be immediately modeled in an animal system, whereas noncoding single nucleotide polymorphisms found in association with a condition may require additional fine mapping and systems biological analysis before pursuing *in vivo* model experiments.

Given the finding of a very high level of locus heterogeneity for both schizophrenia and ASD, efforts are underway to use a variety of approaches to organize disparate genes into more biologically meaningful groups. These include protein-protein interaction analysis and gene ontology approaches that seek to identify molecular pathways of interest. In addition, several recent studies have leveraged genome-wide expression data from typically developing brain (in humans and other species) to gain traction on the question of when and where specific risk genes might show pathophysiological convergence. Studies in both autism and schizophrenia have implicated mid-fetal cortical development (Gulsuner et al. 2013; Parikshak et al. 2013; Willsey et al. 2013). One recent study mapped an ASD-associated co-expression network based on only high-confidence ASD genes to layer 5/6 projection neurons in mid-fetal development (Willsey et al. 2013). All of these "systems biological" approaches are aimed at prioritizing key parameters in model systems studies, including which cell types, signal pathways, and circuits may be of particular interest.

Etiological and symptom heterogeneity has led some to propose that it might be more useful to consider risk genes as distinct clinical entities (i.e., the existence of "autisms" rather than a single "autism"). Others have suggested an analogy to Alzheimer disease, which is considered a single clinical entity based on pathology, despite the fact that only 10% of the cases have been linked to a specific genetic cause. The remaining 90% likely result from a set of changes that take place during aging which, presumably, compromises the ability of neurons to process proteins correctly. Thus, it is taken for granted that a single therapeutic, when it is finally found, will be effective on all patients with Alzheimer disease. This makes a huge difference in the search for therapeutics and explains why knowing the number of classes (in the treatment sense) of diseases is so essential. In the long run, the complexity of psychiatric disorders, both in terms of their waxing and waning symptoms and genetic etiology, suggests that a spectrum of phenotypic and etiologic factors will need to be considered when defining individual disease classes for the development of tractable models and treatment studies.

The distinction between defining a "disease," based on symptoms or pathology, or "diseases," based on genes, could influence selection of models, since these definitions may lead to different predictions about the level (i.e., biochemical, synaptic, circuit, behavior) of pathogenic convergence across etiologies. For example, at one extreme, it could be argued that because the "diseases" are primarily defined by the genetic etiology, the key property required of a model is construct validity (e.g., genetic homology) and the ability to therapeutically target convergent biochemical pathogenic mechanisms. Accordingly, behavioral readouts would be secondary, and a demonstration of face validity for human symptoms need not be prioritized. This view is particularly amenable to using iPS or IN cell-based models to subclassify complex disorders. At the other extreme, one could argue that because the "disease" is primarily defined by the symptoms and pathology, models which achieve the best face validity (e.g., behavioral changes that recapitulate human symptoms) will provide the greatest opportunity for understanding pathogenesis. Accordingly, therapeutic strategies would focus on correcting conserved circuit and behavioral abnormalities, independent of the heterogeneity of genetic disruptions that produced them. This view is particularly useful when the etiology of the disorder is unknown and may require development of model organisms (e.g., marmosets, discussed below) amenable to interrogation of more complex behaviors.

Between these two extremes lies the view that since some of the identified risk genes encode synaptic proteins (e.g., receptor, downstream signaling, and adhesion molecules) with known functions that are largely conserved across species, understanding pathogenesis at the level of synapses will yield the greatest short-term opportunity for therapeutic intervention. The term "synapsopathy" was first coined to describe this view as it applies to autism (Bear et al. 2008; Dölen and Bear 2009) and to distinguish this disorder from those brain diseases where the primary pathology is localized to a specific brain region (e.g., substantia nigra in PD). Despite enthusiasm for the idea, an inherent challenge is that the brain is remarkably heterogeneous in its cellular composition. The wide variety of cell types might thus be the basis for genetic "pathoclisis," a selective vulnerability of subsets of neurons to the effects of mutation. Indeed, recent studies of Huntington disease and autism raise the possibility that genetic pathoclisis will be an important pathogenic mechanism, with profound implications for developing therapeutics (Dölen and Bear 2009; Paul et al. 2014). Moreover, synapsopathy and pathoclisis need not be mutually exclusive; together they might account for symptom heterogeneity despite overlapping clinical presentation. Understanding the relative contribution of each of these mechanisms to the pathogenesis of disease will likely inform decisions about the suitability of modeling synaptic disruptions at the cellular or circuit level, so as to guide future development of tractable interfaces. For example, synapsopathic features might be particularly amenable to ERPs, MMN, gamma oscillations, and ERG measurements (discussed above), whereas TMS, tCS, and brain imaging (see previous discussion) of specific circuits might be more appropriate for interrogating disease symptoms that result from pathoclisis. Of course, a risk of focusing solely on genes, the functions of which are currently somewhat familiar, is that many of the genome-wide significant loci associated with disorders, such as schizophrenia, are noncoding (Schizophrenia Working Group of the Psychiatric Genomics Consortium 2014). Thus we may end up ignoring important pathogenic mechanisms which, in the long run, need to be understood.

Model Systems: Mice

To date, the modeling of identified risk genes has primarily been conducted in mice using constitutive or conditional transgenic systems. This model organism offers the opportunity to interrogate pathogenesis at the level of biochemical, synaptic, circuit, and behavioral mechanisms. Moreover, the availability of a number of other molecular genetic tools—such as BAC-Cre-recombinase driver and BAC-EGFP reporter lines, viral-mediated gene transfer (Callaway 2008; Grimm et al. 2008; Luo et al. 2008; Marie and Malenka 2006; Neve et al. 2005; Salinas et al. 2010), RNA interference (Morris and Mattick 2014), optogenetics (Boyden et al. 2005; Lima and Miesenböck 2005; Lin et al. 2009), genetically encoded voltage (Cao et al. 2013; Dimitrov et al. 2007; Siegel and Isacoff 1997) and calcium sensors (Chen et al. 2014a; Wang et al. 2004) and pharmacosynthetics (Dong et al. 2010; Lee et al. 2013)—has made it possible to address how complex, goal-directed behaviors occur from organized networks of neurons. Many of the plasmids for these molecular manipulations are available from Addgene,[3] a nonprofit, public repository. An in-depth review of the molecular toolkit is beyond the scope of this discussion. Here we will present an overview, with a focus on resources, advantages, and limitations, followed by a handful of examples that have demonstrated the power of these approaches to help understand the pathophysiology of brain disease.

[3] https://www.addgene.org/ (accessed March 3, 2015)

The Gene Expression of the Nervous System Atlas project,[4] in collaboration with the Intramural Program of the National Institute of Mental Health, offers transgenic BAC-EGFP reporter and BAC-Cre recombinase driver lines which allow for cell-specific gene manipulations in the mouse CNS. The aim of this project is to provide the scientific community with reporter and Cre driver lines that will target selected neuronal or glial populations in the brain and spinal cord. An important caveat is that cell type specificity must be confirmed for each line, since BAC transgenics are not generated by targeted mutation; furthermore, cell type specificity in adult brain need not be reflective of expression patterns during development. In addition to this resource, a vast array of constitutive and conditional knockout and knockin mice are available through commercial repositories, such as Jackson Labs.[5] When transgenic mice, particularly conditional knockouts, for the gene of interest are not available, RNA interference is a viable alternative at most synapses. Because various forms of RNA interference are encoded by relatively short sequences, it is feasible, and indeed often necessary, to control for off-target effects using molecular replacement strategies (Alvarez et al. 2006; Jurado et al. 2014).

The introduction of exogenous DNA into neurons by viral transfection has become a standard technique in molecular neurobiology, particularly as our understanding of viral tropism, life cycle, transport, and toxicity has led to the development of increasingly sophisticated recombinant strategies. An illustrated comparison of the viruses commonly used for neuroscience (Table 10.1) highlights the relevant features of each (for in-depth reviewa, see Callaway 2008; Grimm et al. 2008; Luo et al. 2008; Marie and Malenka 2006; Neve et al. 2005; Salinas et al. 2010). This approach can be a powerful adjunct or alternative to the BAC driver lines, particularly when synapse-specific or developmentally restricted expression is required. Nevertheless, *in vivo* viral infection requires labor-intensive stereotaxic injection, except in cases where the vector can cross the blood-brain barrier (e.g., AAV-9; Foust et al. 2009). Finally, this approach is also a promising method under development for therapeutic gene delivery in human patients, with the caveat that viral immunogenicity is often species specific, so not all vectors used in mice are appropriate for humans (Mingozzi and High 2011).

The ability to optically stimulate molecularly specified neurons (termed "optogenetics") has transformed neuroscience (Boyden et al. 2005; Lima and Miesenböck 2005; Lin et al. 2009). Stimulation of neurons with metal or glass electrodes, while still a mainstay, can only resolve individual input pathways when these are arranged in such a way that they can be physically segregated (e.g., Schafer collateral versus perforant path inputs to the CA1 region of the hippocampus). However, this anatomical arrangement is exceptional: in the vast majority of brain regions, inputs are intermingled, and thus stimulation

[4] http://www.gensat.org (accessed March 3, 2015)

[5] http://jaxmice.jax.org/ (accessed March 3, 2015)

Table 10.1 Comparison of commonly used viruses in neuroscience.

Virus	Description	Reference
Recombinant Sindbis (SIN) virus	• Used when rapid, robust gene expression is required • Suffers from precipitous neuronal toxicity, limiting use to short-term expression studies	Chung et al. (2013) Marie and Malenka (2006)
Lentiviral virus (LV)	• Widely used when relatively sparse transfection is desired (e.g., to examine cell autonomous effects) • Appropriate for both acute and long-term expression since LV shows very low cellular toxicity • Because it is a retrovirus, insertion of exogenous DNA into the host genome may disrupt native transcription (positional effect) • Since LV is enveloped, it can be pseudotyped with rabies virus glycoproteins to enable retrograde transport in neurons • Utility of presynaptic uptake and retrograde transport has been exploited most readily with recombinant rabies virus	Klapoetke et al. (2014)
Rabies virus (RbV)	• Is rapidly and robustly expressed • Has relatively slower neurotoxicity compared to SIN and HSV • Because it is an RNA virus, expression cannot be driven by cell type-specific promoters, although alternative strategies have been developed	Callaway (2008) Wickersham et al. (2007)
Canine adeno virus type 2 (CAV-2)	• Is also taken up presynaptically and retrogradely transported • Is a promising alternative to RbV, since it has very low neurotoxicity and is a DNA virus with a relatively high cloning capacity of (36–38 kb)	Hnasko et al. (2006) Salinas et al. (2010)
Adeno-associated virus (AAV)	• Most widely used due to its rapid expression and lack of pathogenicity • New hybrid serotypes have been developed by merging desirable qualities of naturally occurring AAV capsids • AAV-DJ serotype has exceptional neuronal tropism and is particularly well suited for manipulations requiring rapid, robust expression • Limited by relatively restricted cloning capacity (5 kb)	Grimm et al. (2008) Xu et al. (2012)
Herpes simplex virus (HSV)	• Exceptionally high cloning capacity of 150 kb • Efficient transgene expression • Depending on the envelope proteins, potential for anterograde and retrograde transport • Advantages are counterbalanced by neuronal toxicity and immunogenicity	Neve et al. (2005)

of known inputs to a specific cell type is frequently impossible. Furthermore, it is increasingly apparent that many of the assumptions concerning input and output homogeneity, even in well-circumscribed pathways, do not hold (e.g., co-release of transmitters, novel parallel pathways) (Graves et al. 2012; Tritsch et al. 2012; Varga et al. 2009). In addition, optogenetics enables convergence of cellular and behavioral studies since molecularly isolated inputs can be stimulated to both record electrical responses in specific neurons as well as to interrogate the behavioral consequences of evoked responses. This is particularly important in brain regions where the "receptive fields" of the neurons in question are internal states and are not reliably evoked by direct manipulation of the sensory or motor environment. Despite the remarkable advances enabled by the implementation of optogenetics, its current use is restricted by the speed with which the optogenetic proteins can be delivered to membranes. It often takes many weeks or months for adequate expression of optogenetic proteins in axon terminals, thus limiting experiments to late stages in development. Future iterations will likely improve subcellular targeting (e.g., dendritic versus axonal membranes), channel properties (for better temporal fidelity at higher stimulation frequencies), opsin properties (e.g., for better resolution of distinct activation wavelengths), and toxicity due to overexpression.

Genetically encoded voltage (Cao et al. 2013; Dimitrov et al. 2007; Siegel and Isacoff 1997) and calcium (Chen et al. 2014a; Wang et al. 2004) sensors represent a parallel set of emerging technologies, which will do for the recording electrode what optogenetics has done for the stimulating electrode: allow sub-(action potential)-threshold recordings in molecularly specified neurons, both *ex vivo* and *in vivo*. Currently, voltage sensors are limited to *in vitro* approaches in mammalian systems, and the use of genetically encoded calcium indicators *in vivo* is restricted to microscopically accessible brain regions (e.g., the somatosensory cortex). However, ongoing development of brighter fluorophores and endoscopic techniques may likely overcome these hurdles in the near future (Deisseroth and Schnitzer 2013; St-Pierre et al. 2014).

Finally, the G protein-coupled receptor superfamily represents the canonical targets of more than 30% of clinically available pharmacotherapies (across all indications), largely because these molecules are readily druggable, act as modulators of nearly every known physiological process, and have been implicated in the etiology or pathogenesis of numerous disorders. Despite this profile, interrogating the neuronal function of these receptors is complicated by the fact that a single endogenous ligand typically binds multiple receptors, can modulate different signaling pathways through a single G protein-coupled receptor, and are expressed across mixed populations of cells within a given brain region. The advent of second-generation molecularly encoded, pharmacosynthetics, namely the designer receptors exclusively activated by designer drugs, has made deconstructing this functional complexity a tenable goal (Dong et al. 2010; Lee et al. 2013).

While many of these techniques can be used independently in other model organisms, the power of modeling in mice is the opportunity to use a combinatorial approach. For example, AAV (adeno-associated virus)-mediated expression of channelrhodopsin-2 in the striatum of BAC-transgenic Cre driver lines under control of regulatory elements for the dopamine D1 or D2 receptor has enabled direct activation of basal ganglia circuitry implicated in PD. These studies have validated the long-standing hypothesis that two parallel pathways exert bidirectional control over motor behavior and, furthermore, shown that in a mouse model of PD, direct pathway activation rescues movement phenotypes (Kravitz et al. 2010). Another approach has combined viral-mediated expression of pharmacosynthetics with Cre-mediated targeting of neurons which receive inputs from agouti-related protein-expressing neurons in the arcuate nucleus, to interrogate G protein-mediated feeding behaviors relevant to the pathogenesis of insatiability in Prader-Willi syndrome (Atasoy et al. 2012). Others have capitalized on the ability of rabies virus and AAV to infect selectively pre- and postsynaptic neurons, respectively, combining this technology with conditional knockout and knockin reporter mouse lines. Such approaches, for example, allowed the characterization of oxytocin and serotonin receptor-containing inputs to the nucleus accumbens and implicated a novel circuit in the pathogenesis of social deficits seen in autism (Dölen et al. 2013).

Despite the remarkable technological opportunities available in mice, determining the suitability of this organism for modeling disease requires the consideration of genetic similarities and differences between mouse and humans. Sequencing of the mouse and human genome has revealed that in protein-coding regions, there is 97% alignment between orthologs, and in 1:1 orthologs 85% DNA sequence identity. These genes are the most likely to have maintained ancestral function in both species and are therefore most appropriately targeted as disease models (Church et al. 2009). Despite this remarkable homology in protein-coding regions, only 33% of whole genomes align, with 60% DNA sequence identity in aligned regions. This discrepancy is due, in part, to structural variants and segmental duplications (i.e., evolutionarily young and rapidly evolving parts of the genome) in gene regulatory regions, as well as to transcribed microRNA and noncoding RNA sequences (Church et al. 2009). Modeling disease that affects these nonhomologous genomic regions will require alternate systems. For example, patient-derived induced pluripotent stem cells may be appropriate for modeling schizophrenia, since most identified schizophrenia risk variants are regulatory (e.g., cis trans transcription regulators) affecting the time, place, and rate of gene expression. Currently, the success of such an approach will depend on the degree to which the pathogenesis of disease is cell autonomous, although chimeric and organoid systems may be able to overcome this limitation in the future (Anderson and Vanderhaeghen 2014).

In addition to these genetic differences, anatomical differences (e.g., a small medial prefrontal cortex) as well as a limited repertoire of behavioral assays for interrogating complex cognitive function (e.g., episodic learning, language) limit the use of mice for modeling all features of complex neuropsychiatric disease. New genome-editing technologies, such as clustered, regularly interspaced, short palindromic repeats, and associated cas genes, have made it possible to make precise genetic manipulations in many other organisms, including primates (Ran et al. 2013). Considerations such as size, cost, generation time, and ability to breed in captivity will importantly influence the selection of organism for genetic modeling.

The common marmoset, *Callithrix jacchus*, a New World monkey, has significant advantages from a genetic perspective. It is small in size (~400 g), reaches sexual maturity at 12 months, and breeds rapidly in captivity, typically producing two pairs of fraternal twins per year. The neuroanatomy of common marmoset is well described. Like macaques, but unlike rodents, marmosets have a well-developed prefrontal cortex, a critical region for many cognitive functions that are impaired in human psychiatric disorders. Furthermore, marmosets are very social and communicative and can perform some higher cognitive tasks developed for macaque monkeys and humans. The marmoset genome has recently been sequenced, laying the necessary groundwork for genetic manipulations. Moreover, the optogenetic tools described above are also being developed for use in primates (Diester et al. 2011; Galvan et al. 2012).

A number of other organisms offer potential as model systems for understanding the pathogenesis of complex neuropsychiatric disease. For example, SAGE labs[6] has developed several rat models of PD, Alzheimer disease, schizophrenia, and autism, as well Cre lines for specific expression of floxed constructs in dopaminergic neurons. The larger size of rats makes them more amenable to *in vivo* recording as well as an extensive set of well-characterized behavioral assays. Other species of potential interest for future genetic model development are the Etruscan shrew (active touch; Brecht et al. 2011), the prairie vole (pair bonding; Barrett et al. 2013), and the scrub jay (episodic memory; Raby et al. 2007).

Modeling Genes and Environment

The importance of environmental risk factors for neurodevelopmental disorders is well established. Thus, in the long run, it will be essential to interrogate the role of environmental influences in model systems. A major complication, however, is that multiple environmental variables have been identified as potential triggering factors for brain disorders when associated with genetic susceptibility. In schizophrenia, for instance, obstetric/birth complications, loss of parents in early childhood, prenatal infection, physical or psychosocial abuse,

[6] http://www.sageresearchlabs.com/research-models (accessed March 3, 2015)

cannabis use in adolescence, and urbanicity may represent up to 60% of the contributing etiological factors. Many of these factors are difficult to replicate in animals and each one (e.g., urbanicity) consists of many variables, any one of which may critically contribute to the pathogenesis of psychiatric disorders. Nevertheless, attempts have been made to model experimentally some of these factors in animals. For instance, rodent models of prenatal infection using poly I:C treatment (which induces a cytokine-mediated viral-like response) or maltreatment in early life through maternal separation have been found to recapitulate some components of symptomatology. However, to examine gene-environment interactions optimally in experimental animal models, these manipulations need to be combined with animals that express genes, the influences of which on brain function are most susceptible to environmental variables. The analyses of chromatin states and epigenetic alterations across the genome in patients may be one way to gain knowledge on the potential genetic loci affected by nongenomic mechanisms (i.e., DNA methylation, noncoding RNAs). However, this will be a challenging endeavor since epigenetic modifications are often cell type specific.

Conclusions

Interfacing between model systems is critical to approach the daunting complexity of human neuropsychiatric disorders. As our understanding of the etiology and pathogenesis of brain disorders grows, these approaches should be used in an iterative fashion, with the ultimate goal of designing novel therapeutic strategies for the myriad of diseases for which current treatments are of modest utility.

From Cellular and Animal Models to Patients

11

Using a Stem Cell-Based Approach to Model Neurodegenerative Disease

Where Do We Stand?

Lee L. Rubin

Abstract

Interest in improving methods to study human disease is rapidly growing, with the goal of providing effective therapeutics for serious, complex disorders. Following the discovery of cellular reprogramming, much of this interest has focused on the role of induced pluripotent stem cells (iPSCs). The notion is to use patient-derived iPSCs to produce differentiated cells with disease genotypes and phenotypes. These differentiated cells, theoretically available in unlimited numbers, can then be used to analyze disease pathology as well as to identify and test therapeutics in a highly personalized manner. The idea is simple, but execution difficult. This chapter reviews the advantages and complexities of this new approach.

Introduction

For over a decade, the limitations of model systems used to study human disease have been discussed and dissected. Driving this, in part, was the recognition that mice are only capable of representing certain components of human physiology, as can readily be observed in the context of neurodegenerative disease. For example, the standard mouse model used to study amyotrophic lateral sclerosis (ALS) involves significant overexpression of mutated *SOD1* and is problematic for several reasons: *SOD1* mutations account for just a small percentage of ALS cases. In addition, the accelerated course of the mouse version of the disease, which is presumably due partly to *SOD1* overexpression, makes the mouse easier to study, but it does not accurately reproduce the human condition. For Alzheimer disease (AD), the "triple transgenic" mouse engineered to express three separate disease-related mutations seems advantageous because it reproduces, in a reasonable timeframe, much of the pathology and some of the behavioral defects found in AD patients. However, it is again artificial; none of the mutations expressed in the mouse model have been found in individual patients, leaving some investigators to question whether studying this mouse can provide any disease-relevant information.

An even better example may be the set of psychiatric diseases that is presumably more heterogeneous and complicated than either ALS or AD. In numerous, provocative papers, Hyman discussed the "mouse problem," stating in no uncertain terms that neither new information nor new therapeutics have arisen in decades, owing in part to the use of mouse models and their inability to represent complicated behavioral human diseases (Hyman 2014). Differences in cortical development between mouse and humans may be to blame, but the problem is not confined to just neurological and psychiatric disorders. In a recent article, Seok et al. (2013) described how inadequately mouse models of sepsis mimic the human clinical condition.

Of course, as described by Robbins (this volume), mouse models may be used to make increasingly reliable predictions concerning human pathophysiology. However, logically, human disease is best studied using diseased cells from humans. Historically, for neurodegenerative diseases, this has only been possible through postmortem tissue. The disadvantages from this are obvious. Conclusions drawn from such analyses resemble a CSI (crime scenes investigation) type of method: scientists try to surmise what happened during the course of the disease to produce the changes observed in postmortem tissue. The focus is on pathology, rather than on physiology. However, notably, postmortem tissue often lacks the very cells that were affected by the disease, since they have already died. In spinal cord biopsies obtained postmortem from ALS patients, for example, very few motor neurons remain. In AD, neurological scientists are still debating whether amyloid plaques, a pathological hallmark of the disease readily seen in AD patient brains, are anything more than tombstones, denoting that something bad has happened, but not proving that plaque formation causes the disease.

Everyone agrees that we need to be able to study the progression of disease at its earliest stages, and introduce interventions accordingly, but how can this be done? Early intervention is particularly important as recent observations suggest that underlying disease-related changes in familial cases of AD occur over a 20-year period, and this could be characteristic of other degenerative diseases. Tremendous advances in imaging and biomarker analysis are beginning to reveal some of these earlier changes. In addition, genome sequencing is pointing investigators toward disease-related genes, and that type of information has already been extremely useful to cancer biologists. Nonetheless, having *in vitro* human cell-based systems that are capable of recreating diseases from beginning to end would likely contribute substantially to our knowledge of neurodegenerative diseases.

Problems that scientists experience in studying disease become magnified when the research shifts to identifying therapeutics that work. Again, it is reasonably obvious that an imperfect understanding of human disease progression will lead to suboptimal targets for drug discovery. In addition, the limited availability of faithful mouse models, combined with access to postmortem human tissue alone, means that many therapeutics are focused on late, rather than early, events. The inherent disadvantages associated with studying cells already dying are exacerbated by other flaws in the drug discovery system practiced virtually universally in the pharmaceutical industry. The problem can be defined in quite a simple way: too much money (billions of dollars per drug) is spent developing far too few breakthrough treatments. Even worse, more and more money is being spent, more and more screens are being done, but the number of new drugs is not increasing concomitantly. The efficiency of the entire process is shockingly low, in addition to being (perhaps) unavoidably lengthy.

Numerous articles attempting to understand the underlying causes of these issues have been published, and many possibilities have been raised:

- All of the easy-to-discover drugs have already been found.
- Individual target-based approaches are less good than strategies aimed at correcting overall disease phenotype.
- The nature of the chemical matter used for drug screens and drug development is suboptimal.
- Decision making is poor.

Here I focus on a simple, not yet validated, hypothesis: namely, the process of finding safe and effective drugs will improve greatly if drug discovery campaigns place human cells—diseased and control—front and center. This hypothesis is based on the fact that most pharmaceutical projects use highly engineered cells or cells which are not disease relevant (e.g., motor neuron diseases have never been studied in human motor neurons) in screening and *in vitro* efficacy testing. Furthermore, drug safety is established (by FDA requirement) in two nonhuman species so that human safety data are not available preclinically. Finally, there is no current method, at least for the bulk of neurodegenerative diseases, to match specific treatments to individual patients who may most benefit from them. Most neurodegenerative diseases are heterogeneous, reflecting individual mutations (often in an extremely small percentage of patients), environmental components, and/or aging. Thus, it is perfectly possible that at least a subset of patients with particular neurodegenerative diseases could have been treated with specific drugs that are ineffective on the majority of patients enrolled in a trial (Engle and Puppala 2013; Merkle and Eggan 2013; Yu et al. 2013a).

History of the iPSC Approach to the Study and Treatment of Human Disease

In 2002, Jessell and colleagues described a simple protocol to produce spinal cord motor neurons from wild-type mouse embryonic stem (ES) cells using just two small molecules: retinoic acid and an agonist of hedgehog signaling (Wichterle et al. 2002). This method was a successful attempt to approximate embryonic spinal cord development. A surprisingly small number of investigators recognized this discovery as a key enabler of a new approach to studying disease. In essence, this paper showed that it was possible, for the first time, to produce large numbers of motor neurons—billions of them—when, previously, scientists had access to thousands or, in the extreme, hundreds of thousands of those cells. Furthermore, this method could be applied to mouse ES cells expressing various mutant genes, such as those found in ALS and the genetic motor neuron disease spinal muscular atrophy (SMA). In other words, investigators could viably produce control and diseased mouse motor neurons in large enough numbers to carry out drug screens. A few years later, it was shown that motor neurons could also be produced from human ES cells using a similar protocol (Amoroso et al. 2013).

One reason that this concept failed to catch on was that it relied on the use of ES cells. This was particularly controversial in the pharmaceutical world, which tends to be inherently conservative. Another limitation revolved around poor availability of human ES cells, especially those carrying disease genes. Both issues disappeared with the discovery of cellular reprogramming, the generation of induced pluripotent stem cells (iPSCs) from skin and blood, by Yamanaka and colleagues (e.g., Takahashi et al. 2007; Takahashi and Yamanaka 2006; Yamanaka 2007). It became possible—indeed, even practical—to produce iPSCs from tens or hundreds of patients with diseases of interest. It also suggested a conceptually simple way of establishing assay systems focused squarely on differentiated human patient cells recapitulating human disease processes (Rubin 2008). Fully implemented, this approach can be broken down into a series of seemingly straightforward steps:

1. Identify appropriate numbers of patients with disease of interest and obtain skin or blood samples.
2. Produce clones of iPSCs (typically three well-characterized clones) from each patient.
3. Differentiate iPSCs into neuronal type(s) affected by the disease of interest (e.g., motor neurons for ALS and SMA).
4. Establish culture conditions so that relevant phenotypes can be observed.
5. Follow the progression of the "disease in a dish" so that its earliest deviations can be pinpointed.

6. Validate findings from steps 4 and 5 across an array of individual patients to ensure that they are observed in more than a single set of patient cells.
7. Use this information to discover appropriately screenable targets (e.g., receptors, kinases) or phenotypes (e.g., decreased synapse number).
8. Screen appropriate libraries (small molecules, biologics, siRNAs) against the target or phenotype using cultures of diseased neurons.
9. Test best hits (before and after chemical optimization, when appropriate) on neurons produced from multiple patients. This information becomes part of the preclinical *efficacy* package.
10. Test efficacious compounds for safety on individual patient-derived cardiac myocytes and hepatocytes. These results constitute part of the preclinical *safety* package.
11. Test best compounds on groups of patient neurons prior to a Phase I study. Choose most responsive patients for enrollment.

In short, if this concept were to be validated (ultimately in the clinic), the outcomes would be significant. It could lead to an improved understanding of various types of neurodegenerative disease and result in better targets for drug identification. More relevant cells would be available for screening and safer, more efficacious therapeutics could be developed. Our ability to test particular drug candidates on individual patient cells would be enhanced and clinical trials initiated using specific patients who are most likely to benefit from candidate drugs.

Implementation of the Method

In this section, I highlight various considerations related to particular steps that will be required to use iPSCs for drug discovery and other types of studies.

Production of iPSCs

Unfortunately, none of the steps involved in this method is without controversy. Many papers have addressed issues related to cellular reprogramming, some of which are worth emphasizing (Cahan and Daley 2013; Liang and Zhang 2013). The first is that iPSCs are generally clonal: each person is, in essence, represented by the progeny of one cell, and the cell-to-cell variability in gene expression and in mutational burden is well known.

Currently, multiple methods are employed to produce iPSCs. While they may be functionally similar, the reprogramming achieved by all of them is thought to be both incomplete (with cells retaining an epigenetic memory of their tissue of origin) and capable of introducing additional genetic and epigenetic variability. Furthermore, while it has been clear for several years that human ES cell lines differ, iPSCs, although ES-like, are even more variable. Hence, there is an understandable lack of clarity surrounding the question of how many iPSC lines should be used for each individual patient. Similarly, because there is patient-to-patient variability among individuals even with simple monogenic diseases, it is not entirely clear how many patients should be used to represent a disease cohort. This problem is magnified enormously when studying heterogeneous diseases, such as psychiatric disorders.

Differentiation of iPSCs into Neurons

Surprisingly, motor neuron production from human ES cells or iPSCs is still one of the most efficient differentiation systems available. To produce other neuronal types (including cortical neurons, excitatory and inhibitory and midbrain dopaminergic neurons), production techniques are available. Three features characterize most of the differentiation methods, which, as with motor neurons, are generally based on recapitulating mouse embryonic development. First, most are extremely time consuming: producing neurons from stem cells can easily take one to four months, in addition to the three-month period required for iPSC generation. Second, there is surprising variability in the efficiency of the different methods, even when practiced by scientists within individual labs. Third, neurons in the cultures may be too immature to model late onset diseases accurately (discussed further below).

The variability issue is widely recognized. Most cultures described in the literature contain 20% or so of the desired cell product, in addition to residual progenitors, other types of neurons and, eventually, glial cells. Some neurobiologists assert that mixed cultures are preferable since this may reproduce the *in vivo* environment more accurately. However, the variability and heterogeneity of the cultures present challenges for some types of investigation. For example, microarray and RNA sequencing studies that compare diseased and control neural cultures may be compromised by the variability of cellular composition in the cultures. This is also true for some kinds of phenotypic assays (e.g., synapse formation) where results can be affected simply by having variability in percentages of the component cell types. Although it appears that particularly effective cocktails of inducing molecules can partially overcome intrinsic differences among iPSC lines, it may be overly optimistic to imagine that standard differentiation protocols will ever become efficient enough to produce completely uniform populations of individual kinds of neurons. Thus, it will be important to have the ability to purify desired neuronal subtypes. In the near future, this will be possible by using new genome-editing techniques (TALEN, CRISPR) to introduce cell type-specific reporters or by identifying specific sets of surface markers that allow fluorescence-activated cell sorting purification of neuronal populations. Purification of different types of neurons can be followed by remixing them, so that defined cultures with multiple neuronal types can be prepared. While technically feasible, this method may still not capture all of the biology of live brain tissues.

Other Ways to Derive Neurons from Patient Biopsies

Recently, differentiated cells have been obtained from patient fibroblasts using techniques that do not require an iPSC intermediate. These methods are generally referred to as transdifferentiation, direct differentiation, or lineage reprogramming and use the underlying principle of reprogramming (i.e., the dominant action of particular, often small, sets of transcription factors). For instance, expressing high levels of small numbers of transcription factors found in embryonic neurons and their progenitors is sufficient to convert fibroblasts to neurons (Chanda et al. 2013). Reassuringly, induced neurons made from mice expressing neuroligin-3 mutations displayed a synaptic defect similar to those seen in mouse neurons *in situ*. Other interconversions are also possible; for example, astrocytes overexpressing neurogenin-2 become neurons *in vitro* and *in vivo* (Guo et al. 2014). Surprisingly, this technique does not produce more uniform cultures than more standard differentiation methods. However, since it bypasses the reprogramming step, it can produce differentiated cells in weeks, rather than months. One possible disadvantage is that there is no recognizable neural progenitor stage between fibroblasts and neurons, since differentiation does not occur through a normal developmental sequence. Thus, if an investigator desires to produce large numbers of neurons, the starting fibroblasts themselves have to be expanded. At some point, highly expanded fibroblasts, especially those obtained from older patients, may senesce or experience a reduced ability to differentiate. A similar and promising concept has been applied to iPSCs which, following expression of the single transcription factor neurogenin-2 and puromycin-based selection, can produce electrophysiologically active cortical-type neurons in relatively good purity (Zhang et al. 2013).

Functional Validation of Neuronal Identity

What are the best ways to characterize stem cell or fibroblast-derived neurons? On the surface, this may seem like a trivial question. For example, motor neurons differentiated from stem cells express the motor neuron transcription factor Hb9, synthesize acetylcholine, innervate skeletal muscle cells in culture, and send axons to limb muscle when transplanted into chick embryos. However, these motor neurons have a rostral-type identity and must be treated with additional factors to form more caudal motor neurons or neurons found in particular motor pools.

Other types of neurons have been evaluated in similar ways. Which transcription factors and other individual markers (e.g., transporters, neurotransmitter receptors, synthetic enzymes) do they express? Do they have normal membrane potentials, produce synaptic proteins, and form spontaneously active synapses? Often investigators will call neurons "mature" if they express a small number of appropriate markers, fire action potentials, and form synapses. However, these cells are not truly mature since embryonic and fetal neurons express the same markers and also form synapses.

A stricter criterion, one preferred by more rigorous neurobiologists, judges cell functionality after transplantation. For example, Sudhof and colleagues found that induced human neurons were physiologically active when analyzed six weeks after transplantation into newborn mouse brain and were even capable of receiving synaptic input from existing mouse neurons (Zhang et al. 2013). Similarly, Espuny-Camacho et al. (2013) transplanted immature human cortical neurons into neonatal mouse brain and found that after several months they projected extensively and were able to form synapses with the host neurons. A higher-order demonstration of functionality would be similar to one achieved by Macklis and colleagues (Czupryn et al. 2011), who transplanted immature neurons *purified from normal embryonic hypothalamus* into brains of postnatal leptin-deficient mice and achieved behavioral rescue.

Discussions of how to define cell identity are not without basis. For example, recent work indicates that cells derived directly from fibroblasts maintain (by gene expression) a partial fibroblast character even when they show properties of other types of differentiated cells. Interestingly, it appears that cells can be functional, to at least some degree, even with a mixed identity. Perhaps of more concern are other sets of new experiments which seem to demonstrate that all *in vitro* differentiation protocols, which may closely mimic but are not identical to the *in vivo* environment, can only generate cells that are similar (but again not identical) to their "real" counterparts. This may mean that a compromise definition for cellular identify may need to be adopted—one based on functional outcomes, such as recapitulating properties of particular diseases.

How Mature Are Neurons That Are Differentiated from iPSCs (or Fibroblasts)?

From a purist's point of view, transplantation experiments measure only the potential of cells to function *once delivered to an appropriate* in vivo *environment* (i.e., they may not have been truly functional prior to transplantation). This is an important distinction because a variety of studies now suggest that most cell types produced from stem cells are quite immature. Given that *in vitro* differentiation methods are designed essentially as near (or not-so-near) copies of *in vivo* embryonic processes, it is not surprising that the cells produced are immature. However, the fact that most types of cells do not appear to mature appreciably under standard culture conditions is of more concern. Famously, pancreatic β-cells of stem cell origin are poor responders to high glucose; they only develop this capacity after weeks of being injected into a mouse kidney capsule. For researchers interested in modeling late onset neurodegenerative disorders, this is particularly significant. While it may be that immature neurons are still an advance over nonneuronal cells, a strategy for producing more mature cells is needed. Intriguingly, Sasai (2013) and Lancaster et al. (2013) utilized three-dimensional suspension cultures and found that, under those conditions, stem cells display a surprising amount of self-organization, constructing organ-like tissues of different kinds. Still, even the most organ-like individual tissue is still an individual tissue and will not reproduce cell phenotypes that rely upon more complicated interactions among organs.

Using Patient Neurons to Study Neurodegenerative Diseases

What Types of Diseases Can Be Studied?

From the perspective of disease modeling, nervous system disorders can be distributed into different categories: (a) primarily familial versus primarily idiopathic, (b) monogenic versus polygenic for familial forms of disease, (c) early onset versus late onset, and (d) cell autonomous versus noncell autonomous. Most investigators agree that early onset monogenic cell autonomous diseases are the easiest to study, whereas late onset, predominantly idiopathic, diseases are the hardest. Practically speaking, most neurological disorders involve multiple cell types, often including glial cells and immune cells, to some degree. However, many disease models are only expected to reproduce the cell autonomous components readily (although exceptions exist). Furthermore, questions have been raised about the likelihood of *ever* being able to use an iPSC approach for late onset diseases or for idiopathic forms of disease. Of course, a disease currently regarded as idiopathic may simply be caused by a complex set of genes whose expression is altered in a small number of patients, making genomics studies more difficult. Importantly, from the perspective of iPSC-initiated disease modeling, sets of mutations, however complicated, should be preserved in derived neurons.

For each class of disease, there are additional considerations: Which patient types have the highest probability of producing a robust, easy-to-study phenotype? Given patient-to-patient variability and additional complications surrounding the production of iPSCs and neurons, what are the most appropriate controls that will allow firm conclusions to be drawn? For obtaining robust phenotypes, choosing patients with the earliest onset, most severe pathology or most rapid disease progression seems to be a rational choice. Surprisingly, the topic of best controls is somewhat contentious. In the first few years of stem cell-based disease modeling, the standard approach was to compare phenotypes observed in a (very) small number of patient lines, often, but not always, from patients with known mutations, to an equally small (or smaller number) of controls. More recently, a preferred approach has been to start with patient iPSCs and use genome-editing technology to produce isogenic mutation-corrected lines to use as controls. Alternatively, some investigators favor the notion of introducing disease-associated mutations *into* control iPSC lines. While each of these has advantages and disadvantages, using multiple starting lines still seems advisable to get around idiosyncratic features of individual starting cell populations.

What Has Been Learned Thus Far?

Logically, for diseases with well-recognized later stages, it would make sense to reproduce the known pathology and then "rewind the clock" to determine the events that accompany disease onset. Thus, starting with AD patient biopsies, it would be exciting to produce neuronal cultures replete with plaques and tangles and with readily detectable synaptic dysfunction. Similarly using PD patient biopsies, it would be an excellent accomplishment to produce cultures of dopaminergic neurons that show accumulation of α-synuclein-containing Lewy bodies followed by selective neuronal death. To date, this has rarely been accomplished. Below, I summarize key features of some disease modeling studies that have been published in the last few years.

Spinal Muscular Atrophy

One of the seemingly simplest of neurological disorders is SMA, a childhood motor neuron disease caused by a mutation in the gene *survival motor neuron1* (*SMN1*). Several years ago, in one of the first disease modeling papers, Svendsen and colleagues produced an iPSC line from a skin biopsy taken from a child with SMA and showed that derived motor neurons had diminished survival (Ebert et al. 2009). Motor neuron death is, of course, an important feature of SMA. More recently, Corti et al. (2012) found reduced survival of SMA patient-derived motor neurons and showed that the defect could be corrected by genome editing. In both cases, motor neuron death occurred within one to two months or so, suggesting that onset of disease may have accelerated in the cell culture setting. Obviously, if this were to hold generally true, it would be very helpful to the disease modeling community.

More recently, my lab produced iPSCs from multiple SMA patients with mild to severe phenotypes (Rubin et al., unpublished). The first thing we did was to compare several different methods of reprogramming, including Sendai virus, episomal and modified RNA techniques. Reassuringly, results obtained comparing iPSCs, as well as motor neurons, were similar using all of the methods. Of extreme importance, the degree of pathology or abnormality that we observed in individual lines was directly related to the severity of the disease in the children who provided the biopsies. A big surprise was that the iPSCs themselves had a phenotype: in general, they had reduced proliferative capacity and were biased toward premature differentiation. This led us to predict that children with severe cases of SMA will have defects in other tissues. In fact, my laboratory previously showed that there is a similar proliferation/differentiation abnormality in muscle stem cells isolated from a mouse model of SMA (Hayhurst et al. 2012). An increasing number of other observations made either in children with SMA or in a mouse model of severe SMA indicates that there will be other tissue defects, as we are suggesting. This reinforces an important use of iPSC-focused studies: namely, the discovery of currently unknown disease symptomatology.

We also observed changes in motor neuron survival, but with some additional features. First, we found that neurons, in general, and motor neurons, in particular, differentiated faster and with greater efficiency when produced from mouse or human SMA ES cells or iPSCs. However, many of the prematurely born motor neurons died quite rapidly. Intriguingly, SMA is characterized by an acute phase followed by a more chronic phase, so this aspect of disease progression seems to have been maintained in the *in vitro* environment. Furthermore, at the single cell level, individual motor neurons with low protein levels of SMN died before those with high SMN. Conclusions from this work suggest that very early therapeutics for SMA (earlier even than currently thought) have the best chance of keeping the highest number of motor neurons alive and that SMN is a more precise regulator of motor neuron survival than has been previously suspected.

Amyotrophic Lateral Sclerosis

ALS is another motor neuron disease, but it is quite different from SMA. It is primarily idiopathic, but perhaps 20% of the cases can be accounted for by mutations in one of many different genes, including *SOD1*, *TDP43*, and *c9orf72*. It is also late onset and involves both upper (cortical) and lower (spinal cord) motor neurons. Nonetheless, it was one of the earliest diseases modeled using a ES cell approach. Experiments carried out using cultures of mouse spinal cord motor neurons made from wild-type ES cells and from ES cells expressing an *SOD1* mutation supported the idea that the disease has both cell autonomous and noncell autonomous components (Di Giorgio et al. 2007; Nagai et al. 2007). Specifically, motor neurons expressing the mutant gene were more likely to die over a two- to three-week culture period than control motor neurons. However, the death was again accelerated when motor neurons were grown in the presence of mouse astrocytes expressing the *SOD1* mutation (although, admittedly, the astrocytes were prepared as primary cultures). The negative effects of mutant glia could also be observed on human motor neurons. While some of these results had already been suggested by studies using transgenic mice, observations with cultured cells were especially convincing and provided a system that might, in principle, allow characterization of the toxic astrocyte factors. These studies had two important achievements. First, they showed that a feature of ALS (a late onset disease) could be observed when starting from stem cells, and that death was significantly faster than what is observed *in vivo* (mice from which the mutant ES cells were obtained die after about four months). Second, the variables involved in regulating the progression of disease could begin to be dissected. Several laboratories are also attempting to produce upper motor neurons from patient iPSCs, and it will be important to see if they can also duplicate disease phenotypes. Furthermore, there may be autoimmune or inflammatory components in this disorder that can perhaps be studied in iPSC systems.

My laboratory used wild-type and ALS mutant ES-derived motor neurons in a first of its kind screen (Yang et al. 2013). We tested about 5,000 small molecules to identify those that could promote the survival of both wild-type and mutant motor neurons. Our goal was to identify drugs that might act both in familial and sporadic forms of the disease. From our analyses, we identified multiple hit classes as well as a new drug target (MAP4K4) for treating ALS. We then showed that one of our most potent compounds was able to improve the survival of human motor neurons representing several different familial forms of the disease, all produced from iPSCs. By way of comparison, we also tested two drugs that had recently failed in Phase III clinical trials. Neither had been tested on human motor neurons prior to entering the clinic, but both had shown at least some efficacy in the mutant *SOD1* mouse model. One, dexpramipexole, was totally inactive on human motor neurons, while the other, olesoxime, showed inconsistent activity across different lines of motor neurons. This work suggests that (a) it is possible to carry out drug discovery screens at reasonable scale, even using motor neurons; (b) valuable information may be gained by pretesting ALS clinical candidates on human motor neurons; and (c) different forms of ALS may respond differentially to individual compounds, and this system may be able to help choose patients who are most likely to benefit from particular types of treatment.

Psychiatric Disease

More recently, scientists have begun to take on the challenge of using iPSCs to study human psychiatric disease, such as autism spectrum disorders and schizophrenia. Although these conditions may be diagnosed at relatively early ages, they encompass mutations in many different genes, each mutation occurring in a small percentage of patients. Environmental contributions seem likely as well. There is a significant debate as to whether there are multiple different forms of these diseases, such that each form may need to be treated with separate cocktails of drugs. Alternatively, many of the genes known to dysfunction are involved in synaptic transmission; thus, identifying a small set of common treatments may also be possible. This presents a particularly good opportunity to compare a genomics-based analysis to an iPSC-derived phenotypic method of study. In cases of extreme disease complexity, but with an underlying albeit complicated genetic basis, stem cell disease modeling might enable the emergence of robust and consistent disease phenotypes. Mapping these phenotypes back to the disease itself could, however, prove quite challenging.

A few have attempted to do so. Brennand et al. (2011) produced iPSCs from four schizophrenic patients, three of whom had a family history of disease, but none of whom had identified mutations. They used normal iPSCs as controls. Examining heterogeneous cultures of cortical-type neurons, they found that a consistent phenotype arose, one of reduced synaptic connectivity, not necessarily accompanied by changes in synaptic function. These defects could be partially corrected by treating cells with the antipsychotic drug loxapine. They also measured alterations in neurite growth, in the levels of some synaptic components and in the expression of particular gene pathways. They concluded that an iPSC approach might be revealing, even with such a challenging disease.

In a somewhat related study, Muotri and colleagues studied Rett syndrome, a genetic form of autism caused by mutations in the *MeCP2* gene (Marchetto et al. 2010). They compared heterogeneous cortical cultures prepared from iPSCs of four Rett patients to cultures prepared from control patient iPSCs. They obtained evidence of structural and functional alterations in the Rett cultures, which were partially correctable following treatment of cultures with IGF-1. Dolmetsch's laboratory studied Phelan-McDermid syndrome (Shcheglovitov et al. 2013), an autism condition associated with heterozygous deletions of chromosome 22q13.3. Again they produced iPSCs, in this case from two patients, and cortical neurons from the iPSCs. They observed synaptic defects, particularly in glutamatergic neurotransmission. These defects could be corrected by treating the cultures with IGF-1, but also, importantly, by expressing the *SHANK3* gene in neurons. *SHANK3* is one of the genes with previously suspected disease relevance in the deleted chromosomal region. Thus, Shcheglovitov et al. (2013) illustrate the progress that can be made even when studying patients with rather large chromosomal alterations (admittedly under circumstances in which there was a good candidate gene).

Parkinson Disease

PD is a classical late-onset disease characterized by protein aggregation (often in the form of Lewy bodies) and death of dopaminergic neurons. It is also thought that environmental factors, such as exposure to insecticide-like molecules which affect mitochondrial function, may be involved. Several studies have used the iPSCs from PD patients. Most of the studies have used familial cases, especially ones with mutations in the gene for *LRRK2* gene, a kinase known to be mutated in both some familial and some sporadic cases of the disease. Nguyen et al. (2011) produced immature neurons from a patient with a *LRRK2* mutation and showed that they were sensitive to a fairly wide range of stressors. In other words, a disease phenotype was discovered, but the known pathology of PD was not reproduced. Therefore, the relevance of the enhanced stress sensitivity to PD remains unknown. In another study, Liu et al. (2012)

also produced *LRRK2* mutant iPSCs from which they derived neural stem cells, which were themselves passaged multiple times to add a stress component, at which point the cells showed deformed nuclei. The investigators then showed that similarly aberrant nuclear morphology could be produced in neural stem cells produced from human ES cells with the *LRRK2* mutation knocked in by viral-mediated gene targeting. Importantly, the phenotype of patient neural stem cells could be corrected by preparing an isogenic control line or by using a small molecule inhibitor of kinase activity. Thus, there seems little doubt that the phenotype relates to the genotype. The overarching question is: Does the phenotype relate to PD in any meaningful way? To answer this, Liu et al. looked at postmortem tissue derived from a PD patient and found that a percentage of hippocampal neurons also had misshapen nuclei.

Both of these papers described phenotypic differences between patient and controls, mostly in the realm of stress sensitivity. This has actually been seen a number of times in different neurodegenerative disease iPSC studies. Since this cellular behavior does not relate to any known aspects of disease pathology, what is this telling us? It may not explain components of the human disease as it develops in its true CNS setting, but may simply be a consequence of exposing cells to "unnatural" *in vitro* environments. On the other hand, perhaps real insight can be gained here—pointing us in the direction of identifying new and early features of even late onset diseases.

A recent study from the Studer lab is instructive in this regard (Miller et al. 2013a). These investigators tried to accelerate artificially the aging of neurons by expressing progerin, a mutated form of the *LMNA* gene involved in Hutchinson-Guilford progeria syndrome. They did get some encouraging results, pointing in the direction of being able to observe changes in dopaminergic neurons more like those seen in human PD patients, and suggest that late-onset aspects of neurodegenerative disease can be reproduced by accelerating the aging or maturation of neurons.

Using Patient Neurons to Find Therapeutics

Once an adequately controlled phenotype has been identified and confirmed across a suitable number of patients, a screen to find therapeutics can be established. There are several different ways in which this might be done. The most routine, or at least the most conventional, is to identify a druggable target based on a molecular dissection of disease phenotype genes or pathways. Under ideal conditions, the target could be screened in its endogenous cell type, although there are other options. Hits from target-based screens should then be checked for their ability to correct disease phenotypes. Alternatively, the screens themselves could be phenotypic (i.e., aim to correct a defect, such as reduced survival, in the diseased neurons) without even knowledge of an individual target. The advantages and disadvantages of each approach are still being debated, but phenotypic screens, in particular, may be able to capitalize on the strengths of iPSC disease modeling by collapsing changes in multiple genes and processes into a small common set of correctable phenotypes.

Most often, screens like these would be carried out and hits confirmed using a single patient line, presumably one with the easiest to quantify phenotypic change. Confirmed hits would be tested on multiple lines to determine how broadly individual classes of compounds might act. In the case of small molecule drugs, optimization of compounds would proceed as usual. As mentioned above, final compounds of interest can be tested for efficacy on patient neurons. However, they can also be tested for safety on cardiac myocytes and hepatocytes derived from the same patient iPSC lines. Admittedly, this type of testing can only capture toxicological events experienced by isolated single cell types, but future efforts will add additional complexity introduced at the organ level. Compounds of most interest would be those predicted to be effective and safe when applied to the highest percent of diseased patients. Alternatively, particular compound classes may be chosen based on their predicted safety and efficacy toward an identifiable subset of diseased patients.

The Future of the "Disease in a Dish" Idea

A tremendous amount of technical progress has been made in establishing systems and approaches capable of improving our understanding of neurodegenerative disease, leading to more effective treatments and a more efficient pathway into the clinic. Ultimately, the ideas put forth here can only be validated definitively once these concepts are tested on patients. This is, admittedly, a large hurdle. Over the next few years, it seems likely that at least some compounds, including those discovered by conventional means, will be tested on human-diseased neurons *in vitro* prior to testing them *in vivo*. Thereafter, it is likely that targets and compounds identified entirely using the methods described herein will be tested in a clinical setting.

At the moment, one of the major sources of disagreement between stem cell biologists and hard core neuroscientists relates to the lack of complexity of the stem cell-derived cultures, and hence, their potential inability to model neural systems in a meaningful fashion. Which is better: studying circuits and behavior in the mouse or studying simple human systems? It is fairly obvious that both systems are necessary and that both will always need to be evaluated critically. However, it seems highly unlikely that the notion of using human neurons, in the ways described in this brief article, will prove totally without merit. Rather, a likely view of the future is that stem cell biology, systems biology, bioengineering, and genomics will combine to produce more sophisticated functional neuronal circuits, followed by methods that allow for interactions among tissues in different organs, to capture important elements of human disease.

12

From Rodent Behavioral Models to Human Disorders

Trevor W. Robbins

Abstract

To what extent is it feasible to generalize behavior in rodents to complex cognitive functions in humans via rodent models of neurodegenerative diseases or neuropsychiatric disorders? This chapter takes the view that prudent application of rodent models is actually essential to drug development in these areas, but has been hampered by what has been at times their naïve and inaccurate application. Some behavior in rodents may constitute building blocks of more complex cognitive or affective domains in humans. Many of the underlying neural systems, together with their chemical modulation and mechanisms of neuronal plasticity, appear to have been conserved across species. Tests of the validity of models should incorporate their underlying neural substrates to triangulate possible homologies in humans. The ultimate validity of the models depends crucially on the way in which the biological bases of the disorder are modeled, for example, in terms of contributory genetic, neurodevelopmental, experiential, or neuropharmacological factors.

Introduction

For pharmaceutical companies, the huge expense of Phase III trials and relative lack of success of developing effective new compounds in clinical neuroscience, including psychiatry, has raised questions over the utility of animal models of complex neurological or neuropsychiatric human disorders. A reasonable case can be made that the pathway to effective and useful back-translation of such models has not always been facilitated by multicenter clinical trials with heterogeneous patient groups, resulting from sometimes imperfect nosology, coupled with insensitive outcome measures based on clinical scales and compounds of limited efficacy with adverse side effects. Nevertheless, in a climate where some have debated the need for any animal models, we should review what constitutes and limits them, and what aspects may and may not be useful in future attempts at effective translation. This article seeks to synthesize elements of my recent previous reviews with colleagues, which focused on animal models of cognition (Keeler and Robbins 2011) and psychiatric disorders (Fernando and Robbins 2011).

What Is a Model?

This issue has been discussed by many authors who rehearse the several criteria employed to validate animal models: face validity, predictive validity, etiological validity, and construct validity. Most of these are self-explanatory, except perhaps the latter, which refers to theoretical plausibility based on an integrated neurobehavioral explanation of a disorder which thus essentially subsumes the other forms of validity and includes behavioral, neural systems, cellular, and even molecular components.

Essentially, animal models depend on two separate elements: the "disease" (or "plausible perturbation") aspect, in which, for example, the molecular pathology of a disease is reproduced, sometimes via monogenetic inheritance; and the "symptomatic" aspect, in which the symptoms (e.g., behavioral or neurophysiological) are enacted and quantified. Both of these depend to some extent on comparability with humans; molecular pathology may show subtle cross-species differences that bear on the model's validity. However, when the underlying molecular pathology of a condition is well-known (e.g., in terms of the "huntingtin" protein of the autosomal dominant Huntington disease), there is obviously some form of comparability with the human condition, even if other aspects of the phenotype do not approximate the human disorder. The precise causal role of the pathology in the disease or the mechanism by which the pathology causes neurodegeneration may, of course, be more difficult to unravel. More problematic are those examples where (a) there are multiple forms of molecular pathology, as occurs for the amyloid plaques and neurofibrillary (tau) tangles in Alzheimer disease, and when (b) the molecular pathology of human disease does not occur in experimental animals, necessitating strategies for producing, for example, "humanized" disease-bearing mutants. Nevertheless, it is difficult not to believe that animal disease models bearing these forms of pathology will play important roles in elucidating disease mechanisms, as well as screening new therapeutic strategies.

I also suggest that further attention to the "construct" validity of such models may also be important. For example, if a therapeutic agent was found that potently blocked the formation of amyloid, but did not prevent neurodegeneration in relevant brain regions or remediate a dysfunctional outcome, such as a memory deficit or even an impairment in long-term potentiation found in those animals, then one would be more skeptical of its likely therapeutic efficacy. In other words, it may still be desirable to subject such "disease models" to tests of memory appropriate for those brain regions at risk in Alzheimer disease.

In the case of neuropsychiatric disorders such as depression and schizophrenia, we are rather far from identifying the molecular pathology that causally leads to symptoms. It is undoubtedly true that genetic analysis may change this situation in years to come, but it is still likely that multiple genes (generally of small effect) will contribute—most likely additively, or via epistasis—to the complex heterogeneity, for example, of schizophrenia; thus it will be difficult to specify precisely what aspects of the disorders will be a consequence of particular genes or, more likely, the converging products of massive gene clusters. Moreover, there are also the complexities of epigenetic factors to consider, as well as the importance of relevant environmental factors gleaned, for example, from human epidemiological studies. Hence, most of the other "plausible perturbations" used in animal models of psychiatric disorders have a developmental or "stress"-related basis, reflecting perhaps a declining tendency to employ pharmacological interventions, such as intracerebral 6-hydroxydopamine or MPTP (Parkinsonian movement disorder), amphetamine or ketamine/phenylcyclidine (psychosis), or scopolamine (cognitive deficit) to "short-cut" etiological considerations. Of these, the production of striatal dopamine loss to model Parkinson disease has produced perhaps the most successful models; however, they still fail to simulate in rodents the same qualitative motor deficits that are seen in human patients with Parkinson disease, and they are unable to reproduce the chronic progression of the disease.

Environmental factors such as "stress" are also difficult to control, partly because of the likelihood of different forms—physical stressors, social (e.g., isolation or "social defeat") stressors, "learned helplessness"—and the difficulties of modeling specific human stressors (e.g., "urban alienation"). Developmental events also vary widely according to their timing and nature, and it is sometimes unclear whether they impact the brain in a disorder-relevant manner. Moreover, our understanding of how best to combine environmental manipulations with genetic factors to optimize the models is weak.

Thus for psychiatric disorders, the task of modeling becomes far more difficult and inevitably focuses, to a much greater extent, on behavioral and cognitive aspects and their neural underpinning. These aspects themselves hinge controversially on the extent to which animal behavior can be said to bear any reasonable approximation to the complexities of human cognition and upon the interpretation of possible neural homologies. One way of addressing this difficulty is to invoke again the concept of construct validity. If the animal behavior in question can be shown to be the product of homologous brain systems, controlled by similar physiological interactions as occur in humans, and subject to the same sorts of environmental conditions or interventions (e.g., stress, therapeutic drug effects), then it becomes more plausibly a "model" of what occurs in humans.

An inherent limitation of the construct validity approach is the importance and utility of the mouse mutant models for molecular genetics. A very real difficulty has been that, compared with rats and primates, mice have hitherto been far from the ideal species for neuroscientific and behavioral (e.g., based on learning theory and operant methods) investigations. This has partly been for historical reasons, but there are also mundane practical considerations, such as brain size. Hence, currently it is simply not feasible to provide optimal models, if the goal is construct validation. This situation is slowly changing due to excellent innovation in this area; see, for example, the elegant review on the application of mouse models to psychiatry by Arguello and Gogos (2006) as well as new behavioral approaches for the mouse (e.g., Bussey et al. 2012). I predict that it will remain a major obstacle to progress until molecular genetic approaches can also be applied more comprehensively to rats and nonhuman primates.

Another limitation of the construct validation approach is the controversy that often surrounds the identification of neural homology. Although there are major differences in size, and sometimes evolution, of brain structure in the parietal, temporal, and prefrontal cortices between humans and other animals, there is also considerable conservation, encompassing structures of the limbic system, basal ganglia, midbrain, and brain stem. For example, the human "reward" or value systems appear to have very similar neural representation in other animals, including rodents with a focus on the ventral striatum and ventromedial prefrontal cortex (PFC) (Haber and Knutson 2010). The classical ascending neurotransmitter systems such as dopamine, which is often identified with reinforcement and motivational functions in humans and rodents, as well as serotonin, noradrenaline, and acetylcholine all exhibit remarkable phylogenetic conservation (and correspondingly similar functions in behavior). At a different level of investigation, it is intriguing to consider whether the molecular basis of different forms of plasticity (especially learning and memory) involve fundamental differences across species. While this is to some extent an open question, it seems likely that the same molecular "building blocks" hold across species.

There are obvious and often impossible hurdles to surmount in animal models of human behavior, including largely unique human characteristics (e.g., the capacity for language and moral reasoning). Even nonhuman primates fail many tests of "social cognition," such as the tendency to reject unfair offers in variants of the "ultimatum game" (Jensen et al. 2007). Moreover, it is dubious that even monkeys entertain a "theory of mind," at least according to all of the main criteria (Penn and Povinelli 2007). Animals appear to lack the capacity for subjective thought, although this, of course, is in part a function of their inability to express it. Doubtless, with this in mind, many have reasoned that it is impossible for animals to experience such complex symptoms as depression or the hallucinations or delusions associated with psychosis. But this may be taking things slightly too far. Subjective symptoms are undoubtedly an important component of psychiatric disorders, but they can mislead and in fact only form part of the symptomatic picture, which is also somatic and behavioral in nature. The latter point is important as, for example, monkeys chronically treated with amphetamine have been described to exhibit behaviors that would be difficult to explain without invoking the notion "hallucinations" (Nielsen et al. 1983). Moreover, operational measures of concepts such as "reward threshold" (Paterson and Markou 2007), "negative feedback bias" (Bari et al. 2010), and the ability to discriminate between different internal states (e.g., drug cue discrimination learning, which can be generalized to emotional states such as anxiety; Vellucci et al. 1988) provide a basis for inferring altered motivational states in animals in a manner that could be used more routinely in human patients to approximate their subjective status.

A related general stratagem is to "triangulate" animal findings in a way that is relevant for human disorders. A major discovery of recent years has been that fast phasic firing of the midbrain dopamine neurons encodes a "prediction error" that represents the difference between expected and obtained outcome (e.g., reward), thus corresponding to an essential parameter of reinforcement learning (Schultz and Dickinson 2000). Translation of this notion to the functional magnetic resonance imaging (fMRI) setting confirms that similar activity in regions homologous to the midbrain ventral tegmental region is found during causal learning in humans. Moreover, ketamine, the NMDA receptor antagonist and well-known psychotogen, disrupts the formation of prediction errors in a way that predicts its capacity to produce delusional phenomena in human volunteers (Corlett et al. 2007). Therefore, by measuring prediction errors and their disruption in experimental animals (including rodents), it may be feasible to gain predictive data relevant to aspects of psychosis.

There is an increasing trend not to model psychiatric syndromes in their entirety, but to focus instead on modeling well-defined symptoms or symptom clusters. This is especially exemplified by schizophrenia, which has major cognitive and motivational symptoms, in addition to the positive symptoms of psychosis. In fact, there is an argument for considering most of the symptoms of schizophrenia, depression, or any of the other major neuropsychiatric disorders as being "cognitive" or "motivational" in nature one way or another. For example, we have seen above how the fundamental basis of delusions may possibly derive from aberrant prediction errors in learning. However, it is much easier to model fundamental units of cognition (such as learning, working memory, or executive control) that also occur in schizophrenia in experimental animals because of the considerable literature that has accrued on the neural mediation of these processes. In parallel, understanding the neural basis of motivation in humans has gained immeasurably from animal studies of drug-related reward processes in the nucleus accumbens (Wise 2004) and learned fear in the amygdala (LeDoux 2000), to name but two obvious examples.

This realization has informed current initiatives to rework or even replace the current DSM-IV/5 nosology in terms of "dimensional criteria" that depend on underlying neuroscience concepts such as the "reward system" (e.g., Research Domain Criteria; Cuthbert 2014) as well as to the possibility that some common deficits may occur across different psychiatric disorders. In terms of the latter, the new classifications may depend on a mosaic of features, some of which may be similar and some different, thus accounting in part for pervasive comorbidities.

A combined neurobehavioral approach to understanding symptoms may also enable the definition of intermediate endophenotypes that represent intermediate markers between top-level symptoms and bottom-level genetic contributions (Gottesman and Gould 2003) to mental disorders. Endophenotypic markers are believed to be closer to the underlying neuropathology than top-level symptoms of a clinical phenotype. Thus, "neurocognitive endophenotypes" could be used to enhance the power of psychiatric genetic studies, to produce purer cohorts of patients for clinical trials, to tailor potentially effects of new treatments more effectively in accordance with burgeoning pharmacogenomic principles, and to identify prodromal states for which it is possible to treat "before the damage is done" (for mild cognitive impairment is prodromal to Alzheimer disease and at-risk psychosis possibly heralds first episode psychosis).

Modeling Cognition

Cognition refers to the set of processes that manipulate representations in the brain in various ways to produce thinking or behavior. These processes include perception, attention, working memory, episodic and semantic memory, symbolic and propositional functions such as language, and executive control processes which coordinate these modular functions to effect decision making and planning. Such mechanisms interact with motivational and emotional processes, and include social aspects of cognition. Cognitive processes are mediated by different, frequently overlapping neural networks that include wide regions of the neocortex, such as the temporal, parietal and frontal lobes, as well as the subcortical brain. Remarkably, the elements of many of these processes can be studied in experimental animals and have been used or could be used for drug discovery. I will illustrate this by surveying some aspects of cognition; an overview of behavioral constructs, animal models, and clinical applications is given in Table 12.1. If one adopts, as here, a "modular" approach that recognizes the existence of separate aspects of cognition under control by different neural systems, in other words assuming that cognition is not unitary, the logical conclusion is that it has to be assessed with a battery of tests. This "battery" approach mirrors similar batteries for testing cognition in humans, such as the Wechsler Adult IQ Scale (WAIS) which extracts a "general" measure of IQ, the MATRICS battery[1] for schizophrenia, and the CANTAB battery[2] which utilizes some tests that can be given to both humans, including clinical patients, and experimental animals, nonhuman primates or, with suitably modified tests,

---

[1] http://www.matricsinc.org (accessed March 3, 2015)

[2] http://www.cambridgecognition.com/academic/cantabsuite/battery (accessed March 3, 2015)

rodents. Human test batteries, however, differ greatly in how they are administered: the WAIS is mainly a set of paper and pencil tests which incorporates tests of verbal intelligence; the MATRICS battery recapitulates many classical neuropsychological tests under separate cognitive domains; the CANTAB battery employs largely nonverbal tests using a touch-sensitive screen to emulate tests of cognition in experimental animals.

The use of touch screens to test cognitive function in humans and other animals is a relatively recent development. The requirement to make actual contact with discriminative stimuli is a considerable aid to ensure rapid learning, as the task contingencies are rapidly discerned, probably because of the elicitation of fundamental Pavlovian approach tendencies toward the discriminative stimuli. The use of touch screens was originally introduced for the testing of nonhuman primates (Gaffan et al. 1984), but was extended to humans (e.g., via the CANTAB battery) and most recently to rodents (Bussey et al. 1994, 2012) with the aim of enhancing translational potential. A recent application of CANTAB has been to show the relationship between a hierarchy of different behavioral capabilities as a function of gene variants of Dlg mutant mice, with implications for the evolution of cognition, conservation of cognitive "building blocks" in evolutionary terms, and for certain neuropsychiatric disorders (Nithianantharajah et al. 2013).

Whether considering human or animal cognition, such functions are generally componential: impairment in perception, attention, or motivation caused, for example, by sedation or satiety might lead secondarily to learning and memory problems. The ideal test of memory would therefore incorporate basic behavioral controls to show that any effects of a putative cognitive-enhancing compound could not be ascribed to such other factors. General sensory, motor, or motivational factors which may affect cognitive function only indirectly (and thus lead to possible artifacts or mistaken reasoning) must also be tested. Sensorimotor functions, for example, in transgenic mice are often tested in a neurological battery (e.g., Irwin 1968). Primary motivation can be assessed in terms of ingestive behavior (eating or drinking), including high- or low-incentive or high- or low-calorie foods, as well as the propensity to work or exert effort; for example, in terms of instrumental (or operant) behavior directed toward a particular reinforcer, as on a progressive ratio schedule where the amount of work required to gain the reward becomes progressively greater and motivation is measured in terms of the "break-point" (i.e., the number of lever-presses a rat is prepared to make to gain the reinforcer). Unfortunately, the use of similar objective tests for human motivation is poorly developed, although modern paradigms employ relevant parameters of performance to assess this, including measures of latency and force of responding. The better designed tests of cognition incorporate a condition or measures that provide internal controls for motivational effects, and several examples of this will be exemplified below. Overall, it is important to note that although animals may vary widely in their motor capacities or their dependence on different sensory

Table 12.1 Summary of behavioral constructs, animal models, and some clinical applications (see text for details). OCD: obsessive-compulsive disorder; 5-CSRTT: 5-choice serial reaction time task; ADHD: attention-deficit/hyperactivity disorder; PTSD: posttraumatic syndrome disorder; CANTAB PAL: CANTAB paired associate learning.

Behavioral process	Behavioral paradigm	Animal equivalent?	Relevant to which neuropsychiatric or neurological disorder
Perception	Gain control Integration	Yes for gain control (cat, monkey)	Schizophrenia
Attention			
Preattentive gating	Prepulse inhibition	Yes (rodent, monkey)	Schizophrenia, Huntington disease, OCD
Covert attention	Posner's spatial attention	Yes (rodent, monkey)	Parietal lobe damage
Habituation	Latent inhibition	Yes (rodent)	Schizophrenia?
Sustained attention, vigilance	Continuous performance	Yes (e.g., 5-CSRTT) (rodent, monkey)	Schizophrenia, ADHD, dementia
Attentional set-shifting	ID/ED shift test, Wisconsin Card Sort	Yes: digging test (rodent), visual (marmoset, macaque)	Schizophrenia, OCD, frontal lobe injury
Associative learning (Pavlovian, action-outcome, habit learning)	Under development	Yes (rodent)	Psychosis, OCD, addiction, autism

Table 12.1 (continued)

Behavioral process	Behavioral paradigm	Animal equivalent?	Relevant to which neuropsychiatric or neurological disorder
Memory			
Consolidation, reconsolidation, extinction	Under development	Yes (rodent)	Anxiety, addiction, PTSD
Recognition memory, recall, declarative memory	Warrington Faces Hopkins Verbal Memory, REY Auditory-Verbal Test, CANTAB PAL	Yes: delayed matching to sample object recognition, paired associates (rodent, monkey)	Dementia, schizophrenia, temporal lobe/hippocampal damage, depression
Reference memory	Virtual water maze	Yes: Morris water maze, Olton maze (rodent)	Dementia, schizophrenia
Working memory	One-back task; spatial	Yes: spatial delayed response (primate), delayed alternation (rodent)	Dementia, schizophrenia, basal ganglia disorders
Executive functions (e.g., cognitive inhibitory control, planning)	Stop-signal task, Stroop test, Tower of London	Yes: stop-signal (rodent, monkey); Stroop but less developed (rodent) No: Tower of London (rodent)	ADHD, schizophrenia, OCD, binge eating, bipolar disorder, basal ganglia disorders, frontal dementia, frontal lobe injury
Cognitive-motivational interface: "hot" cognition	Iowa Gambling Task Delayed discounting Response to faces, words or feedback	Yes (rodent) Yes (rodent, monkey) Only to feedback (rodent, monkey)	Depression, addiction
Social cognition	Trust game, Ultimatum game, "theory of mind"	Not readily modeled in animals	Social interaction/recognition, autism, negative symptoms in schizophrenia

modalities, this most likely does not affect the fundamental principles of learning theory, such as the Rescorla-Wagner rule (an equation that captures the concept of learning through the "surprise" generated by a deviation of actual from expected outcomes, defined as prediction errors) or cognitive representation. Thus, it is possible to generalize findings on cognition from animals to humans despite basic differences in sensorimotor performance.

Testing Specific Aspects of Cognition from Rodents to Humans (and Vice Versa)

Perception

Perception arises from "online" representation of the world by the sensory systems and is ultimately used to make decisions about how to perform. Perceptual capacity is often tested in experimental animals and humans by what appear to be similar methods. Thus, perceptual assessment in rats generally involves the capacity for discrimination, whereby responding in the presence of one stimulus is rewarded or reinforced (e.g., with food for food-restricted rats) while the other is not. Stimuli are presented randomly across two locations to avoid the task being solved incorrectly by the animal on the basis of spatial factors. For humans, perceptual testing is less often accompanied by obvious reinforcing feedback (e.g., by points, money, or praise), although such factors would come into play in the case of training or rehabilitation. Human psychophysical methods, such as titrations of stimulus threshold, determination of contrast sensitivity functions, and application of signal detection theory to separate perceptual sensitivity from motivational bias (as in pain perception), can all be readily applied in rodents. However, an overriding consideration is, of course, that the dominant modality in humans and other primates is visual, whereas in rodents it is olfactory, which clearly limits the utility of rodent models involving perceptual ability.

In the CNTRICS (cognitive neuroscience treatment research to improve cognition in schizophrenia) process, two constructs in perception were identified as being possibly relevant to perceptual disturbance on schizophrenia: gain control and visual integration (Green et al. 2009). The former is defined as processes enabling sensory systems to adapt and optimize their responses to stimuli within a surrounding context. Integration is defined as processes enabling local attributes (e.g., of a scene) to be encoded globally. Of the two, only gain control is readily modeled in experimental animals. The contrast-contrast effect (CCE) task derives from the common illusion that contrast sensitivity is modulated by the properties of adjacent or surrounding stimuli. CCE has been well characterized in terms of its psychophysical and neural underpinning, and has been studied in macaque monkeys and cats (for a review, see Green et al. 2009).

Attention

Attention generally depends on fluctuations in asymptotic test performance that cannot be attributed to changes in motivational state (see Lustig et al. 2013; Robbins 2014). Several forms of attention identified in humans can be modeled in experimental animals: selective attention (focusing on one input or feature, while ignoring the rest), sustained attention (maintaining attention over a long period), vigilance (detecting rare inputs), and divided attention (maintaining attention to more than one input or task). Attentional deficits occur in such disorders as attention-deficit/hyperactivity disorder (ADHD), schizophrenia, addiction, and mania, as well as after brain damage. It is important to take possible attentional deficits into account when considering possible changes in memory and learning; without appropriate input, these processes cannot easily occur.

Preattentive gating. Attention involves both conscious (explicit, overt) and unconscious (implicit, covert) aspects. Moreover, "preattentive" filtering or "gating" processes have been postulated that are also assumed to be unconscious. Within this domain, the prepulse inhibition (PPI) paradigm is the most well-developed test because of its enormous translational validity across species, including rodents, and has been of considerable use in experimental studies of schizophrenia, where implicit measures of cognition are useful. (PPI has also been suggested to be an example of perceptual gain control, see above.) PPI is a modulation of the startle response to loud noises or other intense stimuli. This modulation is an inhibition of the reflexive startle response (frequently whole body displacement) produced by the presentation of a brief surrogate stimulus (generally of the same modality but of much reduced intensity) that occurs immediately prior to the startle stimulus. PPI is impaired by dopamine D2 agonists and remediated by D2 receptor antagonists; it is also sensitive to a number of other pharmacological manipulations (Geyer et al. 2001). PPI is widely used in testing genetic mutants of relevance to schizophrenia because of its objective nature, reliability, and convenience (mass testing being feasible as long as there is suitable auditory insulation). However, impairments in PPI occur in other disorders besides schizophrenia and could possibly be epiphenomenal to the symptoms, although an attempt at "construct" validity suggests that failure of "sensory gating" is a fundamental cause of schizophrenia, presumably subsuming both positive and cognitive symptoms.

Covert attention. A covert form of attention has been described in humans to occur when attention is cued spatially in advance to a particular location by the brief presentation at that location of another stimulus. Attentional capture by such a stimulus occurs apparently automatically and unconsciously. The process can be demonstrated when this cueing is misleading and the subject has to disengage that process and respond to a stimulus that is presented elsewhere. This form of attention has been attributed to mechanisms within the human parietal cortex (Posner and Petersen 1990), but similar processes can be characterized in experimental animals, including rats (Ward and Brown 1996).

Habituation and latent inhibition. Habituation, another attention-like process, can occur during learning when repeated stimuli have no consequence, leading to a waning of response. This reduction is readily measured in animals and humans (often using psychophysiological recordings in the latter case, such as skin conductance or heart rate). However, an additional consequence, that of latent inhibition, can be more persistent and long-lasting. Latent inhibition refers to the retardation of learning about a stimulus that occurs, if it has never previously predicted reinforcement (Weiner 2003). A major hypothesis of latent inhibition is that it reflects the subject simply "ignoring" the stimulus by an active process of inhibition. Latent inhibition (or a related phenomenon termed learned irrelevance) has some advantages as a test of attention in that impairments may be expressed subsequently as improved learning, thus ruling out motivational impairments. Although latent inhibition has translational potential, it is sometimes difficult, for practical reasons, to be sure that what is measured in humans as latent inhibition corresponds to the same process in rodents.

Sustained attention and vigilance. Continuous performance tests measure the ability to sustain attention, as well as vigilance, and usually reveal impairments in disorders such as schizophrenia or ADHD. A simple analogue of this task for rodents is the 5-choice serial reaction time task (5-CSRTT), based on a paradigm used to assess attention in human volunteers in a variety of experimental situations, including stress, distracting white noise, and following drug treatment (Robbins 2002). The 5-CSRTT measures the accuracy (errors of commission) and latency of detecting visual targets, as well as errors of omission and impulsive responding (i.e., responding prior to target onset). The latency to collect food pellets provides a control measure of motivation. The difficulty of the task can be enhanced in various ways, including shortening of the duration of the visual target, varying its rate of presentation and temporal predictability, as well as the occurrence of defined distractors, such as bursts of white noise interpolated into the intertrial interval. This task has been widely employed in rats, and more recently mice (as well as in nonhuman primates), to measure effects of drugs, regional brain lesions, and manipulations of the central neurotransmitters or genetic mutations. The 5-CSRTT has been used to measure beneficial effects on response accuracy of some putative "cognitive-enhancing" drugs, such as dopamine D1 agonists, as well as to characterize impulsive behavior that predicts the escalation of compulsive cocaine-seeking behavior in rats. A human "4-choice" version of the task has recently been shown to reproduce deficits in impulsive responding in patients who exhibit methamphetamine abuse (Voon et al. 2014) or humans receiving dietary tryptophan depletion, which mimics the impulsive behavior produced in rats on this task with extensive forebrain serotonin depletion (Worbe et al.

2014). The CANTAB "5-choice" version has been used to demonstrate attentional improvements in patients with Alzheimer disease treated with the anticholinesterase tacrine (Sahakian et al. 1993). This demonstration was borne partly out of data showing that cholinergic agents could improve attention in rats with lesions of the cholinergic basal forebrain, and more recently, in "triple transgenic" mutant mouse models of Alzheimer disease.

There are several variants of the standard 5-CSRTT: the primary one requires the rat to make an observing response into a central location to detect a peripheral target, and has been used to quantify the "attentional neglect" that can occur after unilateral manipulations of corticostriatal brain regions (Carli et al. 1985). Another, rather different, operant test requires the cross-modal integration of auditory and visual stimuli (McGaugh and Roozendaal 2008; Lustig et al. 2013) and includes an important control that animals must detect not only the presence of the stimulus but also its absence. Recent developments in touch-screen technology have enabled the development of a continuous performance task that is very similar to that used in humans, requiring responses to specified visual targets (see Bussey et al. 2012).

Attentional set shifting. Attentional set-shifting tests that are sensitive to frontal lobe damage in humans, such as the Wisconsin Card Sorting Test, involve the formation and shifting of attentional "sets" and the capacity to avoid a prepotent response to one aspect of a stimulus in order to respond to another (Keeler and Robbins 2011). This latter aspect of inhibitory control over prepotent tendencies has resulted in these tests being employed as tests of "executive function" (see below). Such tests have proven highly "translational" in that they have been applied to a range of experimental animals (mice, rats, marmoset monkeys, rhesus monkeys) and humans in ways that appear to reflect neural homologies. Thus, attentional set shifting appears to implicate the prefrontal cortex (lateral PFC in primates, including humans, though medial PFC in rodents) on the basis of both lesion and neuroimaging studies, whereas reversal learning appears to implicate the orbitofrontal cortex from studies of humans, marmosets, rats, and mice (see review in Keeler and Robbins 2011).

The attentional set-shifting task uses compound visual stimuli, which vary in at least two perceptual dimensions: visual shapes and superimposed lines for primates (Dias et al. 1996); texture and odor stimuli in different modalities for rodents (Birrell and Brown 2000). Humans, nonhuman primates, or rodents (rats or mice) are trained to attend to one dimension (on the basis of positive feedback or reinforcement) and to ignore the other. Training is accomplished through tests of reversal (where the two exemplars within a perceptual dimension have their reinforcement contingencies reversed so that what was previously correct is now incorrect and vice versa) and intradimensional shifting (where novel exemplars are introduced, but the same dimension, e.g., lines or shapes, is reinforced). These training stages serve to focus the animal on particular stimulus features and to ignore the others, with the result that they have a prepotent tendency to respond to the trained dimension (e.g., shapes). Finally, an extradimensional shift (ed-shift) is programmed in which novel exemplars are again introduced, but now the previously irrelevant dimension is reinforced. This latter stage is analogous to the category shift on the Wisconsin Card Sorting Test. In the rodent version (being available for mice as well as rats), the test is implemented using olfactory cues and texture in a "digging for food" test paradigm (Birrell and Brown 2000). Performance across the various stages is qualitatively comparable to that seen in primates: the ed-shift and reversal learning are the most sensitive stages to drug effects, performance at other stages usually being employed as internal controls. There are now various versions of these tests of "cognitive flexibility" that use similar logic for shifts between, for example, responding to body turns or to space on a cross maze or alternatively, attending to discrete (e.g., visual) cues versus contextual cues on a maze. These tests do not use different stimuli at each test and so are also confounded by any response to interference.

The attentional set-shifting and reversal tasks have found many applications in neuropsychiatry. Thus, impairments in ed-shifting have been found not only in patients with obsessive-compulsive disorder (OCD) but also their first degree relatives, suggesting that the capacity to shift attention in this way is an endophenotype for OCD (Chamberlain et al. 2008). By contrast, although ed-shifting is also impaired in schizophrenia, it is apparently not impaired in the unaffected siblings of patients (Ceaser et al. 2008). Nevertheless, the attentional set-shifting paradigm has been shown to be sensitive to cognitive-enhancing effects of modafinil in patients with schizophrenia (Turner et al. 2004), and this has allowed an impressive, and rare, case of "back-translation" using the rodent version of this paradigm. In this study, a common "perturbation" to model schizophrenia was employed: subchronic treatment with the NMDA receptor antagonist phencyclidine, which impairs relatively selectively the ed-shift. This deficit was rescued by treatment of the rats with acute modafinil, thus mimicking the human finding (Goetghebeur and Dias 2008).

Associative Learning

Both Pavlovian and instrumental (operant) conditioning have cognitive aspects: the former enables the causal prediction of events in the world leading to expectancy and the latter affords control over environmental contingencies. Disruptions of different aspects of Pavlovian and instrumental learning, alone or in combination, probably underlie all of the major forms of neuropsychiatric disorder, including drug addiction, anxiety, depression, and schizophrenia. In the case of drug addiction, Everitt and Robbins (2005) have proposed that it represents a transition from instrumental action-outcome goal-directed learning (i.e., intravenous drug self-administration) to stimulus-response habit learning, performed in a compulsive fashion. Intravenous drug self-administration arguably contributes to one of the best rodent models of human behavior that accurately predicts drug abuse liability of different compounds. Of course, it is not tantamount to addiction itself; however, a more compulsive element, specified in DSM criteria for addiction, can be discerned when rats work under a second-order schedule to earn infusions of cocaine, as well as occasional electric shocks. A small proportion of animals, rather similar to that proportion of humans who become addicted following cocaine use, continue to self-administer the drug despite these aversive consequences (Belin et al. 2008).

Habit learning can be defined operationally through tests of "outcome devaluation" (such as selective satiety for a particular food), when habitual behavior continues irrespective of the goal. These principles were worked out in studies of rats, but the general principles, as well as the underlying neural substrates, appear to generalize remarkably to normal human learning (Balleine and O'Doherty 2010).

The principles of aversive conditioning are also highly relevant to the understanding and treatment of psychiatric disorders. In the case of anxiety (see Fernando and Robbins 2011), the dominant influence has been Mowrer's two-factor theory, developed mainly in rodents, which suggests that phobic anxiety develops as a consequence of both Pavlovian and instrumental (avoidance) conditioning (Mackintosh 1983). Behavioral therapy for phobias depends on a principle of exposure (repeatedly presenting the anxiogenic stimulus in the absence of adverse consequences), which is synonymous with the procedure of extinction in experimental animals. Recently, the realization that extinction is probably a form of additional inhibitory learning has led to the successful combination of exposure therapy (extinction) with the cognitive enhancer d-cycloserine (DCS) to treat patients with vertigo. This development was motivated by extensive experiments in rodents that showed analogous effects (Davis et al. 2006)—a particularly impressive example of translation from rodent studies to human patients. In fact, DCS augmentation has now been used effectively in several other forms of anxiety disorder (Hoffman et al. 2013). However, the efficacy of this behavioral pharmacological interaction may be considerably limited by the difficulty of translating the highly controlled aspects of the laboratory to the real world.

The detection of instrumental contingencies (e.g., the arbitrary act of lever pressing that leads to food delivery for a rat) can be thought of as a higher-order cognitive process which plays an important part in our ability to make the sequence of voluntary actions that constitute goal-directed behavior. Learned helplessness is a theory of how experience of loss of control over environmental contingencies can lead to depressogenic behavior. Such loss of control can also be produced by disrupting top-down connections from the rat medial PFC to such regions as the serotoininergic raphé nuclei (Amat et al. 2005). Although there is some doubt now that learned helplessness is a paradigm specific to human depression, it is clear that tests such as the forced swim test in a tank of inescapable water ("behavioral despair") and the tail suspension test (in mice) do not simulate the main aspects of depression. These paradigms are somewhat crude examples of the helplessness concept, and are nevertheless much used in influential neurobiological studies of the molecular basis of depression, doubtless because of their convenience and utility to predict antidepressant drug efficacy (Nestler et al. 2002). However, the lack of precise definition of stimulus-response contingencies (e.g., the exact adaptive significance of "floating" or planing without success in the forced swim test) renders these tests of dubious construct validity and thus their ultimate lack of success unsurprising.

Memory

Memory can be divided into many subprocesses. A basic, but increasingly questioned, distinction is between relatively transient short-term memory (which may include "working memory") and long-term, more permanent, memory. Memory "traces" are thus hypothesized to undergo "consolidation" into long-term memory. The distinction made in human long-term memory by Tulving (1983), between "episodic" (generally autobiographical, the "what, where and when" of memory) and "semantic" memory (memory for meaning), is perhaps more difficult to investigate in experimental animals, such as rats and monkeys, because of linguistic encoding. Nevertheless, it is apparent that animals, including rodents, have long-term memories for salient stimuli of motivational significance.

Consolidation, reconsolidation, and extinction. A major contribution of rodent studies has been the post-trial or post-training paradigm introduced and developed extensively by McGaugh (e.g., McGaugh and Roozendaal 2008). Here, what is generally a single trial or training session is followed immediately by a drug treatment that can either be amnestic (e.g., protein synthesis inhibitors) or promnestic (e.g., amphetamine). Retention is tested on a subsequent trial, perhaps 24 hours or 3 days later. The post-training manipulation is designed to influence consolidation, either beneficially or adversely, and is often most efficacious when administered directly into the amygdala, a likely site of consolidation of simple cue-related aversive and appetitive memories. The procedure most often used is of aversive memory: the rodent is punished for stepping down from a platform or through a door by presentation of electric foot shock. Memory is expressed on the retention trial by a longer response latency to step down or through the door.

The advantage of this design is that the drug cannot affect memory indirectly by its actions on perceptual, attentional, or motivational mechanisms, as it is administered when these no longer impinge on learning. Controls are necessary with longer post-trial treatments to check that the drug effects are not affecting retention proactively (e.g., by being active at the time of retention and thus affecting memory retrieval). Studies of the consolidation of appetitive memory are also feasible, but are used less often because of the unreliability of one trial appetitive learning. The post-trial paradigm has not (yet) translated particularly well to studies on human memory, although it has been useful in highlighting possible effects on emotional memory, relevant to such syndromes as posttraumatic syndrome disorder (PTSD). The discovery of a related process of reconsolidation (Nader et al. 2000), however, may have greater applicability to the clinic. After initial conditioning, presentation of a conditioned stimulus alone in the absence of the unconditioned reinforcer may enable reconsolidation, a period of memory destabilization when the memory trace regains its vulnerable status through new protein synthesis and when it has been shown to be feasible to target this phase with pharmacological agents (e.g., NMDA receptor antagonists, β-blockers) to produce selective amnesia (or enhancement). The reconsolidation process is mirrored by the active process of extinction learning, which occurs when the reinforcer (e.g., shock or food) is omitted (see also above). It is too early to be sure that these rodent studies of reconsolidation will have therapeutic significance for disorders such as PTSD, anxiety, and addiction, but this is currently an active area of research (see Nader and Hardt 2009; Milton and Everitt 2012).

Recognition memory, recall, and declarative memory. Recognition memory refers to the ability to detect familiarity and, for humans at least, to reminisce about previous experiences involving objects, people, or places. A commonly used task is that of object recognition (devised by Ennaceur and Delacour 1988) in which a rodent (or monkey) explores a novel object during a sample trial and is then given a choice between this familiar object and a novel object, in terms of the amount of time it allocates to exploring both objects. Lesser exploration of one object indicates greater familiarity, and hence recognition of it (in a restricted sense that does not include the subjective elements). The test can also be adapted to measure "social recognition" by using experimental animals as the "object." Recognition memory is generally manifested over long delays, up to 24 h, although it can be tested at much shorter intervals and has been shown to depend on structures such as the rodent perirhinal cortex, rather than the hippocampus (Murray et al. 2007). Recognition memory tasks generally employ stimuli only once, so the test is "trial unique." If the same set of objects were to be used over many trials (as occurs in the spatial delayed alternation or spatial delayed response task, see below), this would produce considerable proactive interference. The test, therefore, becomes one of recency memory (how *recently* the stimulus has been experienced) rather than one of recognition memory (how *familiar* is the stimulus). The test also becomes one more of frontal rather than temporal lobe function (Chiba et al. 1997). As a rapid and easily implemented test, object recognition memory is perhaps the most used of all rodent assays of memory in the screening of putative cognitive-enhancing drugs, although it has been employed in several different variants (Dere et al. 2007). Its translational properties, in terms of human tests of recognition memory, are clear, particularly when equivalent touch-screen versions of visual recognition memory are employed.

Recognition memory is, however, a less-sensitive test of memory than either cued or free recall, in which the memory has to be generated from long-term memory store. Unlike recall, recognition is not particularly sensitive to hippocampal damage, nor is it the earliest manifestation of Alzheimer disease, where amnesia for episodic memories is more evident.

Rodent (as well as primate) data indicate the hippocampus to be implicated in forms of associative memory, in animals involving space or other contexts; for example, remembering the location of objects (Murray et al. 2007). These forms of memory are often referred to as reference memory tasks in the animal literature. Recognition and recall correspond to what Squire (1992) defined as "declarative memory," which is distinct from "procedural memory" (memory for "how" or "skill"). Although procedural memory is not discussed in detail in this article, it could be tested in rodents in motor-learning situations, such as the rotor-rod test or as memory for "habits."

Reference memory. Reference memory is a form of long-term memory referring to rodent task requirements that stay constant over trials. This definition was introduced by Olton when rats were required to remember the constant location of food-baited arms in an 8-armed radial maze (Olton and Paras 1979). It can also be applied to the Morris water maze, a notable assay of hippocampal function, in which rodents are required over a number of learning trials to learn the location of a hidden platform in order to escape from a tank of water (Morris et al. 1982). The rodent is allowed to swim the maze from different starting points; thus successful learning depends on the construction of a "cognitive map" to navigate the environment. Although this task may appear to lack translational validity for testing, for example, memory in patients with Alzheimer disease or schizophrenia, promising "virtual reality" tests of spatial navigation are currently being employed to assess human memory (Hanlon et al. 2006).

A related approach has been to "back-translate" from human to animal studies. The CANTAB paired associates learning (PAL) task has some of the attributes of episodic recall ("what and where" learning), as humans are required to learn and remember the different locations of several abstract visual objects over short delays. This task is sensitive to deficits in patients with mild cognitive impairment (likely prodromal Alzheimer disease) and some patients with schizophrenia. It has been shown to be sensitive to impairments in patients with mild cognitive impairment three years before formal diagnosis (Swainson et al. 2001). Recently, the main aspects of this test for rodents have been simulated in a visuospatial learning task performed on a touch-sensitive screen (Talpos et al. 2009) and may be useful in evaluating effects of new drugs for Alzheimer disease. The translational utility of the human and animal versions of the test is emphasized through deleterious effects of hippocampal dysfunction on performance across species as well as by the results of functional neuroimaging in humans with mild cognitive impairment.

The Morris water maze and the CANTAB PAL task effectively test the capacity of "what and where" memory, but they do not quite capture what is meant by episodic memory, which also requires the tagging of that memory to a particular time. Until recently, it has been assumed that episodic memory is uniquely human, involving the "mental time travel" of subjective reminiscence. Demonstrations of "what, where, and when" memory have appeared in the literature, beginning with food-caching birds (Clayton and Dickinson 1998), but also including rats (e.g., Eacott et al. 2005) and nonhuman primates (Martin-Ordas et al. 2010). This is clearly an important growth area in translational neuroscience, especially in terms of modeling the earliest manifestations of Alzheimer disease.

Working memory. Working memory refers to the active use of short-term information for the purpose of constructing representations of the world and guiding behavior (Baddeley 1986). Working memory is often termed "online" memory; memory traces are activated during planning and long-term memory retrieval. Working memory is at the interface between perceptual processes and the formation of long-term memory, and is often associated with so-called "executive processes." A major component of working memory is responsible for response selection and coordination of the outputs of different short-term memory buffers. In operational terms in animal studies, the use of the term "working memory" in the Olton maze refers to the requirements of that memory test procedure in which rodents are required to visit each of the eight arms once, and once only, to retrieve a maximum of eight pellets (contrasting with reference memory, see above). The rodents have to remember only where they have recently been, and this memory is irrelevant to performance on subsequent test days. It can be argued that this form of "working memory" is not quite the same as that defined by human memory theorists, such as Baddeley, where there is a coordination of different, modality-specific short-term memory buffers for use in various tasks such as planning, linguistic discourse, and logical reasoning. However, it does seem to overlap the human form of working memory in some important respects.

Olton's working memory tasks are related to the classical spatial delayed response and delayed alternation tasks that have been used to establish the role of the primate PFC in working memory (Goldman-Rakic 1996). This involvement of the PFC has been important for modeling cognitive deficits associated with schizophrenia, as working memory is impaired very early in the course of schizophrenia, possibly even the prodromal state (Wood et al. 2003). For the purpose of screening drug effects, delayed alternation in rodents is easily implemented in a maze or operant chamber, where it is often referred to as "delayed nonmatching to position." Nonmatching is an easier task for rodents than matching because of their preexisting foraging tendency to alternate spatial choices. The operant version of the task allows the systematic variation of delay intervals, which can extend from 0–60 s. A "delay-dependent" effect in such a task is generally taken as evidence of a specific memory effect, independent, for example, of attention (Dunnett 1985).

However, for that inference to be valid, it is necessary for performance on the task at 0 s to be shown not to be similarly susceptible when the perceptual difficulty of the task is enhanced. An artifact that is difficult to surmount in the operant task is mediating responses which allow the rodent to adopt postures or positions minimizing the memory requirements of the task. One way to overcome this problem is to use sensitive touch screens to record responding, which can more precisely vary the spatial requirements of the memory tasks, as in the CANTAB battery for humans and nonhuman primates (Bussey et al. 2012).

Executive Functions

Executive functions are control processes that serve to optimize performance (e.g., in terms of earned rewards or reinforcers) by coordinating the various components of complex cognitive functions. Frequently (though not exclusively) associated with the functioning of the PFC, executive function is impaired in a variety of neuropsychiatric disorders, including ADHD and schizophrenia, and is thus a target for pharmaceutical therapy. Executive functions include some aspects of cognition already covered: cognitive flexibility in the face of changing environmental circumstances (attentional set shifting and reversal learning), the control of working memory, and the capacity to resolve conflicts between competing actions or predispositions. The latter may include what is termed "cognitive control" or "inhibitory response control," disruptions of which can lead to impulsive and compulsive behavior, relevant to such syndromes as OCD and ADHD.

Impulsive-compulsive disorders are one area in which animal models of human disorders fare surprisingly well. For example, there are several tests of impulsivity in rodents that have their equivalents in humans:

1. The stop signal reaction time (SSRT) test (Eagle and Robbins 2003) is a sophisticated form of the so-called Go/NoGo task, and is often used to measure motor impulsivity, e.g., in ADHD (Solanto et al. 2001). The SSRT test estimates the time it takes to cancel an initiated response. A stop signal is presented on about 20% of trials to indicate not to respond on that trial. This stop signal is interpolated at varying times (of the order of fractions of a second) after the onset of a Go cue; when it is presented at longer delays, it is correspondingly more difficult to cancel the response, as reflected by increased stop errors and by longer SSRTs. The SSRT task assumes that there is a "race" between a "go" process and an independent "stop" process; whichever "wins" determines the outcome of the trial. Pharmacological results with this

test show excellent translatability between human and rat studies. For example, the relatively selective noradrenaline reuptake blocker atomoxetine improves SSRT in both humans (Chamberlain et al. 2006) and rats (Robinson et al. 2008; Bari et al. 2009), whereas the selective serotonin reuptake inhibitor citalopram has no effect in either species (Chamberlain et al. 2006; Bari et al. 2009).

2. Delayed discounting of reward (also referred to as delayed gratification or delay aversion) occurs when an individual chooses between a small, immediate (or more certain) reward and a larger (or less certain) delayed one. The "impulsive choice" is to select the small, immediate or the small, certain reward. The assumption is that rewards are discounted by time or probability, and that a hyperbolic discounting function can predict an individual's choice. One of the parameters of this function (k) is essentially a measure of the "steepness" of discounting, or of impulsive responding. This task can be used readily in animals (Mazur and Herrnstein 1988; Evenden 1999) as well as humans (e.g., patients with ADHD or drug abusers). Measures of this form of impulsivity ("impulsive choice") do not always correlate in the same individual with measures of motor impulsivity (e.g., SSRT). This was the case for a large population of children with ADHD, resulting in a suggestion that ADHD reflected a spectrum disorder with different forms of impulsivity (Solanto et al. 2001). Structures such as the nucleus accumbens and orbitofrontal cortex are implicated in mediating delayed discounting in both rats (e.g., Cardinal et al. 2001; Winstanley et al. 2004; Bezzina et al. 2008) and in humans (McClure et al. 2004).
3. Impulsivity can also be measured in tests of visual attention, such as 5-CSRTT (discussed above; see also Robbins 2002), and in tests requiring timing, such as the differential reinforcement of low rates of responding schedule. Premature, defined as impulsive, responses on the visual attentional task have been shown to predict compulsive cocaine seeking in rats (Dalley et al. 2007) and may capture a similar vulnerability in human stimulant drug abusers (Ersche et al. 2012; Voon et al. 2014).

Impulsive behavior can thus be seen as a loss of inhibitory response control, which is a key feature of dysexecutive syndromes. It is, however, important to realize that impulsivity is not the only consequence of such a loss. Compulsive behavior may also be related to impulsive responding, but the difference between them is that compulsive behavior persists abnormally whereas impulsive behavior is frequently premature as a consequence of an inability to wait. Both are invariably associated with adverse consequences (e.g., more negatively valued events or a loss of positive reinforcements or rewards). Compulsive behavior may be modeled, for example, at several levels of response organization: motor stereotypy, rigidity of attention set (see above), and persistent responding to the formerly reinforced stimulus during reversal discrimination learning when the previously nonrewarded stimulus now becomes correct or during extinction, when reward is omitted completely. OCD is the clinically prototypical compulsive disorder and is thus associated with impairments in extradimensional shifting (Chamberlain et al. 2006). In terms of brain activation, the performance of reversal learning (Chamberlain et al. 2008) helps to validate these rather general behavioral expressions of compulsive behavior as possible (neurobehavioral) endophenotypes for OCD. The parallel is heightened by the finding that serotonin depletion in the marmoset orbitofrontal cortex similarly impairs reversal learning (Clarke et al. 2004), possibly consistent with the use of SSRI (selective serotonin reuptake inhibitor) pharmacotherapy in OCD.

The Cognitive-Motivational Interface: "Hot" Cognition

The concept of "top-down" control by PFC executive mechanisms over striatal (and other subcortical) substrates can be extended to aspects of cognition and behavior other than impulsivity and compulsivity. The notion that the PFC is implicated in "emotional regulation" suggests that the PFC and associated structures (such as the anterior cingulate) have roles in moderating activity in structures which control emotional behavior and learning, such as the amygdala, with implications for clinical anxiety and depression. Observations of depressed patients suggest they have exaggerated (and indeed "catastrophic") reactions to negative feedback which impact their cognitive functioning (Elliott et al. 1996; Murphy et al. 1999). In the context of a probabilistic reversal learning task, where the correct choice is only rewarded on the majority (e.g., 80%) of occasions (with negative feedback following a minority, e.g., 20%, of trials), depressed patients often make inappropriate shifts in response choice on the next trial in response to spurious negative feedback—a tendency accompanied by a reduced PFC-induced deactivation of the amygdala (Taylor Tavares et al. 2008). It is possible that this rather complex paradigm can actually be modeled in rats; Bari et al. (2010) recently showed that manipulations which reduce serotoninergic function mimic some of the effects seen in depressed patients on "lose-shift" behavior.

Several other novel behavioral approaches to measuring affective bias in rodents have recently been introduced in both operant (Harding et al. 2004; Enkel et al. 2010; Anderson et al. 2013) and semi-naturalistic settings (Stuart et al. 2013). Enkel et al. (2010) trained rats to press a lever to receive a food reward associated with one tone, and to press another lever in response to a different tone to avoid punishment by electric foot shock. In an "ambiguous cue test," responses to tones with frequencies intermediate to the trained tones were taken to indicate expectation of a positive or negative event. In fact, a negative response bias was found in congenitally helpless rats, a genetic animal model of depression, and treatment with a pharmacological treatment mimicking stress-related changes, biased rats away from positive responding.

Anderson et al. (2013) used a similar paradigm to show that chronic citalopram induced a shift away from negative bias.

In another novel behavioral approach to measuring affective bias, rats encounter two independent positive experiences: the association between a food pellet reward and distinctive food bowls in which the pellets could be found by digging (Stuart et al. 2013). These experiences are gained on separate days under either neutral conditions or during a pharmacological or affective state manipulation (in this way it resembles the classical conditioned place preference procedure also employed for measuring reward preference). Affective bias is then quantified using a preference test where both previously rewarded substrates are presented together and the rat's choices recorded. The absolute value of the experience is kept consistent, and all other factors are counterbalanced so that any bias at recall can be attributed solely to the treatment. Similar to studies in healthy human volunteers observing emotional faces (e.g., Pringle et al. 2011), Stuart et al. (2013) observed significant positive affective biases following acute treatment with typical (fluoxetine, citalopram, reboxetine, venlafaxine, clomipramine) and atypical antidepressants (agomelatine, mirtazapine), and significant negative affective biases following treatment with drugs associated with inducing negative affective states in humans (FG7142, rimonabant, 13-cis retinoic acid). They also observed that acute psychosocial stress and environmental enrichment induced significant negative and positive affective biases, respectively, and involved memory consolidation. Thus, this is an excellent example for showing how motivational and cognitive processes may interact to generate apparent changes in affect in rodents.

Higher-Order Cognition, Including Social Cognition

I have shown that it may be feasible to decompose higher-order decision-making tasks in infra-human animals, as such components of tests—such as the Iowa gambling task (e.g., Zeeb et al. 2009), associative (trial-and-error) learning, reversal learning, delayed discounting and inhibitory control—all contribute to deficits seen in frontal and neuropsychiatric patients on this classic neuropsychological test. However, it may prove intractable to model higher-order planning tasks, at least in rodents, unless it can be shown that rats are capable of anticipating future needs and subordinating them to current ones. In the same context, it may be too much to ask that aspects of social cognition can be modeled in rodents, although elements of the detection of others' intentions (the so-called "theory of mind" functions) may be present in apes, though not in rhesus monkeys (Penn and Povinelli 2007). This altered efficiency is a severe problem because of the pervasive nature of social cognitive deficits, most notably in autism and schizophrenia; in the MATRICS battery, social cognition is one of seven main domains of deficit in schizophrenia. Tests of social behavior and interaction in animals may be useful through ethological observation (Moy et al. 2004), but they cannot hope to capture the complexity of human social cognition. As mentioned above, social recognition has been employed as an appropriate test of certain social factors. For this to be considered a test of social (as distinct from more general information) processing, it is necessary to show that any effects are restricted to the social domain and to contrast them with a lack of effects, for example, in tests of visual, tactile or olfactory recognition memory.

Conclusions

Modeling specific aspects of motivation and cognition in rodents can bear some approximation to the greater complexity observed in human patients. As cognition is not a unitary construct, it is necessary to focus on specific aspects, for example, of memory that are highlighted by the patient's deficits. It is often important to test cognition in humans and other animals in ways that are as similar as possible, even using similar types of stimulus material and response mode where feasible. Although testing effects of drugs on intact (i.e., "normal") experimental animals can be informative, as is also the case with healthy human volunteers, the ultimate tests involve clinical trials and thus have to be paralleled by animals that have been manipulated in some way so that the potential of any cognitive-enhancing effect can be best evaluated.

It is not possible to survey here all of the relevant "animal models," subsuming "disease models" that have been employed for the different disorders, but it should be clear that the choice of relevant "perturbation" of normal system functioning is often crucial to the success of the model. In the next period, I anticipate that it will become increasingly important to compare effects of different treatments on different animal models, and to take increasing notice of "back-translation" as the critical test of validity. It is also emphasized that dependent variables measured should not necessarily simply be behavioral ones; the power and validity of a model need to be also based on concomitant neurobiological indexes, such as electrophysiological activity, neuroendocrine or neurochemical change, or genetic expression, so as to define the significance of changes in cognitive performance in the more general context of brain functioning. In addition to theoretical sophistication in the behavioral, neural, and genetic domains, there is also the increasingly important issue of methodological rigor in experimental design. Replicability and reliability, for example, across multiple laboratories or test centers are increasingly becoming important criteria that should be combined with other examples of good practice, such as random assignment of subjects to groups, unbiased blind assessments, and adequate statistical power.

First column (top to bottom): Ilka Diester, Bernd Sommer, Steve Hyman, Isabelle Mansuy, Trevor Robbins, Marius Wernig, Lennart Mucke
Second column: Franz Hefti, Akira Sawa, Trevor Robbins, Alvaro Pascual-Leone, Gül Dölen, Ilka Diester, Bernd Sommer,
Third column: Isabelle Mansuy, Marius Wernig, Lee Rubin, Karoly Nikolich, Franz Hefti, Akira Sawa, Lee Rubin

13

Bridging the Gap between Patients and Models

Ilka Diester, Franz Hefti, Isabelle Mansuy, Alvaro Pascual-Leone, Trevor W. Robbins, Lee L. Rubin, Akira Sawa, Marius Wernig, Gül Dölen, Steven E. Hyman, Lennart Mucke, Karoly Nikolich, and Bernd Sommer

Abstract

Classically, research into human disease tends to be done in a top-down or bottom-up manner, starting from symptoms or genes, respectively. While bottom-up approaches may work well in oncology, and might advance understanding of monogenic neuropsychiatric diseases, successful application for complex, multifactorial disorders is more difficult and has resulted in many translational failures. This chapter investigates the existing obstacles and explores options to overcome them. Complex diseases need to be dissected into measureable, manageable factors and investigated in a comparable, compatible assembly of model systems to test hypotheses, concepts, and ultimately drug candidates or other therapeutic interventions. While some of these factors might best be investigated top down, a bottom-up approach might be more effective for others. Both approaches may only be successful up to a specific point. Thus, the two must be linked and a bidirectional approach pursued. Inclusion of patients is essential as are behavioral readouts, since disease-associated dysfunctions or symptoms are often behavioral in nature. To connect models and humans, behavioral readouts need ideally to be linked to evolutionary conserved neural substrates. Some anchor points already exist and new promising ones, such as induced pluripotent stem cells (iPSCs), are emerging. Recent developments may speed up translation of research into clinical applications (e.g., faster drug screens in a patient-specific manner). When positioning different models, it is important to characterize their predictive power diligently, to emphasize their scientific rigor, and to not overstate their application potential. Finally, to effect faster transition from research to clinical applications, organizational structures are needed to foster interdisciplinary research and collaborations between academia and industry. A "third space" concept is proposed to conduct early proof of principle studies (Phase 0 and I). To increase the success rate in clinical development so as to provide actual benefit for patients, proactive interaction is needed between all organizational entities involved in drug development and therapeutic discovery (e.g., academia, guidance agencies, biotech, device and pharmaceutical companies, regulatory agencies, and funding agencies).

Learning from the Past: Unidirectional Approaches

To outline discovery strategies clearly and to systematize arguments, we categorized research processes into bottom-up and top-down approaches (for a definition of terms used in this chapter, see Appendix 13.1). The bottom-up approach starts with genes, whereas top-down procedures begin with a syndrome or phenotype. The two streams likely overlap at the level of molecular pathology, although in some instances overlap at other levels is also possible. Depending on whether pathological changes are found from analyses of brain tissue or predicted and found from genetic studies, these approaches should be considered top-down or bottom-up, respectively (Figure 13.1).

Top-Down Approaches

The top-down approach takes some features or attributes of a disease or disorder from a detailed phenotypic analysis and uses these either to predict possible drug targets and new therapeutic concepts or to test the efficacy of a drug discovered by serendipity. These disease/disorder phenotypes should be

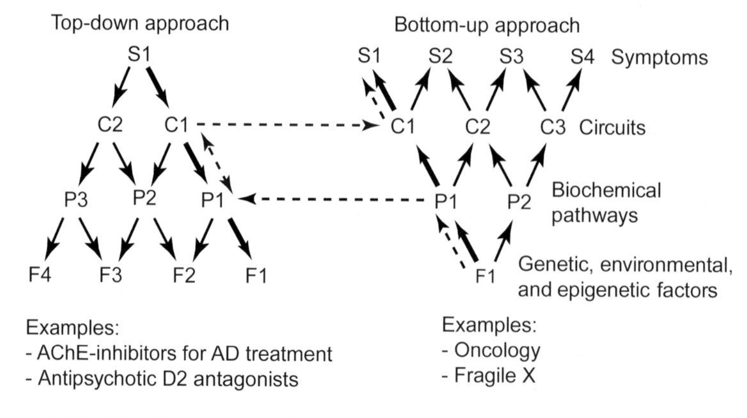

Figure 13.1 Top-down versus bottom-up approaches in research. Top-down approaches begin at the level of symptoms (the usual starting point is one symptom, S), and move down to genetic or environmental factors (F). Bottom-up approaches work their way up from a genetic factor to identify the corresponding symptom. Naturally, the approaches have the most likelihood of success at the level one step away from the starting point: circuits (C) for top-down approaches and biochemical pathways (P) for bottom-up approaches. The most difficult step is the gap between biochemical pathways and circuits. Here, knowledge derived from basic research can serve as a bridge. Broken lines indicate a desirable research approach to combine the fields; see Appendix 13.2 for further discussion in regard to Parkinson disease.

investigated in detail at the level of molecular or cellular pathology as well as neural system/circuit dysfunction, cognition and behavior, and epidemiological factors. The top-down approach has worked quite well in diseases characterized by a dominant symptom and relatively direct delineation to underlying pathways and pathology, such as Parkinson disease (PD). For PD, motor dysfunction could be associated with degeneration of the nigrostriatal dopamine (DA) pathways, which appeared sufficient to explain much of the movement deficits of this neurodegenerative disease. Therefore, it made sense to attempt replacement of depleted DA as a major therapeutic concept; hence the introduction of L-DOPA therapy and DA receptor agonist medications. However, cognitive deficits in PD patients, which have been attributed to other, extrastriatal pathology, were found to be largely independent of DA, and are thus not treatable with DA replacement (Appendix 13.2). For a detailed review of a historic perceptive as well as contemporary top-down models for aspects of schizophrenia, see Appendices 13.3 and 13.4.

Applying modern analysis methodologies to top-down approaches today may start with molecular profiling of patient biospecimens—molecular substrates from patients' surrogate tissues (e.g., cerebrospinal fluid or olfactory cells obtained through quick and safe nasal biopsy). This allows a set of targets to be identified directly in humans. Many neurodegenerative disorders occur as a result of the combination of genetic and environmental factors. Unbiased proteomic and/or candidate molecular approaches can be applied to identify targets of interest. In contrast to genetic sequences, which contribute to the identification of targets associated with trait change, the molecular analysis of biospecimens provides the opportunity to identify both trait- and state-associated targets (see Appendix 13.1). This approach is considered to be the most effective when applied during early disease stages or even with subjects in the prodromal stage. Molecular profiling can also be combined with molecular imaging, such as positron emission tomography (PET) and magnetic resonance spectroscopy (MRS). One of the most promising examples is the identification of potentially predictive markers for Alzheimer disease (AD), such as pathological changes in amyloid PET imaging and alteration of Aβ42 and phospho-tau in cerebrospinal fluid (Lista et al. 2014; Blennow et al. 2012).

Bottom-Up Approaches

To date, oncology provides some of the best examples of success for the bottom-up approach. Specific subtypes of cancer are identified by their genetic mutations and corresponding antibodies or other antagonists are generated to treat them. As a generalization, all cancers share a common phenotype: excessive proliferation. However, the drivers of the phenotype (e.g., mutated kinases) vary from one cancer subtype to another. This means that there are two broad classes of potential treatments: standard chemotherapeutics which block cell proliferation in a relatively nonspecific manner or act on tumor-specific pathways that regulate proliferation. In one of the initial examples of this approach, mutations of the HER2 gene were identified as causative genes in some forms of breast cancer. Trastuzumab (Herceptin) is an anti-HER2 antibody developed to treat this specific condition (Sliwkowski and Mellman 2013). Several further antibodies followed the same paradigm, and the cancer field is progressing rapidly with this approach. Companion diagnostics are developed in parallel to the development of therapeutics. Continuing progress with this approach is expected to result in highly individualized but highly effective treatment of cancers.

In contrast to the oncology field, there have been no clearly positive examples for the bottom-up approach in the area of neurologic and psychiatric diseases. For instance, despite the discovery of a dominant disease gene in Huntington disease two decades ago, no effective therapy has thus far been developed (Ross et al. 2014). The most advanced example is provided by fragile X syndrome and the link to mGluR5 receptors. Fragile X syndrome, a monogenic cause of autism for which the identification of an expansion of CGG repeats in FRM1 gene leading to altered FRM1 functions (mRNA transport), helped identify mGluR5 as a potential therapeutic target (mGluR5 antagonists can rescue some of the defects resulting from FRM1 deficiency in knockout mice; see Appendix 13.5). Although this identified a promising therapeutic concept, translation to human disease has thus far been unsuccessful, illustrating the challenge of bridging model and human systems.

Summary of Unidirectional Approaches

Although unidirectional approaches have proven successful in disease areas such as oncology and a few promising examples have emerged for diseases of the nervous system, the majority of CNS disorders are characterized by their complexity and multifactorial etiology. Diseases such as schizophrenia, AD, or bipolar disorder are syndromes composed of multiple symptoms, which imply the contribution of several genes to different degrees, including complex interactions between these genes and environmental factors. The combined implication of genetic and nongenetic (epigenetic) factors is one of the major obstacles in the understanding of these diseases. For example, schizophrenia may involve up to 10% of all known genes, and the different forms of the disease may implicate different genes or genetic loci in distinct patients or families. Further, environmental factors (e.g., early life stress or nutritional deficiencies) are known to play a contributory role. Thus, schizophrenia is not only multigenic but also highly diverse genetically, involving epigenetic factors. A unidirectional approach might not be best suited for tackling such complex diseases. In comparison to cancers, the genetics appear to be more complex, with each disease having multiple drivers. In addition, neurological and neuropsychiatric diseases are associated with an unknown number of phenotypes.

Understanding how to divide each disease optimally into discrete therapeutic categories would represent a significant advance in the field.

The Bidirectional Proposition

It seems clear that neither a strict bottom-up nor a top-down strategy alone suffices to approach complex neuropsychiatric disorders. Instead, both approaches are equally important to pursue in parallel and interactive ways. While the ultimate goal of both approaches is identical, on the short term both differ principally in their output to new therapeutic concepts. Top-down approaches are closer to patients and their symptoms; thus, they tend to lead to symptomatic rather than curative treatment concepts. On the other hand, understanding the genetic basis of syndromes or disease groups might not immediately lead to an understanding of how symptoms are produced, but even without such knowledge, genetics may guide the discovery path toward curative rather than symptomatic treatments. An integration of genetic and epigenetic factors will be essential in this respect.

One might consider, for instance, generating animal models with mutations or copy number variations similar to those identified in patients, alone or together, and in combination with an environmental manipulation. One could also consider using induced pluripotent stem cells (iPSCs). Nonetheless, open questions and limitations remain: the type of cells to use (those directly involved in the pathology are not accessible), variability of cultures, no recapitulation of environmental factors, and associated epigenetic processes. One could employ biospecimens (e.g., cerebrospinal fluid, olfactory cells, or blood) to identify transcriptomic and/or proteomic alterations associated with the disease. In addition, functional (e.g., symptomatic) readouts remain essential to the ultimate demonstration of clinical efficacy. These examples illustrate the need for a bidirectional approach and highlight the limitations of a strict separation between bottom-up and top-down approaches.

Deconstruction of Neural Diseases in Measurable Manifestations

Neural diseases are complex constructs, and patients manifest variable combinations of symptoms and disabilities. Thus, neural disorders have to be deconstructed into measurable manifestations (symptoms, behaviors, disabilities, or functional disturbances) that can be mapped onto specific neural substrates or neural dysfunctions (Figure 13.2). Importantly, each given symptom may map onto more than one disease, and the combinations of symptoms in a given disease may vary significantly. This can give rise to phenotypic variability even though the neural substrate, or the causative gene or genes, may be the same. Thus, better results may be exacted if we aim to treat specific behavioral or cognitive alterations and their underlying neuronal substrates, rather than

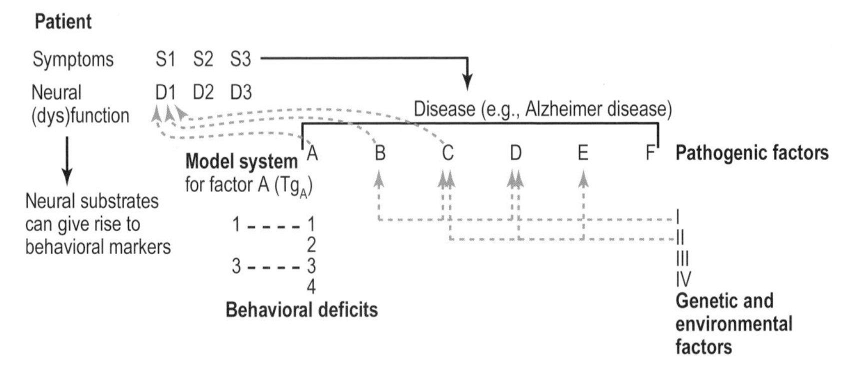

Figure 13.2 Mapping manifestations of complex polygenic human neural diseases to models. A patient might show symptoms (S1, S2, S3), and associated with each symptom is one neural dysfunction (D1, D2, D3). Together the symptoms build the disease. The disease is associated with several pathogenic factors (A–F), and each pathogenic factor results potentially in behavioral deficits (1–4). Model systems (like a transgenic mouse) might show only part of the behavioral deficits (here, 1 and 3), even though they fully express the pathogenic factor. Pathogenic factors result from the interplay between mutated genes and particular environmental factors (I–IV). One gene can be associated with several pathogenic factors, and one pathogenic factor can be dependent on more than one of the disease genes. Finally, each pathogenic factor can contribute to more than one neural dysfunction, and each neural dysfunction can be associated with more than one pathogenic factor.

address disease disabilities or symptoms in a holistic manner. The complexity of pathogenic factors in polygenic diseases leads to a multiplicative space of factors which, if modeled completely, would lead to an exploding number of possible combinations. Principal component analysis and related statistical techniques, such as discriminant cluster analysis, can potentially be used on large data sets to determine the interrelationships between, for example, disease/disorder behavioral/cognitive/physiological symptoms, pathologies, and neural circuit/network dysfunctions to determine their interrelationships and how many orthogonal (i.e., distinct) clusters or factors (i.e., phenotypes) make up the disorder.

Evolutionary Conservation as a Bridging Point between Human and Model Systems

Neural dysfunctions can serve as targets for the development of therapeutic concepts and inform models. This conceptualization is consistent with the new recommendations for clinical trials from the U.S. National Institutes of Mental Health (NIMH) which emphasize the need to have patient-centric outcomes (e.g., improvement of quality of life or suppression of a given symptom or disability), while demonstrating the link and specificity of the tested intervention with a specified neural target or underlying neural dysfunction. Neural dysfunctions can be measured using a variety of methods, including neuroimaging as well as neuropsychological, neurophysiological, and neurochemical techniques. Importantly, the introduction of methods into the process of therapeutic concept discovery will facilitate the rigorous testing of interventions in humans and so enable the bridge from humans to models, whether whole animal, cellular, or computational.

The expectation is that neural substrates and dysfunctions will map better from human to model systems than symptoms or manifestations of disease. The mapping of neural substrates from humans to a suitable model will enable the identification of the pathogenic factor or factors that account for a given dysfunction. In experimental model systems, it is then possible to move from the pathogenic factors to the responsible genes or gene products.

The bridge from humans to models is bidirectional (Figure 13.3). Eventually, after a therapeutic concept is identified, translation from the model to the human

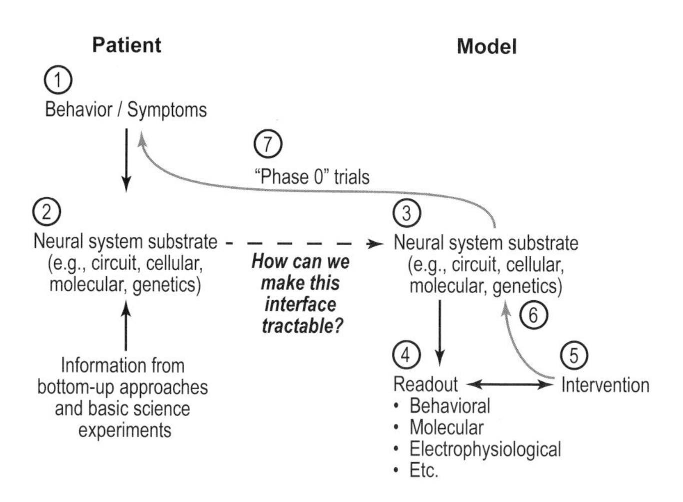

Figure 13.3 Process of bridging animal models to patients and back. (1) Symptoms are identified in patients. (2) A neural system substrate associated with this symptom is then identified. Importantly, this can only be done based on information from previous basic science experiments and disease-oriented, bottom-up research. (3) An evolutionarily conserved neural system substrate is identified in the best-suited model system, again informed by basic research. (4) An appropriate readout is chosen. (5) This readout is used to evaluate the success of the intervention, which ultimately works on the neural system substrate (6). (7) Findings should then be translated to humans in a proof of principle manner using identical or at least similar readouts. Results obtained in humans should then be back-translated to the model for further refinement. The critical step is the interface between patient and model system.

and back will be necessary. We argue that an experimental medicine approach is critical here, as it permits rigorous testing of candidate therapeutic concepts before embarking on traditional clinical trial phases. This approach would assess the ability of the hypothesized mechanism of action of the therapeutic concept to alter the neural substrate and normalize the neural dysfunction that could lead to an improvement in symptoms or disability. As formulated by the NIMH[1]: "Rather than testing the intervention in a traditional efficacy outcome trial, the experimental medicine trial involves objective measures of target engagement and effects on brain function as initial evaluation points." Such approaches may include dose finding studies and a comparison of different methods of intervention delivery. The results of such experimental medicine assessments ought to represent an early "Go/No-Go" decision point; demonstration that a given therapeutic concept adequately engages and modifies the neural dysfunction is necessary to ask whether it can affect clinical symptoms.

The selection of inappropriate neural dysfunctions and their use as the bridge between human and model systems, along with the lack of experimental medicine approaches, may account for past failures to translate therapeutic concepts to humans and achieve greater success in clinical neurotherapeutics. The suitable point to bridge human and model systems is at the level of evolutionary conservation. Three points are of relevance here:

1. The level of conservation may vary with different model systems.
2. Conservation will likely decline in the following order: gene, gene product, molecular mechanism, cellular function, neural circuit, behavior, symptom.
3. A model will obviously be progressively closer to the symptoms of human disease in the opposite order: behavior, neural circuit, cellular function, molecular mechanism, gene product, gene. This emphasizes the importance of behavioral measurements.

The Importance of Behavior for Investigating Mental Disorders

While the approach of evolutionary conservation may suggest the importance of the neural system substrate over behavior, and thus imply that behavioral measures are less important, there are good arguments to include behavioral studies in this process, especially when there are no overt pathological hallmarks of a disease available:

1. Drugs generated using the bottom-up approach for most (if not all) diseases/disorders of the nervous system will eventually have to be tested

[1] http://www.nimh.nih.gov/funding/opportunities-announcements/clinical-trials-foas/nimh-clinical-trials-funding-opportunity-announcements-applicant-faqs.shtml (accessed March 3, 2015)

in clinical trials. Here, the main outcome measures, however crude, are overtly behavioral (or cognitive) in nature and are generally based on clinical measures of behavior, including functional outcome.

2. Clinical trials can be improved by combining better-defined behavioral measures with measures of the integrity of neural circuitry (e.g., through neuroimaging as well as electroencephalography and magnetoencephalography) to be able to link the neural changes to specific symptoms (i.e., behavioral changes).
3. To understand how treatment impacts a mental disorder—for example, depression—sensitive measures are needed to assess the degree to which a patient is suffering from (a) undue bias to think negatively, (b) a diminished hedonic response that derives from a primitive approach/reward or reinforcement system, or (c) both. These measures, then, need to be related to the specific neural circuitry deficits (from which different biomarkers may emerge). Currently, our understanding of neural circuitry functions is not sufficiently advanced to use reverse inference logic. For example, the hypothesis "this circuit is damaged so it *must* be a reward problem" must be tested and updated continually. Further, in many cases it is necessary to "challenge" the neural system behaviorally in order to identify a circuit dysfunction, such as in provocation studies in obsessive-compulsive disorders or relapse cues in addiction, including attempts at remediation.
4. To enhance the predictive power of animal models and tests, these need to be related to the measures of Phase I and III clinical studies. Thus, where a "reward system" deficit is invoked, there should be an animal model or test equivalent, ideally behavioral as well as neural (see previous point). Even if it were not necessary to use the behavioral measure, it would greatly increase the confidence in interpreting any neural change as having an important functional consequence, thus also raising the predictive power of Phase I and II outcomes for Phase III.
5. Automated behavioral measures can be made to be very reliable and with small error variance, suitable for detecting the effects of drugs or genetic manipulations, even when clear equivalents to humans (e.g., for human language) do not exist.
6. Many examples of complex behavior can be reduced to evolutionary building blocks of the more sophisticated functions in humans, such as goal-directed behavior, working memory, spatial attention (for an example of cholinergic treatment of AD, see Appendix 13.6), and performance of clinical neuropsychological tasks such as the Wisconsin Card Sorting Test (a test of cognitive flexibility sensitive to damage or dysfunction of the human frontal lobes). The latter can be reduced to tests of discrimination learning and can be used in parallel in humans, including patients, and animals including monkeys, rats and mice (see Robbins, this volume). To optimize the tests across these species,

different sensory modalities may have to be employed; that is, different visual features for primates, smell and touch for rodents. However, the basic structure of the behavioral tasks, and hence the functions governing cognitive flexibility, are similar if not identical (e.g., reversal learning depends on the orbitofrontal cortex). Moreover, for some tasks, such as the stop signal reaction time (SSRT) test, there may be comparable pharmacology across species (see Robbins, this volume). These examples provide considerable confidence that the same type of function, mediated by the same type of circuit, is being measured across species.

7. Finally, one can begin to relate (by experiment) basic processes of learning discovered in animals (to humans), and aberrations of these processes (to patients) in a way that may relate to their symptoms and help explain them. An example of this (logically "triangulating") approach is the prediction error deficits in causal learning (i.e., how events can be predicted or expected) that relate to functioning of frontostriatal circuits and can be mapped using electrophysiological or neuroimaging methods (e.g., the BOLD response using functional MRI). NMDA receptor antagonists, such as ketamine, impair learning by disrupting these prediction errors, and the degree of disruption in humans predicts perceptual impairments which contribute to delusions in human volunteers. Similar disruptions are observed in first episode patients in schizophrenia. Given that the NMDA receptor antagonist similarly impairs the readout of prediction errors during learning in experimental animals, these disruptions in animals can, in principle, act as "biomarkers" of more complex delusional phenomena in humans.

Summary of Bidirectional Proposition

Translation from patients to models, or from a model to the patient, should never be approached solely in a top-down or bottom-up fashion, but rather always through a bidirectional approach. This requires us to identify the point(s) of evolutionary conservation, be it on the level of neural substrate or (in certain cases) behavior. Fortunately, much work has been done and published in this realm, often coming from basic research approaches, once again emphasizing the importance of this research line. To allow easier access to this information, communication and efforts between researchers must be enhanced, especially between clinicians and basic scientists, but also between academia, industry, and governmental and other agencies. (This point is taken up below, where a "third space" is proposed to accomplish this goal.) Finally, instead of attempting to cover the complete disease spectrum all at once, a clear focus on specific, tractable aspects of disease comparable between models and species will help accelerate the process.

Choosing the Right Model System and Evaluating Predictive Power

According to our idea of evolutionary conservation, models should faithfully reflect a given neural substrate or dysfunction. A neural substrate may be a neural circuit, a cellular element, a molecular target, a gene, or a gene product. It is important to recognize, however, that one or more pathogenic factors may be relevant for a given neural substrate/dysfunction. The experimentally measured model readout should be a faithful reflection of the impact of a tested intervention onto the neural substrate. Ideally, the most proximal marker of the neural substrate (e.g., circuit activity, cellular firing pattern, molecular or gene expression) offers robust and faithful metrics. Thus, if a given therapeutic concept engages the neural substrate in the same manner in both the model and the human, one could argue that a sufficient level of predictability will be achieved. Nevertheless, behavior should not be neglected as it can be extremely important and helpful in linking the neural substrate to an overtly measurable malfunction, ultimately bridging proximal efficacy measures from early clinical trials into Phase III.

Rodent Model

Although animal models are sometimes said to have "failed" in predicting drug efficacy, closer examination shows that these "failures" have not necessarily been failures; instead, they stem from false prior expectations, overstatements, and inappropriate extrapolation or overgeneralization of findings. For example, a positive drug effect in a simple screen of passive avoidance learning in a mouse or rat may be taken as optimistic evidence of a general cognitive enhancement, even though the effect in such a test may arise simply from an effect on a special form of emotional learning. Care must be taken when extrapolating to human memory. On the other hand, drug failure may have been due to factors other than misinterpretation of the animal data, including a poorly designed clinical trial that did not include measures to track preclinical translation. Back-translation from such trials is, in practice, rarely employed to check the validity of the animal model objectively. Overall, the predictive power of a rodent model should be assessed in the context of the entire clinical development cascade, taking reference to related tests in humans that use similar readouts and target neural substrates of evolutionary conservation. These are rarely conducted in Phase III but rather much earlier to obtain proof of mechanism or principle. Thus, the predictive power of such tests in early human trials needs to be aligned to those used in Phase III. An informative story about cholinergic influence on AD and the predictive power of rodent models is given in Appendix 13.6.

Transgenic Mice

The etiology of neurodegenerative and neuropsychiatric disorders is complex and multifactorial. Using experimental animal models has been more helpful in dissecting than simulating this complexity. Such models have provided a large number of useful insights into the pathogenic activities of individual factors suspected of contributing to the development of these or other neurological conditions.

For example, overexpressing familial AD-linked human amyloid precursor proteins (hAPPs) in neurons elevates amyloid beta (Aβ) levels in brain and is sufficient to cause a range of AD-like disease manifestations in these models, including impairments in learning and memory, behavioral alterations, synaptodendritic rarefaction, synaptic and neural network dysfunction, formation of amyloid plaques, neuritic dystrophy, astrocytosis, microgliosis, and vasculopathy (LaFerla and Green 2012). Several molecular alterations first observed in such models have been subsequently identified in humans with AD; for example, depletions of calbindin in dentate granule cells and of specific voltage-gated sodium channel subunits in parvalbumin-positive GABAergic interneurons in the neocortex (Verret et al. 2012; Vossel et al. 2013).

Because several other factors likely contribute to the pathogenesis of AD, including apolipoprotein (apo) E4, tau, α-synuclein, inflammatory mediators, and vascular alterations, it is not surprising that hAPP transgenic mice do not replicate all aspects of this complex human condition (LaFerla and Green 2012). Similarly, it is unreasonable to expect that testing drugs in such models will be sufficient to predict their efficacy in the much more complicated human condition. Notwithstanding, some therapeutic interventions have extrapolated quite well from hAPP mice to patients with AD or amnestic mild cognitive impairments, including the removal of amyloid plaques by anti-Aβ antibodies and the reversal of abnormal network activity by the antiepileptic drug levetiracetam (Vossel et al. 2013).

Compound experimental models combining different genetic and/or pharmacological manipulations can be used to simulate part of the complexity observed in the human conditions. In the AD field, such models have revealed interesting co-pathogenic interactions between Aβ and tau, Aβ and apoE4, apoE4 and tau, and Aβ and α-synuclein. Naturally, species barriers must always be kept in mind and make it critical to validate results obtained in animal models in human subjects. Whether iPSC-derived cell culture models can form a useful bridge in this process and whether they may be able to predict therapeutic efficacies more reliably than animal models remains to be determined.

To arrive at a therapeutic concept, it is important to quantify the predictive power of a given experimental state. For transgenic mice, increased rigor in preclinical trials that test the robustness of an effect plays a pivotal role. It should become standard (even if expensive) to include more independent biological cohorts and different age groups. Further, the species comparison can give confidence in a result.

Although the phenotype of a given transgenic mouse model may indicate large overlap to the human disease, careful consideration of the best strategy to generate and analyze these mouse models remains key. Although AD mice carrying triple mutations in APP, PS, and tau replicate most of the pathological features of the disease, they might not be perfectly relevant: the introduced genetic lesions do not appear together in patients; the transgenic, mutated tau is not even associated with familial AD but rather with another form of dementia, namely frontotemporal lobe dementia (Hutton et al. 1998). In addition, artificial promoters are introduced into each gene, and thus the interactions between gene products may not be captured appropriately. In regard to assays, protein disposition, synaptic function (slice), and cognitive functions (*in vivo*), and their interactions need to be studied. Finally, the contribution of environmental factors and their interaction with genetic factors need to be integrated into the models to recapitulate fully the etiopathology of the disease.

Nonhuman Primates

According to the idea of evolutionary conservation, models should faithfully reflect a given neural substrate/dysfunction. Due to their evolutionary similarities, nonhuman primates are more closely related to humans than other animals, thus leading to the greatest overlap in neural substrates and behavior. Nonhuman primates are, therefore, a prime (in some cases the only) model for brain structures and functions which are specific for primates.

One explicit example is the connection between motor cortex and spinal cord, which is uniquely organized in primates. In rodents, there are no direct connections between corticospinal neurons and the cervical motor neurons that innervate forelimb muscles—brainstem pathways and spinal interneurons relay cortical input to motor neurons. In nonhuman primates and humans, direct corticospinal connections with motor neurons have evolved, together with an increase in the size and number of the corticospinal fibers. This development correlates with an improvement in dexterity, particularly in the ability to perform finger-thumb precision grip (Lemon 2008; Courtine et al. 2007). This is reflected in the field of neurodegeneration or neuroregeneration, where nerve cell regrowth studies in monkeys have led to key discoveries that were ultimately channeled into a new treatment concept. In human adults, the neurite outgrowth inhibitor ("Nogo") protein prevents the healing of damaged nerves following a spinal cord injury (Schwab 2004). Experiments on monkeys have demonstrated that antibodies developed to counter "Nogo" following spinal cord injuries led to significant functional improvements through nerve growth (Freund et al. 2006). Currently, "Nogo" antibodies are being used in clinical studies on patients with spinal cord injuries. Another example from the neurodegenerative realm is the MPTP (1-methyl-4-phenyl-1,2,3,6-tetrahydropyridine) primate model used to decode the pathology of PD and to develop therapies (Capitanio and Emborg 2008). Nonhuman primates exposed to MPTP develop symptoms similar to PD, whereas rodents show a less specific response to MPTP. Studies on MPTP-treated nonhuman primates have enabled detailed investigation into the pathological consequences of PD, and thus a range of therapeutic approaches have become possible, such as the targeted administration of DA agonists and use of deep brain stimulation (Jenner 2003). Specifically, deep brain stimulation is a new and effective procedure for treating patients with movement disorders that would have been barely conceivable without research on primates (Rosin et al. 2011). Further, investigations on nonhuman primates are important for research of aging because the closer genetic relationship to humans produces a highly similar aging phenotype (Roth et al. 2004).

In highly cognitive or social contexts that rely on prefrontal cortex structures that are unique in primates, the nonhuman primate model is irreplaceable. Here, nonhuman primates allow for deeper insights into the neural substrates underlying these higher-order brain functions (Watson and Platt 2012). Many neuropsychiatric diseases (e.g., schizophrenia or attention-deficit/hyperactivity disorder) are accompanied by brain dysfunction in the frontal lobe (Robbins and Arnsten 2009; Nelson and Winslow 2009). Here, research on rhesus monkeys has been decisively important for decoding the mode of operation of the frontal lobe, which is specific to primates (Castner et al. 2000). A significant driving power for developing drug-free treatment for psychiatric diseases is expected from this research branch (Goldman-Rakic et al. 2004).

While transgenic nonhuman primates are an exciting prospect (Niu et al. 2014; Sasaki et al. 2009), there are also other more established techniques for correlative and causal investigation of the primate brain. Electrophysiological characterization of brain functions and neural correlates of behavior, as well as electrical and optogenetic manipulations for causal probing of the same, promise to widen our knowledge substantially (Wang et al. 2012; Diester et al. 2011). In particular, optogenetic techniques that allow for axon-specific manipulations could help disentangle complex pathways, thus elucidating the role of subcircuits for a specific behavior. When electrical stimulation is employed, we cannot be certain that only a local area or one single pathway is affected. Optogenetic manipulations, by contrast, allow for this very precise stimulation. By stimulating or inhibiting axons, while measuring simultaneously the impact on neural activity as well as on behavior, it is now possible to measure the impact of a pathway very specifically (Deisseroth 2014). Such a deeper understanding of the neural underpinnings of an observed behavior or symptom will also allow evolutionary conserved bridging points to be identified. Further, noninvasive brain stimulation approaches that can be applied to animal models, as well as to humans, offer the appeal of potentially serving as consistent biomarkers in models and patients. For example, transcranial magnetic stimulation can be used in such a manner (Freitas et al. 2013; Fox et al. 2012; Shafi et al. 2012; McClintock et al. 2011). Research on nonhuman primates is subject to justified ethical concerns about working with highly differentiated and sensitive animals. There is consensus that experimentation with these species needs to be restricted to areas where lower animal species provide insufficient information, and that they need to follow highest standards of animal welfare.

Induced Pluripotent Stem Cells and Induced Neuronal Cells

Current Possibilities and Limitations of Stem Cell Technologies

There is tremendous excitement about recent progress in cellular reprogramming and stem cell biology. Skin fibroblasts can be converted into iPSCs or induced neuronal cells (Vierbuchen et al. 2010; Mitsui et al. 2003). iPSCs can be differentiated in a second step into neurons, glia, and, in principle, any other cell type of the body. Fibroblasts can also be directly converted into various cell lineages, such as cardiomyocytes, hepatocytes, oligodendrocytes, and neural progenitor cells (Graf 2011; Vierbuchen and Wernig 2011). From a disease modeling point of view, these reprogramming methods are attractive because for the first time disease-relevant cell types (e.g., functional neurons) can be generated from the patients' own cells containing the patients' authentic genetic material—a "dream" for disease modelers since such a complete genetic representation cannot be accomplished in model organisms given the vastly different genomic organization in other species. This is particularly relevant, given the complex polygenetic basis of the major neuropsychiatric and neurodegenerative diseases. For these reasons, the community is extremely excited about the potentially transformative new cell models that are on the horizon to enrich our current portfolio of disease models.

There are several principal applications of these reprogramming techniques. In the context of neurological and psychiatric diseases, patient-specific neurons could be used (a) to divide each disorder into phenotypic categories (such as reduced number of glutamatergic synapses), which is vitally important since many genes may alter the same "druggable" process; (b) to establish disease-relevant screens that identify the best therapeutic classes for each phenotypic category; (c) to test each therapeutic candidate across large numbers of patient neurons; and (d) to choose patients most likely to benefit from specific drugs for clinical testing and, eventually, for treatment. Finally, a few relevant iPSC-derived cell types, such as cardiomyocytes and hepatocytes, could be pre-screened from hundreds of patients for toxic side effects. Another principle application of this technology is cell transplantation, where the idea is to interfere positively with disease processes (Figure 13.4).

As with every nascent technology, there are currently important limitations to this methodology. First, at the moment, protocols exist that claim similarities to cortical, spinal motor, and midbrain DA neurons, but they typically have

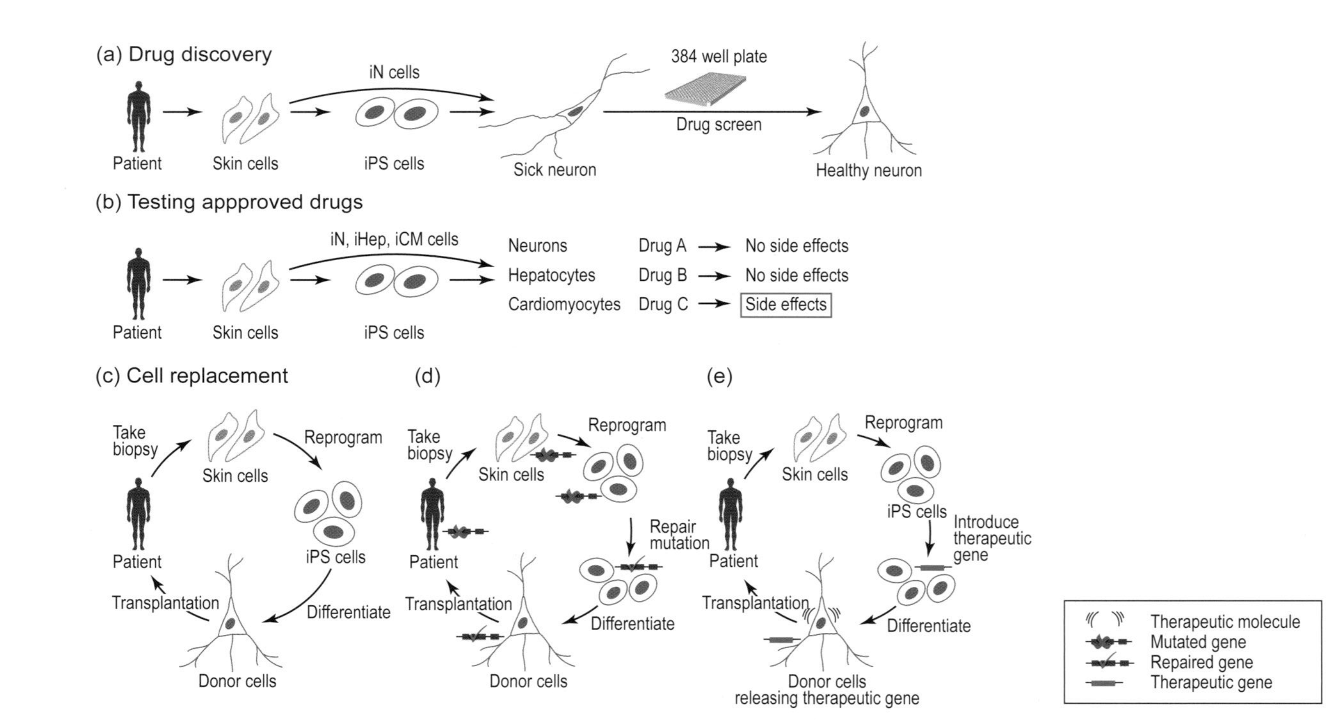

Figure 13.4 Three putative applications of induced pluripotent stem (iPS) cells. (a) Drug discovery: Neurons with disease phenotypes are used in screens to identify effective drug candidates. (b) Testing approved (or investigational new) drugs: Prior to clinical administration, tissue and cell types of primary patient-specific neurons, cardiomyocytes, and hepatocytes are used to find the safest and most effective drugs for each individual patient. (c) Cell replacement: Patient-specific cells can be transplanted to replace missing neurons, correct genetic defects, or provide a local source of growth or neurotrophic factors. Induced neuronal (iN) cells; inner cell mass (iCM) cells; induction of functional hepatocyte-like (iHep) cells.

low yields and represent mixtures of various undefined cells (Ming et al. 2011). Given the possibility that specific pathological defects might occur only at specific types of neurons or synapses, it will be important to have good control over these parameters. Differentiation protocols will have to be devised to generate pure or at least defined combinations of specific neuronal and glial subtypes. Second, a general issue in the iPSC field is that differentiated progeny are typically immature. This is true for neurons, glia, cardiomyocytes, hematopoietic cells, and hepatocytes (Ming et al. 2011; Sahakian and Jaenisch 2009). This is not surprising, since human fetal development takes nine months, as opposed to three weeks for a rodent. It will have to be determined what time frame can be tolerated for feasible disease models. The maturation state of brain cells is a concern because many neurodegenerative diseases occur in the adult brain and late in life. Therefore, even the production of fully mature cells might not be sufficient; we may need to produce "aged" cells. An extremely interesting approach to accomplish this goal is to introduce genes known to cause premature aging. It seems that at least certain cell biological effects typical of aging can be recapitulated in this fashion (Miller et al. 2013a). Unfortunately, however, the maturation problem has not been overcome by this approach, despite the observation of cell biological aging effects. Finally, human cells cultured *in vitro* cannot readily reconstitute complex neural circuitry as seen in the brain. This may require a three-dimensional organization of the cultured cells; thus the cultured iPSC-derived neuron system is more equivalent to dissociated brain cultures and is fundamentally different from the standard brain slice preparation performed in rodent or in rare epilepsy-associated human brain resections. Still, promising progress has been made in the generation of three-dimensional retina and corticoid spheres, which take advantage of the remarkable ability of embryonic stem cells and iPSCs to self-assemble (Kadoshima et al. 2013; Lancaster et al. 2013; Nakano et al. 2012). At this time, early developmental processes can be recapitulated in such organoid cultures, but the lack of vascularization imposes size limitations and thus may restrict the degree of maturation. We note, however, that even the most advanced human cellular and organoid model will be restricted to issues between gene and neural circuit. The connection between circuits and behavior is not accessible by a culture system, although new avenues are under discussion on how to overcome this hurdle (see next section).

There are already a couple of examples in which iPSCs have shown their immense potential, suggesting that they can play a role in disease modeling (see Rubin, this volume). For example, in spinal muscular atrophy (SMA), a genetic motor neuron disease with early onset, several papers have shown that iPSC-derived motor neurons carrying the SMA mutation can recapitulate many of the known disease processes. In addition, experiments using these cells have predicted other aspects of disease pathology that await verification in SMA patients. Another example is amyotrophic lateral sclerosis, a motor neuron disease with late onset and many subcategories (genetic and idiopathic). *In vitro* drug testing is only rarely carried out using motor neurons. One mouse model is available that highly overexpresses a mutated *SOD1* gene. These mice might capture only ~2% of the patient population. Thus, not surprisingly, drug tests in mice have not been predictive of clinical responses in patients. So, instead of generating more transgenic mice which again might capture only a small fraction of patients, iPSCs might offer a valuable alternative. Since they can be generated from large numbers of patients with various backgrounds, therapeutic concepts and drug candidates could be pretested in these highly individualized cultures before going into clinic.

All therapeutics need to be effective *and* safe. An important aspect of iPSC-based disease modeling is that it permits the generation, in a patient-specific way, of multiple cell types. For example, for evaluating motor neuron disease drug candidates, three cell types might be particularly relevant: motor neurons (for efficacy), cardiac myocytes, and hepatocytes (for safety).

To evaluate any new strategy using iPSCs, screens with already known drugs and careful observations of the effects are necessary. In the future, iPSCs might be used as an important add-on to more traditional approaches. They could play a role at the very start of drug discovery efforts and/or they could be valuable in stratifying compounds for clinical testing. They could also play a vital role in choosing patients most likely to benefit from particular treatments (see Figure 13.4 for three putative applications of iPSCs).

Current findings with iPSCs are exciting, and we want to point out that handling large numbers of iPSCs is possible. However, it will be difficult to capture entirely environmental factors in this system. In this case, the use of surrogate tissues to identify more homogeneous patient populations is useful. For example, olfactory cells from the neural epithelium, obtained through quick and safe biopsies, maintain a reasonable extent of neural molecular profiles.

New Horizons for Disease Modeling: Mouse-Human Brain Chimeras

As discussed, currently used (typically mouse) models of human disease traits offer benefits as well as important shortcomings. Some scientists might provocatively state that mouse models have essentially no predictive value to test drug efficacy in humans. While everyone recognizes that any model is naturally imperfect, certain aspects of disease have been well represented by these models and, indeed, yielded predictive power, such as the specific effects of cholinergic treatments (see Appendix 13.6). In principle, the utilization of human cells is obviously preferable. Thus a logical step is the utilization of human iPSCs or induced neuronal cell systems. However, phenotypic analysis is limited *in vitro*. Obviously, for circuit function and behavioral studies, intact brains and organisms are required. Moreover, there are important cell biological limitations with these cells, most notably cell type specification and functional maturation. Current *in vitro* technology can, at best, supplement mouse models but it cannot replace them (Figure 13.4). Thus, there is an urgent need to develop *in vivo* models utilizing human cells.

A general solution would be to generate mouse-human brain chimeras consisting of mouse and patient-derived brain cells. In that setting, essentially all shortcomings discussed would be addressed: (a) patient-derived brain cells could be studied *in vivo*; (b) the brain represents the ideal environment for neuronal maturation replacing long-term cultures; (c) regular participation in embryonic development would ensure the generation of authentic neuronal and glial subtypes with appropriate projections into other brain areas.

This could be achieved in several ways. First, one could take advantage of the pluripotent nature of iPSCs. Mouse iPSCs introduced into a blastocyst will participate in normal embryonic development, yielding chimeric mice made up of host blastocyst-derived and iPSC-derived cells. By analogy, human pluripotent cells could similarly participate in embryonic development, resulting in organs that contain both mouse and human cells. Preliminary attempts at this approach appear promising (Gafni et al. 2013). Second, human neuroectodermal cells derived from human iPS cells could be introduced into the early developing mouse brain. Third, transplantation of glial-restricted precursor cells into the neonatal mouse brain would generate chimeric mouse brains containing mostly human glia and mouse neurons (Han et al. 2013b). Naturally, such experiments would have to be executed under close ethical supervision.

Summary of Model Systems

Any model is successful if it does not pretend to be a model of the "disease" but rather of a given pathogenic factor. Models need to reflect, as faithfully as possible, the human neural substrate that best characterizes the symptoms of disease caused by the pathogenic factor being scrutinized.

Increasing Translational Success

Over the past decade, rapid progress in basic neuroscience has brought many enthusiastic claims of advances in effective drugs and other therapies to treat neurologic and psychiatric diseases. Commensurate with this has been a large number of significant failures in Phase III studies. As a result, the neuroscience field has obtained a reputation for being unpredictable, ultra-risky, and extremely expensive. There is an impression of a prevailing culture of easy promises from basic scientists; careless moves from Phase II to Phase III, in spite of marginal Phase II data; and a lack of adequate "probability of success" predictions. For these reasons, many biopharmaceutical companies have shifted efforts away from neuroscience toward oncology and immunology.

The translation of data from models to humans, and the development of therapeutics to approval, is a complex and regulated process—one that involves many diverse organizations and players. Figure 13.5 attempts to summarize

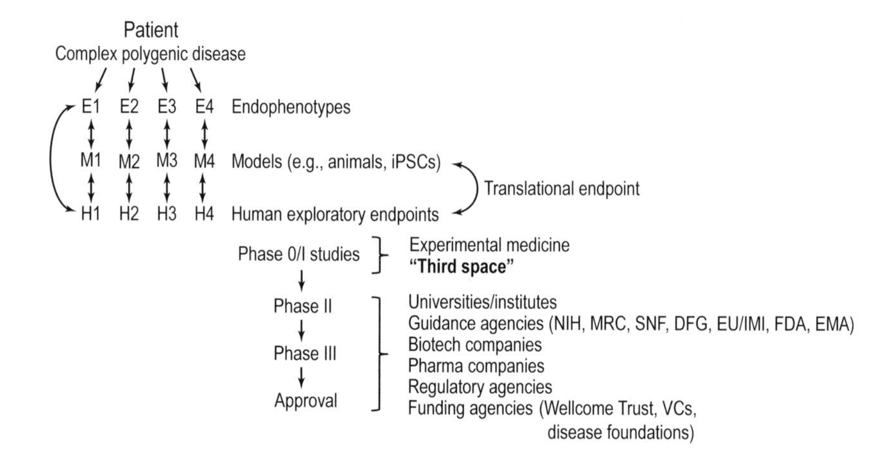

Figure 13.5 Translational sciences and therapeutics development: process and participating organizations. Treatment approaches that emerge from bottom-up and top-down approaches are typically validated at the preclinical level in various animal and *in vitro* models. The models (M) reflect specific endophenotypes (E) of a complex polygenic disease. Exploratory clinical endpoints (H) corresponding to the outcome measures greatly facilitate early clinical development and detection of efficacy in humans. Optimal translational endpoints are identical in humans and models and are measured with the same or closely related technologies. The general area of "experimental medicine" exceeds traditional departmental structures, which compromises translational efforts. Clinical development (in particular, Phase II and III studies) involves multiple organizational structures from academia, government, industry, charities, and financing. These efforts are often poorly aligned and integrated. NIH: National Institute of Health (U.S.); MRC: Medical Research Council (U.K.); SNF: Schweizer Nationalfonds (Switzerland); DFG: Deutsche Forschungsgemeinschaft (Germany); IMI: Innovative Medicines Initiative (EU); FDA: Food and Drug Administration (U.S.); EMA: European Medicines Agency; VC: venture capital.

the process as well as the involved organizational entities for the subsequent discussion.

A Dedicated Home for Translational Sciences: The "Third Space" Concept

Academic scientists have become increasingly engaged in translational biology and medicine. Once promising neural or molecular targets are discovered, many academic laboratories strive to reach a point where they have (a) chemical entities or interventions that reliably engage the target; (b) useful assays to document, characterize, and measure the engagement; and (c) enough target validation data to be able to transfer the technology to a biotechnology or pharmaceutical company. Indeed, both government funders and patient groups encourage academic scientists to pursue these goals and even provide some of the necessary resources for different stages of the process (e.g., the National Institutes of Health, the National Center for Advancing Translational Science, the Michael J. Fox Foundation, and The Wellcome Trust).

Despite this support, however, most academic centers lack the necessary infrastructure (e.g., well-curated, accessible chemical libraries or engineering laboratories for prototyping) to realize effective early stage discovery and develop novel therapeutic concepts. In addition, most academic biologists and neuroscientists do not know how to conceptualize the various steps in the process, how to set up and conduct screens at the appropriate scale, or how to judge whether the process (e.g., chemistry or device development) is proceeding well. Trained personnel (e.g., medicinal chemists, engineers, project managers) are often lacking, as is a broad base of necessary expertise inherent in translational work. Importantly, the expertise needed for "experimental medicine" is not contained within traditional departmental structures of academia; thus, as attempts are made to conduct translational efforts, problems arise all too easily. An effective infrastructure needs to be guided by individuals with industry experience—people who are able to work closely with the academic experts championing the project.

Efforts to create such an infrastructure and develop the necessary expertise within academia are underway. One example is the "Reactor" program in Harvard Catalyst.[2] Launched in 2013, Reactor aims (a) to develop innovative technologies and methodologies and assist their clinical implementation, and (b) to bridge the chasm between discovery of basic biomedical observations and their clinical applications through the provision of resources, mentoring, and expertise. Reactor convenes cross-disciplinary, cross-institutional teams and communities, providing support to expedite clinical and translational research. It facilitates skill development, team building, mentoring, funding, and identifies external resources to assist investigators and teams in the translation of clinical discoveries. Reactor's ultimate goal is to leverage these discoveries to impact patient care through the development, testing, and adoption of novel prevention strategies, diagnostics, therapeutics, and biomarkers. Other similar organizational experiments are underway. In all of these efforts, the ability to address and respond to all challenges and obstacles, both those that are known and those which will emerge, will be of paramount importance.

There are serious and hard obstacles to overcome if a therapeutics facility is to be created within academia. Extensive intellectual and technical platforms are needed and diverse expertise (ranging from intellectual leadership to business development officers, to chemists, engineers, etc.) must be secured. Moreover, there is an inherent conflict between intellectual property and business considerations: results may not be publishable within expected

[2] http://catalyst.harvard.edu/programs/reactor/ (accessed March 3, 2015)

academic time horizons. Here, arrangements reached by universities engaged in classified research (e.g., MIT's Lincoln Laboratories,[3] Draper Laboratory,[4] and the Applied Physics Laboratory at Johns Hopkins University[5]) can offer guidance.

Instead of anchoring a therapeutics facility within academia, controlled by a single university and a single company, we propose that a separate, dedicated home be created to optimize translational science. External to the structures inherent in academia and industry, this "third space" (Figure 13.6) would provide a fertile environment for experts from both fields. Staffed by scientists (and not graduate students or post docs interested in academic careers), this "third space" would be dedicated to early phase human experimental trials and the identification of human endophenotypes for translation. One could envisage the development of a "third space" by a regional consortium of universities and hospitals, which would share in its governance and funding as well as benefit from its pooled expertise. Critical issues to establish at the onset include intellectual property rules, review mechanisms to judge effectiveness,

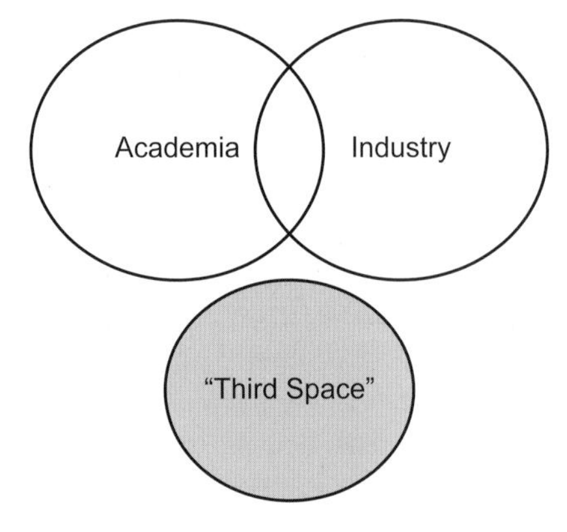

"Third Space" for experimental medicine
- Staff scientists (no graduate students)
- Early phase human experimental trials
- Identification of human endophenotypes for translation
- Philanthropy to launch
- Financial sustainability will be a challenge

Figure 13.6 The "third space" is a separate, dedicated area external to academia and industry that brings experts from both fields together on neutral ground. Staffed by scientists (not graduate students or post docs interested in an academic career), early phase human experimental trials would be conducted and human endophenotypes identified for translation. Charitable organizations could help in its launch, but long-term financial sustainability would require careful planning.

[3] https://www.ll.mit.edu/ (accessed March 3, 2015)

[4] http://www.draper.com/ (accessed March 3, 2015)

[5] http://www.jhuapl.edu/ (accessed March 3, 2015)

and realistic business plans. Charitable organizations could help in its launch, and partial support could be provided by agencies such as the National Center for Advancing Translational Science. Careful thought and planning would be necessary, however, to ensure long-term financial sustainability.

Late-Stage Clinical Studies toward Approval and Patient Benefit

Beyond translational medicine, bridging models and Phase 0/Phase I clinical studies, advancement of a drug candidate into Phase II, and Phase III clinical studies necessitates the involvement of biotechnology and pharmaceutical companies, regulatory authorities—the U.S. Food and Drug Administration (FDA), the European Medicines Agency (EMA)—as well as funding agencies and investor groups. There is no generally accepted format for these interactions. There are positive examples of precompetitive efforts between academic groups and pharmaceutical companies, the Alzheimer Disease Neuroimaging Initiative being the most visible one. Obviously, negative examples of such interactions exist as well, often caused by an inappropriate level of bureaucracy, leading to ineffective collaborations. Small biotech companies and associated venture capital groups play an important role in drug development. Typically, they operate outside of the established frameworks of university–pharmaceutical company interactions, interacting in a more informal networking way with universities, pharmaceutical companies, and regulatory authorities. In recent years, charitable organizations and advocacy groups for specific diseases have become important contributors as providers of information, contacts, and funding. The development of effective therapeutics would benefit from closer integration of and communication among all involved parties. To this end, we offer specific recommendations, in the following section, to improve the success in translational biology and medicine, and to assist in subsequent development toward approval and actual benefits to the patients.

Specific Recommendations

Collaborations between Scientific and Clinical Researchers from Different Fields

The gap between scientific and clinical researchers from different fields is growing due to external pressure. Fragmentation of institutions and of communities (e.g., epilepsy, stroke) limit communication; thus failures are repeated across individual fields. Here, we need to draw lessons from cancer research where, although a breakdown is made into cancer types, cancer is perceived as one big research topic because all cancer types share one important property: excessive proliferation. In neuroscience, diseases are discussed separately, and, as a result, the public perceives research fields into these diseases as unrelated and individually small. In oncology, the situation is simpler, due to a single driving factor, increased proliferation, even if there are specific drivers for individual types (e.g., in bone cancer). In neurology, by contrast, the single readout is different: we lack an equivalent to "proliferation" as the one uniting factor. In psychiatry, a major obstacle is the lack of subjective biomarkers corresponding to disease and symptom. Thus, major efforts on biomarker identification in psychiatric disorders are expected. Since manifestations of patients with psychiatric disorders change, careful distinction of state and trait markers is required in biomarker exploration studies. Although still preliminary, change in C reactive protein in manic symptoms may provide a good example of a possible state marker (Dickerson et al. 2007).

To overcome this issue, the National Institutes of Health have initiated a network that is not tied to a specific disease, to enable better learning between diseases (only Phase II trials are included to ensure better communication and make studies comparable). In Germany, at the *Forschungszentrum für neurodegenerative Erkrankungen* (German Center for Neurodegenerative Diseases), clinicians and scientists jointly conduct translational research. More national research hospitals need to be established so that science does not compete with patient care.

Collaborations between Academia and Industry

It would be beneficial to have institutions which support academia–industry collaborations that are capable of addressing all neural diseases instead of many specialized ones. Within such institutes, academia and industry would work together, and it would be possible to carry out Phase 0 trials (as was done in cancer research). Early human efficacy studies need to be done in genetically identifiable populations. Tailor-made fits between academic and industry scientists might be best suited for such successful collaborations. Interactions between academia and industry are often hampered by business development and intellectual property concerns, as well as a lack of understanding by universities of how to negotiate with industry. One possible solution would be to create an entity aimed at assisting effective communication. To this end, a translational group has been established at the National Institute of Health that combines management and consulting experts. Another solution might be to establish a mobile translational unit, staffed with personnel from industry, chemistry consultancy, and lawyers. When scientists find a good target, they could obtain advice from the mobile unit on how best to move forward. The European Innovative Medicines Initiative aims to improve the development of innovative treatments. Launched under the European Commission's Sixth Framework Program for Research, it assembles relevant stakeholders such as academia, biotech companies, regulators, patient organizations, etc., and is led by the European Federation of Pharmaceutical Industries and Associations. Specific objectives include overcoming the many, significant hurdles inherent in drug development (e.g., the difficulty in predicting safety and efficacy of potential drug candidates). Overarching strategic objectives are approached by formulating specific calls for proposals to form public-private partnership consortia in research in Europe; EU funding is provided to academic partners and is matched by equal in-kind contributions from participating industrial partners. As such, the Innovative Medicines Initiative provides a suitable platform that facilitates and promotes interactions and collaborations between academia and industry on joint work concerning pre-competitive aspects relevant to drug discovery utilizing mutual capabilities of partners. Although there are still obstacles to overcome, such as reducing bureaucracy, and the fact that the performance of individual consortia highly depends on the partners' willingness to subordinate their specific interests to joint objectives, the initiative may eventually prove useful in introducing a more cooperative spirit between academia and industry.

Collaborations between Multiple Industry Partners

A good working example for pre-competitive collaboration between pharmaceutical companies is the Dominantly Inherited Alzheimer Network, which includes ten participating industry partners. In such a consortium, it is very important for all partners to focus on one goal and build milestones based on objective measures. Successful pre-competitive collaborations are conceived to focus on target finding and validation, biomarker and target-engagement measurements, rather than drug development.

Industrial collaborations typically involve profitable, large pharmaceutical companies and exclude small biotech companies and related experts. Biotech companies and associated venture capital investors constitute an innovative, highly talented, and informal group. They operate largely through social networks and, in the United States, are particularly strong in the Boston and San Francisco Bay areas. Venture capital groups are often driven by experienced physician-scientists, a group of people that could contribute more significantly in the overall search for effective treatments.

Biotech companies have been highly effective in taking discoveries at universities to the application level, often in otherwise ignored diseases. Typically they bring a new concept to the level of Phase I and Phase II studies. If successful, the companies are often absorbed by large pharmaceutical companies which then take the treatment to approval. In neuroscience, examples include AC-Immune (antibodies for AD), Rinat Neuroscience (anti-NGF antibodies for pain), Avid Radiopharmaceuticals (PET imaging agent for plaques in AD), and EnVivo Pharmaceuticals (nAChR agonist to treat cognitive deficits in neurodegenerative and psychiatric conditions). It seems worthwhile to attempt to integrate biotech and venture captial groups more effectively into the discussions and interactions between universities, governmental agencies, and large pharmaceutical companies. Small biotech companies typically do not have the funds to buy into pre-competitive pharmaceutical efforts. A separation between profitable and nonprofitable companies may be an approach to solve this problem.

In this context, several foundations should be mentioned: the SMA Foundation, the CHDI Foundation, the Michael J. Fox Foundation, the Simons Foundation, and the Accelerate Brain Cancer Foundation (ABC^2). These foundations have become effective contributors, supporters, and funding sources for specific disease areas. For example, the SMA Foundation, together with academics, biotech and pharmaceutical companies, has made a major contribution to the promotion of credible therapies to treat motor neuron degeneration; ABC^2 multiplied the number of clinical trials for brain tumor and, similarly, other foundations have boosted discoveries and therapeutic approaches for a variety of brain diseases.

Over the next decade, we envision an increasing number of collaborations between pharmaceutical companies, biotech companies, academia, investors, foundations, as well as private sources to promote high-quality science-based therapies.

Translational Approaches in Psychiatry Usually Fail: What Can Be Done?

Several reasons have contributed to the frequent failures observed in translational approaches taken in psychiatry. It is very difficult to motivate psychiatrists to take part in a translational research track, in contrast to the situation in neurology. Most of the psychiatry resident programs in the United States do not include training in scientific investigation, and offer only limited exposure to basic study design (in particular, hypothesis-driven research design), statistical analysis, and interpretation of results. The current environment of reduced funding for scientific pursuits as well as the increased demand for revenues generated from clinical responsibilities in psychiatry, places time and financial constraints on young clinicians, often discouraging a long-term commitment to research. Meanwhile, basic science researchers who wish to study psychiatric illness are limited in opportunities to learn about the complex and heterogeneous manifestations and courses of psychiatric diseases. While basic scientists approach neurological diseases in a correct manner by utilizing molecular pathological hallmarks in the pathology, there are no objective markers available in psychiatric disorders as yet. It is difficult to gain insight into psychiatric disorders without direct patient exposure; knowledge from textbooks and journals can mislead basic scientists as they build working hypotheses to study psychiatric disorders.

The "dual gap" in training of early clinicians and young basic science researchers has hampered progress in translational psychiatry. In response, Johns Hopkins Medicine recently initiated a workshop entitled "Mind the Gap," creating a forum in which young researchers and resident physicians discuss research topics directly raised from clinical questions. This workshop proved beneficial for all groups. Shared research interests voiced by basic scientists and clinician-scientists benefit from the guidance and resources provided by senior faculty members (mentors) (Posporelis et al. 2014).

More emphasis should be placed on training and mentoring psychiatrists so as to provide clinical scientists a broader and deeper knowledge in neuropsychiatry. This could help bridge the somewhat arbitrary separation between neurology and psychiatry. Neurological disorders frequently display psychiatric manifestations, such as depression and psychosis. Studying these manifestations is crucial to patient care management as well as to a better understanding of disease mechanisms (Robinson et al. 1984; Cooper and Ovsiew 2013; Schwarz et al. 2012). There are some promising attempts to fill this need, but more initiatives are needed. For example, the Sidney R. Baer Jr. Foundation Fellowship in Clinical Neurosciences is a joint effort between the neurology and psychiatry departments at Beth Israel Deaconess Medical Center and Harvard Medical School. Designed as a three-year fellowship, it includes both clinical and research mentoring and training. Fellows are matched with two clinical mentors (one from the psychiatry faculty, one from the neurology faculty at Harvard Medical School) as well as two research mentors (one an active clinical/translational scientist, one a basic researcher). Seed funding and appropriate mentoring and support is provided so that a research project can be developed during the fellowship. This provides solid clinical foundation in clinical neurosciences as well as training across the research-clinical divide.

Increasing Interactions with Regulatory Agencies

Regulatory agencies are often seen as impediments and excessively conservative. Positive examples, however, prove that this is generally not true. Very often, the cautious and critical reactions on the part of the regulatory agencies reflect the necessary focus on safety issues, but lack familiarity with the science behind the new drug development programs. It is thus beneficial to include regulatory agencies as early as possible into the discussion. In the United States, pre-investigational new drug meetings provide a format for such early discussions, and the European Medicines Agency has comparable mechanisms to obtain early feedback. Invitations to conferences on translational medicine extended to staff from regulatory agencies will help provide early familiarity with emerging new approaches.

The recent discussions on regulatory paths to approval for drugs in AD and PD illustrate this interactive approach. In particular, the development of PET-imaging agents for the amyloid pathology in AD provides the most prominent example for successful implementation of the proposed approach. Direct discussions between academic groups, biopharmaceutical companies, and the Food and Drug Administration led to a novel and generally accepted path for the development of PET radiopharmaceuticals to visualize pathological deposits in neurodegenerative disease. The approach was then summarized in a highly visible paper published by FDA officials (Yang et al. 2012a).

Improving the Transition between Phase II and Phase III

While previously discussed approaches and employment of appropriate models target primarily an improved success rate in the transition from Phase I to Phase II, drug development for nervous system disorders still confronts a high attrition rate in Phase III. Over recent years, success rates have declined rather than improved (Arrowsmith 2011). Sufficiently large patient populations are required to exact the regulatory endpoints implicit in Phase III trials, thus necessitating a major financial investment from drug companies. However, the high failure rate experienced late in the value chain imposes a major commercial risk for most corporations. Rigid application of translational models and experimental medicine approaches, using appropriate surrogate markers, may enable a front-loaded testing of more therapeutic concepts, thus leading to faster failing of candidates and a more rigid review of Phase II data, followed by a more diligent selection of principles being promoted to Phase III.

One important improvement for the transition between Phase II and Phase III trials would be to create well-characterized patient groups. This general, important challenge impacts both translational research and clinical neuroscience. Although the creation of patient data bases with specific quantitative measures is time-consuming, expensive, and not directly supporting academic advancement, it is imperative if we are to enable effective patient-based research. A "third space" concept, similar to the notions developed above, is to establish and maintain patient repositories. Such efforts could be part of experimental clinical neurosciences initiatives and might be jointly funded by governmental agencies, industry consortia, and foundations. In addition, academic health centers and many national health care systems are currently investing substantial funds to establish electronic medical records; these could be further developed to allow semantic query functions and enable patient selection and even randomization in research. Agreement on basic assessment tools and the establishment of suitable test batteries, careful validation, and quality control will facilitate success in late-stage clinical trials.

Summary of Recommendations to Increase Translational and Development Success

Due to significant failures in late-stage drug development and unfulfilled promises, drug discovery and development for neurologic and psychiatric diseases is considered extremely risky, and many biopharmaceutical companies have shifted their efforts away from these disorders. Recommendations to ameliorate the situation include various organizational and behavioral measures. In particular, we propose the creation of "third space" organizational structures to unite academic and industrial groups involved in experimental biology and medical research. Furthermore, to increase the success rate in clinical development and provide actual benefit for patients, we propose that proactive interactions be supported between all organizational entities involved in drug development and therapeutic discovery (e.g., academia, guidance agencies, biotech, device and pharma companies, regulatory agencies, and funding agencies).

Appendix 13.1

Biomarker: a characteristic that is objectively measured and evaluated as an indicator of normal biological processes, pathogenic processes, or pharmacologic responses to a therapeutic intervention (definition created by the National Institutes of Health Biomarkers Definitions Working Group). Biomarkers that fulfill specific criteria can be endophenotypes.

Bottom-up research: research approach that starts from genetic or environmental factors.

Endophenotype: term used to parse symptoms into more stable phenotypes with a clear genetic connection. In psychiatric genetics, an endophenotype is used to bridge the gap between high-level symptom presentation and low-level genetic variability ("intermediate phenotype"). Endophenotypes are a subclass of biomarkers. For a biomarker to qualify as an endophenotype, the following criteria must be fulfilled: (a) the endophenotype must be associated with illness in the population; (b) it precedes clinical symptoms; (c) it is primarily state-independent (i.e., it manifests in an individual whether or not illness is active); (d) within families, endophenotype and illness co-segregate, and endophenotypes can be manifested on various levels (see Figure 13.7). In summary, endophenotypes can provide better links to genetics, purer samples of patients for clinical trials, and early markers of vulnerability to disease or disorder.

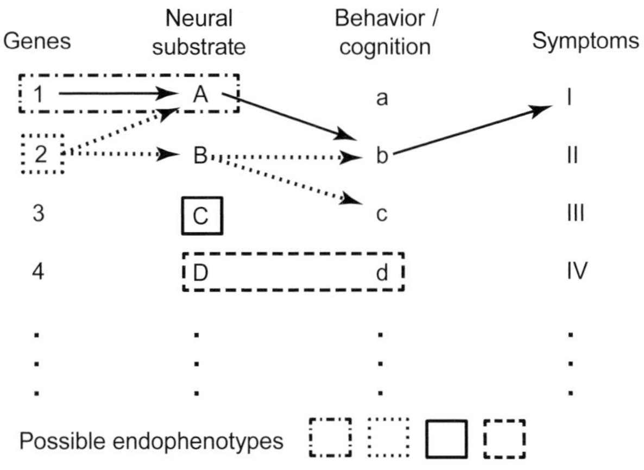

Figure 13.7 Visualization of the different versions of endophenotypes. An endophenotype can either be an altered gene, an altered neural substrate, an altered behavior, or combinations of the different options. Importantly, an endophenotype precedes clinical symptoms.

Phenotype: composite of an organism's observable characteristics or traits, such as its morphology, development, biochemical or physiological properties, and behavior. A phenotype results from the expression of an organism's genes as well as the influence of environmental factors and the interactions between the two.
Phase 0 trials: Exploratory "first-in-human" studies to test activity in a small number of subjects.
Phase I trials: Establish safety, dose range, and adverse effects in small groups of subjects.
Phase II trials: Establish efficacy of the drug, usually against placebo with larger groups of subjects.
Phase III trials: Confirmation of safety and efficacy in the to-be-treated patient population with large groups of subjects.
Prodromal stage: early stage or symptoms of disease before characteristic symptoms appear.
State marker: a state marker reflects the status of clinical manifestations in patients.
Top-down research: research approach that starts from syndrome or behavioral phenotype.
Trait marker: A trait marker represents the properties of the behavioral and biological processes that play an antecedent, possibly causal, role in the pathophysiology of the psychiatric disorder.

Appendix 13.2 Top-Down and Bottom-Up Approaches to Parkinson Disease

Current understanding of PD, the existing therapy, and the ongoing efforts for improved therapy reflect fruitful bridging between top-down and bottom-up approaches. Neuropathological and neurochemical investigations of *post mortem* brain tissue revealed significant degeneration of the ascending dopaminergic system originating in the midbrain. These discoveries, made approximately sixty years ago, led directly to DA replacement therapy and the approval of the first drug treatment, L-DOPA, the precursor of DA in combination with an inhibitor of metabolism outside of the brain. This combination drug and various direct agonists on DA receptors are the mainstay of current Parkinson therapy (Carlsson 2002).

Lewy bodies, pathological inclusions in the brains of PD patients, were first described one hundred years ago, long before the recognition of the dopaminergic deficit. Little progress in understanding was made until about fifteen years ago, when a mutation causing familial PD was found to be located in the gene coding for α-synuclein, a protein aggregated in the Lewy bodies (Polymeropoulos 2000). The discovery of several other disease genes followed, which allowed investigators to create hypothetical pathological pathways integrating the various genetic discoveries. Furthermore, several strains of transgenic animals carrying various single or combined disease genes have been created. Initially, even though some of the mice exhibited Lewy body-like deposits in the brain, none of them showed a dopaminergic neuron deficit, which would be expected in an advanced animal model of PD (Bezard and Przedborski 2011). More recently, however, strains of transgenic mice with dopaminergic neuron deficits have emerged, thus providing a bridge to the original neuropathological discoveries.

The bridge between bottom-up and top-down approaches has brought integrated hypotheses of disease mechanisms, involving several identified disease genes and advanced transgenic animal models, which exhibit dopaminergic neuron degeneration as well as Lewy body pathology. This progress supports the ongoing enthusiasm in the field; in particular, that optimal drug targets can be identified and that potential drugs and treatments can be tested in animals with high predictive power.

Appendix 13.3 Historic Perspective on Top-Down Models for Aspects of Schizophrenia

In schizophrenia, the disorder phenotype is even more complex, with positive symptoms (e.g., hallucination, delusions), negative symptoms (e.g., apathy, social withdrawal), and cognitive deficits (e.g., in working memory). Models were driven by the need to understand the initial, serendipitous discovery of the antipsychotic effects of chlorpromazine and the subsequent motivation to introduce more efficacious compounds with fewer side effects. These led to the hypothesis that a DA overactivity state in schizophrenia is, at least partly, responsible for the positive symptoms (e.g., hallucinations and delusions) of the disorder, and the isolation of a specific DA receptor (D2) responsible for the antipsychotic effects. This was consistent with the phenomenon of amphetamine paranoid psychosis often found after bingeing on high doses of stimulant drugs, such as amphetamines. Initially, new drugs, such as the butyrophenones, were synthesized and screened against amphetamine-induced hyperactivity or stereotyped behavior in rats treated with amphetamines, which increases synaptic levels of DA in such brain regions as the nucleus accumbens and the dorsal striatum (i.e., caudate putamen). Amphetamine locomotor hyperactivity has been shown to be associated with the nucleus accumbens, whereas amphetamine-induced stereotyped behavior is linked with DA-dependent functions of the caudate putamen, suggesting abnormalities of DA regulation in those areas, broadly shown later to be correct (e.g., Howes et al. 2009). The predictive power of these pharmacological assays was excellent, although they were subsequently superseded by actual measures relating clinical potency to measures of D2 DA receptor binding. However, although the D2 antipsychotics were actually one of the major achievements in this area over the last fifty years, they clearly fail to treat both of the other two major facets of schizophrenia effectively—the negative symptoms (encompassing apathy and social withdrawal) and cognitive deficits—which in group terms are quite broad. However, this masks underlying heterogeneity in terms of a predominance of memory impairments, often associated with the hippocampal malfunction or "executive" impairments associated with prefrontal cortex dysfunction. In fact, one prominent monkey study by Castner et al. (2004) showed that chronic treatment of rhesus monkeys with the D2 antipsychotic haloperidol produced an apparent downregulation of the D1 receptor and a significant impairment on a spatial working memory task that was remediated by treatment with a D1 DA receptor agonist. Clozapine (an atypical neuroleptic with many additional modes of action besides the D2 receptor antagonism that is presumably responsible for its antipsychotic effect) emerged after an intense program of drug development, which included animal model studies to differentiate its mode of action from that of the typical antipsychotic compounds. However, its action on negative symptoms and cognitive deficits is generally regarded as weak.

Investigators thus began to query the genesis of the DA overactivity state, which through PET studies has recently been shown definitively to occur in the caudate nucleus, even in the unmedicated and prodromal state prior to first episode schizophrenia (Howes et al. 2009). An early view (Weinberger 1987; Murray and Lewis 1987) was the neurodevelopmental hypothesis, which suggested an early cortical (including hippocampus) deficit that led to (a) dysregulation of midbrain DA cells and (b) ancillary effects (e.g., impaired cognitive function). This early developmental cortical dysregulation hypothetically included changes in glutamate receptors (e.g., NMDA), leading to a glutamatergic hypothesis of schizophrenia fueled by evidence that certain NMDA antagonists, such as ketamine and phencyclidine, in humans produced a characteristic psychosis not dissimilar to that observed in schizophrenia. This led to an entirely different strategy of administering drugs affecting glutamatergic mechanisms, including the mGlu2/3 receptors (the Lilly mGluR2/3 agonist pomaglumetad; Patil et al. 2007), which has not held up on subsequent clinical trials.

This approach has also been used to generate appropriate animal models based on acute or chronic treatment with glutamatergic antagonists, such as the NMDA receptor antagonist phenycyclidine or (less successfully) ketamine (Moghaddam 2013). Whether the glutamate hypothesis eventually bears fruit is unknown; however, the mGluR2/3 agonist might work better early in the course of schizophrenia, when MRS measures of glutamate indicate in many patients a hyperglutamatergic state in the cortex, which presumably would be diminished through treatment with an mGluR2/3 agonist. Even if the drug does not work in chronic schizophrenia, it may protect first-episode patients from progressing to a chronic state by preventing or reducing initial episodes of schizophrenia.

Appendix 13.4 A Brief Survey of Contemporary Top-Down Models in Schizophrenia

Given the evidence that schizophrenia is a neurodevelopmental disorder (although there is also evidence for a progressive decline in many patients after initial psychotic episodes), the animal model approach has been to manipulate brain development in such a way that it mimics some of the main features of schizophrenia. Various means have been used to achieve this, inspired by early studies on schizophrenia which show enlarged lateral ventricles, indicating possible loss of brain tissue. The latter has been confirmed through structural scanning to include loss of brain volume, especially in the temporal lobe, and changes in the hippocampus. The other main anatomical changes in schizophrenia include a loss of cortical parvalbumin-containing interneurons or Chandelier cells as well as cortical thinning and GAD67 protein changes in the prefrontal cortex (Moore et al. 2006).

Many of these changes as well as dysregulation of midbrain DA neuron activity can be simulated by affecting the hippocampus during development. Neonatal lesioning of the hippocampus (Lipska and Weinberger 2000) produces many of these changes as well as interesting behavioral effects of possible relevance, including impairments in cognitive function (e.g., spatial working memory) and social interaction. Indeed, it is possible that this early, self-imposed social isolation itself produces some of the neural changes, given the role of social play in neural plasticity and brain development, as would the effects of isolation rearing, which by itself can lead to several relevant impairments including upregulation of midbrain DA system function (Hall et al. 1998).

Another top-down model is the MAM-E17 model (Moore et al. 2006). MAM is a mitotic neurotoxin which essentially retards neural development. When administered to pregnant dams at the E17 stage, its effects are limited to forebrain structures such as the hippocampus (which is especially affected) and neocortex. The brains are smaller, as is the hippocampus itself, presumably as a consequence of cell death and rearrangement during maldevelopment. There are also changes in the prefrontal cortex and an upregulation of the mesolimbic DA system projecting to the nucleus accumbens. Animals have some cognitive deficits (e.g., associated with hippocampus and prefrontal cortex). Reversal learning in MAM-treated rats (also known to be impaired in schizophrenia) is rescued by an mGluR5 PAM, consistent with the loss of mGLuR5 receptors in the prefrontal cortex (Gastambide et al. 2012).

The earlier model based on neonatal hippocampal lesions reproduces behavior that often looks similar to that of the MAM17 model (Lipska and Weinberger 2000). An interesting aspect of the neonatal lesion model, possibly analogous to early schizophrenia, is that its initial symptoms are social (self-isolation); these may exacerbate development of additional symptoms by depriving the animal of social play and hence possible hippocampal-cortical connectivity.

Environmental factors and their modeling: Early on, it was recognized that the etiopathogenesis of schizophrenia involves environmental factors. While genome-wide association studies have revealed many genetic risk variants (22 loci with up to 8,300 independent single nucleotide polymorphisms, copy number variants; Ripke et al. 2013; Kirov et al. 2014) with low effect, twin studies have highlighted a missing heritability (Sullivan et al. 2003). Further, detrimental environmental conditions such as physical and psychosocial abuse, parental loss during early childhood, chronic social stress and social defeat, history of maternal infection during pregnancy, stress due to birth complications, and urbanicity or cannabis use in adolescence are recognized as critical risk factors for the disease.

Based on these observations, environmental models have been developed in experimental animals. Models of maternal inflammation induced by polyI:C injection during gestation in rat or mice have reproduced some of the negative and cognitive symptoms (Kneeland and Fatemi 2013). Isolation rearing of rat pups after weaning impairs working memory and sensorimotor gating in the prepulse inhibition (PPI) of the acoustic startle response, as observed in schizophrenic patients, in adulthood (Weiss et al. 2001). Consistent with the DA hypothesis of schizophrenia, isolation rearing also alters burst firing of DA neurons in the ventral tegmental area, DA neurotransmission in PFC and nucleus accumbens, and increases DA release and receptor sensitivity in the accumbens and striatum (Hall et al. 1998). Further, several models have combined manipulations in double-hit paradigms. In rat, neonatal MK801 or phencyclidine injection and subsequent rearing in social isolation from weaning on impair PPI response (Lim et al. 2012; Gaskin et al. 2014). In mice, the combination of prenatal inflammation by poly I:C and postnatal chronic mild stress recapitulates PPI deficits and increases behavioral hypersensitivity to amphetamine (Giovanoli et al. 2013).

Comparison of genetic models with pharmacological and environmental models is useful even if each of these models has limitations. Genetic models are partial, an MAM model may be superior for modeling frontal hippocampal interactions, while the neonatal lesion may be the best model for DA upregulation. Environmental models are complex. Overall, there are no convincing models of cognitive deficits or of how cannabis use in adolescence may lead to symptoms. Other environmental factors such as birth complications and urbanicity are difficult, if at all possible, to model.

Appendix 13.5 Background Information to Fragile X

In the case of fragile X syndrome, a monogenic cause of autism, CGG repeat expansion leads to methylation and transcriptional silencing of the *FMR1* gene (Verkerk et al. 1991), which encodes the fragile X mental retardation protein. Although this gene is homologous to the mouse *FMR1* gene, transcriptional silencing does not occur with the introduction of CGG repeat expansion. Thus initial mouse models attempted to recapitulate the human mutation by knocking out the *FMR1* gene (Dutch-Belgian Fragile X Consortium 1994).

Because synaptic plasticity is the foundation of most theories of learning and memory in the brain, a turning point for the field came in 2002, when studies showed that protein synthesis-dependent, mGluR5-mediated, long-term depression in the hippocampus was *exaggerated* in the *FMR1* knockout mice (Huber et al. 2002). Recognition of a number of parallels between phenotypic features of the disease and predicted or known consequences of (over)activation of GpI mGluRs, led to the "the mGluR theory of fragile X" (Bear et al. 2004), which was formally tested by establishing *FMR1* knockout phenotypes relevant to the disorder, and examining these in the context of mGluR5 knockdown (Dölen et al. 2007).

Because this manipulation corrected phenotypes in a number of brain regions at the biochemical, synaptic, circuit, and behavioral levels, these results led to the first speculation that autism might be a "synapsopathy" (Dölen and Bear 2009)—a term coined to distinguish fragile X syndrome from neuropsychiatric disorders such as AD and PD, where the primary pathophysiology appeared to be localized to a particular brain region (Bear et al. 2008). The therapeutic potential of mGluR5 was subsequently further validated by pharmacologic manipulations across development that could address concerns of pathophysiological hysteresis; these have suggested that even late intervention can correct behavioral phenotypes in mice (Chen et al. 2014b; Pop et al. 2014; Michalon et al. 2014; Gantois et al. 2013).

Despite these advances, failure to appreciate the difference between how the mutation is achieved in animal models and human patients have contributed to misalignment of preclinical to clinical studies. Specifically, in a recent Phase III trial, when fragile X patients were treated as a homogenous group, mGluR5-modifying drugs failed to show significant therapeutic effect. However, transcriptional silencing in human patients is not digital; instead, the size of the repeat expansion is inversely correlated to expression of the fragile X mental retardation protein, which is directly correlated to autistic symptoms. Furthermore, when data were stratified by repeat expansion size, patients with the most severe disease presentation showed significant improvement.

Appendix 13.6 Historical Perspective for the Cholinergic Hypothesis of Alzheimer Disease

An interesting historical perspective is provided by the cholinergic story in AD (Everitt and Robbins 1997). Early pathological work established a correlation between intellectual deficit and cholinergic markers in the neocortex of patients with the disease. Preclinical studies sought to investigate the causal role of these deficits using pharmacological approaches in humans (e.g., testing effects on cognition of the anticholinergic drug scopolamine) and in experimental animals using lesions of varying selectivity of the cholinergic neurons that had been shown to degenerate quite early in AD. Several studies converged to show relatively consistent deficits in attention type functions in rats (and also monkeys) with such lesions, with less obvious effects on memory and many aspects of learning. These deficits were remediated by procholinergic drugs, including acetylcholine esterase inhibitors and nicotine. A parallel clinical trial of an acetylcholine esterase inhibitor, which had overall significant clinical efficacy, only showed improvement in the same attentional paradigm configured for patients—but not in concurrent tests of memory—in parallel with changes on a clinical scale that highlighted enhanced alertness (Sahakian et al. 1993). Therefore, these translational data from the animal model to the patients' data more or less correspond to the current clinical judgment that acetylcholinesterase inhibitors are very far from being a drug to "cure" all of the symptoms of AD, but may have a mild beneficial effect in certain restricted domains of functioning. Two subsequent observations are relevant: (a) the same behavioral effects of acetylcholinesterease inhibitors have recently been shown in the "triple transgenic" mouse model of AD (Romberg et al. 2011) and (b) this drug treatment may well prove to be more efficacious in neurodegenerative diseases such as Lewy body dementia and PD dementia (Emre et al. 2004), where there is evidence of often greater cholinergic deficit than in AD.

Epilogue

14

The Revolution Has Begun

Steven E. Hyman and Karoly Nikolich

The results of this Forum give reason to hope that new therapies for nervous system disorders are not only possible but also realistic. This optimism, visible in the vigorous discussions captured in this book, is grounded both in an acute understanding of the fundamental challenges that exist and the synergies that grew out of the process. Much remains to be accomplished if promising ideas, solutions, and paradigms are to be realized in academia and industry. Thus given the importance of innovative ideas and experimentation, we have no desire to reach premature closure on precise approaches to either neurodevelopmental or neurodegenerative disorders. There was, however, broad agreement that genetics and a greater emphasis on human biology offer essential springboards for effective new approaches to pathophysiology and therapies.

After decades of stagnation marked, inter alia, by all too many disappointing failures in costly, late-stage clinical trials, what is the basis for our optimism? Central to this is the remarkable advent of new tools and technologies during the last decade which are truly revolutionizing our ability to study the nervous system in unprecedented ways (Hyman 2012; Pankevich et al. 2014). The resulting sense of promise that we perceive comes with the recognition that a revolution in translational neuroscience has already begun. Our use of the term "revolution" is not accidental: the participants of the Forum agreed that for brain disorders, successful discovery and development of disease-modifying therapies require most certainly a revolution, not merely incremental progress realized through past approaches. We (the attendees and coauthors of this volume) are already participating, to a great degree, in this much needed transformation; this Forum provided valuable space to reflect on what has happened and to discuss, and occasionally argue, about what needs to be done to accelerate progress. Those from industry made clear that they understood the high prevalence of many brain disorders, the vast unmet medical needs, and the enormous global disease burden associated with these diseases. Equally, however, they stressed that unless the new biology discussed at the Forum produced more than incremental advances, it would be difficult to justify significant investments in brain and nervous system disorders when other areas of research (e.g., cancer) were awash in targets, had clear paths to target validation, and possessed biomarkers to facilitate clinical trials. In addition, we also recognized that lessons from other therapeutic areas must be adopted by translational neuroscience if the vast challenges ahead are to be met.

Why has progress been so slow? If we are truly at the beginning of a successful revolution, and not merely involved in another false start, it is critical to reach an understanding of what has gone wrong in recent decades and to identify important obstacles that remain:

1. Phenotyping and disease classification pose major challenges, given the dearth of biomarkers. Some disease definitions (e.g., those in the DSM classification of mental disorders) should have been considered heuristics or "early drafts," yet were permitted to become reified, thus forming a procrustean bed that damaged both translational and clinical investigation (Hyman 2010).
2. Common forms of neurodevelopmental and neurodegenerative disorders are greatly influenced by genetics, but their genetic architecture has proven far more complex than initially suspected, thus dooming older approaches such as candidate genes and linkage analyses. It is critical for the field, as a whole, to embrace successful unbiased genome-scale approaches to genetics.
3. The human brain is generally inviolable in life. Although indirect approaches (e.g., electroencephalography and diverse noninvasive imaging modalities) have contributed much to studies of disease, they do not readily yield molecular or pathophysiological information. Only recently have new ligands for positron emission tomography, along with measures of cerebrospinal fluid, made it possible to stratify patients with dementia for clinical trials (Jack and Holtzman 2013). Given the agreed centrality of human biology, these observations were seen to argue for attention on two fronts: the ability (a) to engineer diverse human neurons and glia *in vitro* and even to model circuits *in vitro* and (b) to advance studies of patients with, for example, new phenotyping methods inspired by genetic findings.
4. For many pathophysiological studies, brain scientists must rely on postmortem tissue, with all its challenges. This stands in marked contrast to oncology where, for instance, excisional biopsies provide living tissue for both clinical purposes and scientific study. This suggests that postmortem studies of brain should be supplemented with *in vitro* studies of human neurons, while recognizing that these approaches are likely to have complementary limitations.
5. Brain disorders are generally not cell autonomous, adding an additional layer of complexity in terms of the need to study synapses and circuits.
6. Animal models have often proven problematic for both neurodevelopmental and neurodegenerative disorders because of their evolutionary distance from humans. Many cell types and circuits in the human brain

that are involved in neurodevelopmental disorders (e.g., in prefrontal cortex) are evolutionarily quite recent and differ markedly from ubiquitous rodent models. Attention to evolution will be critical in thinking about animal models in the future, as will the recognition that the questions posed for animal models are important but limited.

From intense discussions of the foregoing issues, a conceptual roadmap emerged to hasten the hoped-for revolution in pathophysiology and therapeutics (see Figure 14.1). This roadmap outlines an iterative approach, alluded to throughout this volume, aimed at grounding target identification, target validation, and clinical trials on solid scientific foundations based on known mechanisms of disease. As noted, given the emergence of powerful new tools and technologies over the last decade, a well-validated pathophysiology has become possible.

What are the tools and technologies that make a revolution possible? Participants at the Forum agreed that genetics offers foundational advances for such a revolution (McCarroll and Hyman 2013). For neuropsychiatric disorders such as schizophrenia, autism, and bipolar disorder, which lack a distinctive, analyzable neuropathology, genetic insights may prove to be the only opportunity we have to gain insight into pathophysiology. Even for Alzheimer disease, in which amyloid plaques and neurofibrillary tangles have yielded important biochemical clues, genetics is helping to identify critical processes involved in pathophysiology (Lambert et al. 2013).

The discovery of common genome-wide significant loci associated with disease as well as rare variants that influence pathogenesis would not have been possible without a technological revolution, which was arguably the most important product of the Human Genome Project. This yielded inexpensive microarrays to identify common variation, efficient and inexpensive DNA sequencing, and powerful new computational tools. It is staggering to realize that the cost of DNA sequencing has declined over the last decade by approximately one millionfold. Realization of the potential of these technologies has also led to significant changes in how science is organized: given the need for large sample sizes, extensive and durable global coalitions have been formed to investigate human genetics.

One frequently discussed concern regarding the promise of genetics has been the failure to devise effective therapies for monogenic diseases of the nervous system, such as Huntington disease and familial forms of amyotrophic lateral sclerosis, even decades after the genes were discovered. The multiple alleles of small effect that confer risk for common non-Mendelian disorders, such as schizophrenia, most cases of autism, and nonfamilial neurodegenerative disorders, should pose a far greater challenge. No one at the Forum argued that it would be simple to exploit genetic information arising from polygenic disorders, yet as noted, there was significant optimism for the intermediate and long term. Lessons derived from older attempts to develop therapies for

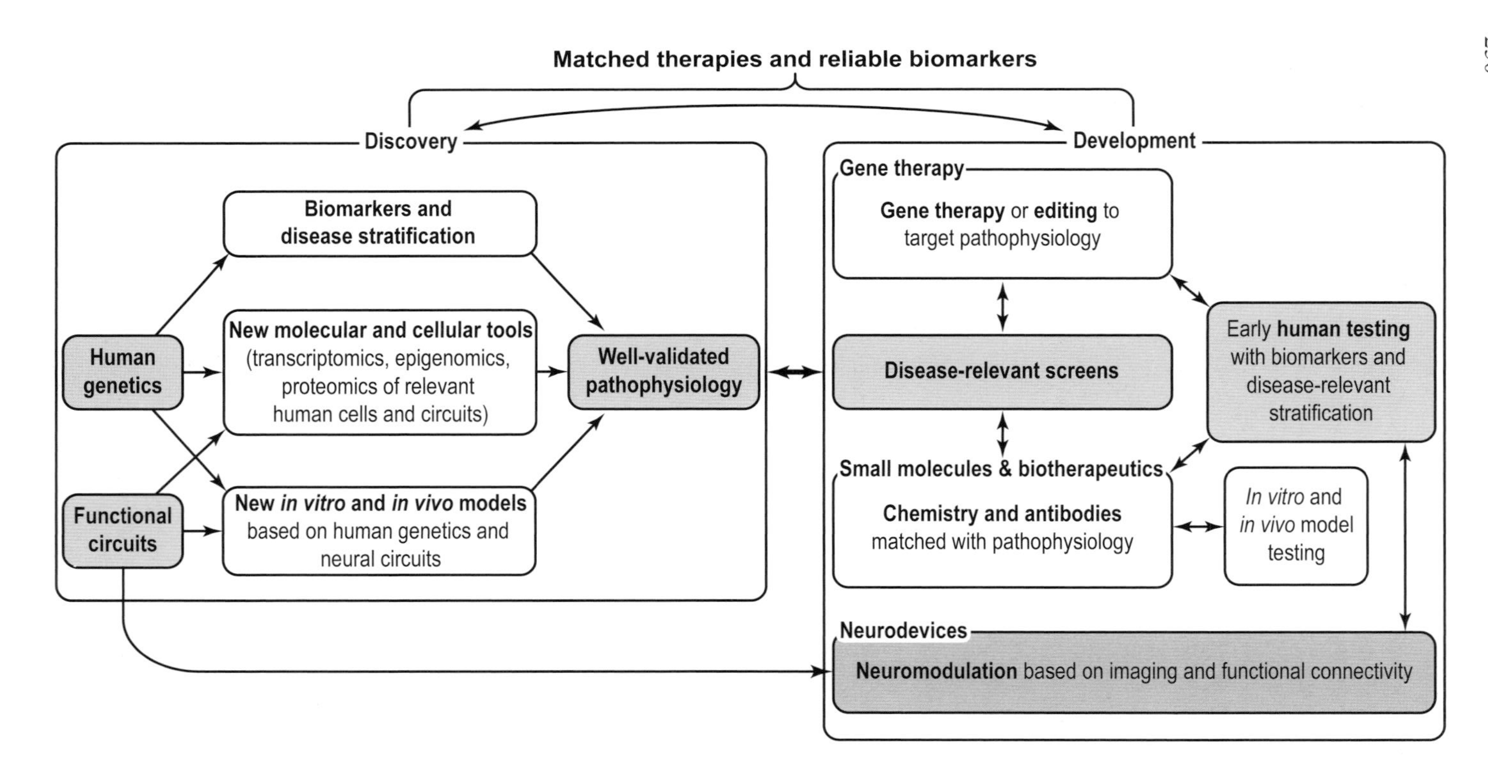

Figure 14.1 Conceptual roadmap outlining new approaches to target identification, target validation, and clinical trials grounded in human genetics and human biology and thus putatively on mechanisms of disease.

monogenic disorders will be valuable for all genetically grounded efforts in nervous system therapeutics. Specifically, there was recognition of the special difficulties posed by toxic gain of function mutations, such as those underlying Huntington disease and some forms of amyotrophic lateral sclerosis, and at present greater likelihood of being able to silence such genes in the nervous system, albeit with significant challenges remaining for delivery of the therapeutic agent. It was also noted that limitations of mouse models are increasingly being appreciated, thus it is likely that animal models will be used to answer questions for which they will be informative rather than misleading. In addition, the advent of stem cell technologies and other approaches to reprogramming cells into neurons should benefit translational neuroscience with respect to both monogenic and polygenic disorders. For example, screens using stem-cell derived motor neurons have been used successfully for a monogenic form of amyotrophic lateral sclerosis (e.g., Yang et al. 2013).

For polygenic brain disorders, it will clearly be challenging to put genetic information to work in the service of pathophysiology and therapies; however, in principle, there is a feasible path. This matter received attention from Forum participants who work on both neurodevelopmental and neurodegenerative disorders, but was a central focus for the former, given the lack of alternative sources of molecular clues. Geneticists at the Forum addressed the widespread fallacy that alleles of small effect could have little biological utility. Whatever its penetrance, any allele shown to have genome-wide significance for an association implicates the gene within which it is a variant in the pathophysiological process—or if the association is to a regulatory sequence, the gene (or genes) to which it is linked. Genes, in turn, implicate pathways or protein complexes in disease processes. Once the directionality of the risk alleles are determined and the relevant genes placed among protein interactors, potential therapeutics can be targeted to increase or decrease, as appropriate, the activity of that pathway or protein complex (McCarroll and Hyman 2013). What matters is not the effect size of the allele that points to the relevant gene, and thus the relevant pathway; what matters is that the association with disease be established with certainty. Far too much effort has been wasted chasing down biological effects that ultimately proved irrelevant because they were inferred from "candidate gene approaches" that were underpowered or otherwise poorly designed.

Beyond the need for continuing the discovery of risk-associated loci, most of which still need fine mapping, and identification of risk-associated variants through DNA sequencing, additional molecular information must still be gleaned if genetics results are to prove fully useful. Unlike metabolism, cardiovascular disease, or cancer in which critical biochemical pathways are well described, risk-associated genes identified in studies of schizophrenia and autism implicate protein complexes that require better definition in relevant cell types, such as the postsynaptic specialization of excitatory neurons. The computational interactomes that do exist and the proteomic results derived from highly mixed cell populations will have to be supplanted by more refined proteomic experiments, which might benefit from comparing results from postmortem brain with results from *in vitro* preparations in which large numbers of identical cells of a single neural type can be purified.

As risk alleles, risk genes, and risk pathways are identified, they can be studied in dissociated human neurons and glia *in vitro*, in small circuits reconstituted from dispersed cells, and in brain organoids (Lancaster et al. 2013). However, most nervous system disorders involve the dysfunction of specific neural circuits. Important new technologies have started to provide exceptional insights into short- and long-range circuits, pathways of the brain. The human and mouse connectome projects have reached major milestones and have provided first blueprints of the brain's major connections at both small and large scales (Smith et al. 2013). Optogenetics (Boyden et al. 2005) has opened up precise new ways of mapping functional circuitry in neural or other excitable cells. Using light-activated ion channels, it has become possible to activate and inhibit, with unprecedented precision, specific neural circuits in live, free-moving animals involved in thought, learning, memory, emotions, and disease-associated behaviors (Asrican et al. 2013; Steinberg et al. 2014). Optogenetics has offered insights into how specific circuits determine behaviors, from which high precision models have been developed to examine a range of behaviors and behavioral perturbations (Deisseroth 2014).

A powerful approach is offered by the combination of genetics and optogenetics. Animal models can be developed based on genetic mutations identified in patients. Such models offer a unique opportunity to identify, and subsequently modify, circuits that underlie behavioral abnormalities. These studies allow targets in neuronal circuits to be identified which, when manipulated, can repair the dysfunction. The resulting knowledge offers a new basis for translation: the malfunctioning circuits in the rodent models can be tested in human brains using imaging and various stimulation techniques. Furthermore, single cell genomics in specific disease-associated circuits allows high granularity for candidate genes that encode targets capable of modulation by small molecule drugs (Riley et al. 2014). Such approaches may ultimately allow drug discovery and development to become robust and reliable. The path toward effective therapeutics appears to be based on specific brain circuits that mediate symptoms of brain disease. Optogenetics has become routinely applied to study circuits involved in anxiety, anhedonia, feeding disorders, learning and memory, as well as many others.

In addition, deep brain stimulation, best developed for movement disorders (Benabid et al. 2009) but also under investigation for depression and obsessive-compulsive disorder, has opened up new opportunities to activate or inhibit certain neural circuits in patients that can be further studied with neuroimaging and other technologies (Ostergard and Miller 2014).

Attention was also given to how a new, well-validated pathophysiology could be translated into effective therapies. This will involve improvements in chemistry to address targets previously thought impossible or exceptionally difficult to "drug" as well as continued improvements in getting both small molecules and antibodies across the blood-brain barrier. Gene therapies have become, after a long period of gestation, increasingly safe, and thus may provide additional therapeutic tools.

Commensurate to the scientific enquiry inherent in our conceptual roadmap (Figure 14.1), academic institutions, foundations, biotech and pharmaceutical companies, and regulators need to work together to deliver candidate treatments more rapidly to patients. In addition, improved organization is needed in how basic and translational science is conducted if there is to be a true renaissance in developing therapeutics for nervous system disorders.

In preparing for the Forum, the program advisory committee recognized that translational neuroscience was at a critical juncture—one that could lead down two paths: extinction or revolution. The unique challenges and numerous failures to treat complex and often slowly developing diseases of pathological neural circuits have been disheartening. Yet, profound progress has been made, through discoveries and techniques, to better our understanding of the underlying pathophysiological mechanisms of nervous system diseases. The revolution has indeed begun, and the results of this Forum give reason to hope that new therapies for nervous system disorders are both realistic and possible.

Bibliography

Note: Numbers in square brackets denote the chapter in which an entry is cited.

Adler, L., E. Pachtman, R. Franks, et al. 1982. Neurophysiological Evidence for a Defect in Neuronal Mechanisms Involved in Sensory Gating in Schizophrenia. *Biol. Psychiatry* **17**:639–654. [10]

Albin, R. L., A. B. Young, and J. B. Penney. 1989. The Functional Anatomy of Basal Ganglia Disorders. *Trends Neurosci.* **12**:366–375. [9]

Alvarez, V. A., D. A. Ridenour, and B. L. Sabatini. 2006. Retraction of Synapses and Dendritic Spines Induced by Off-Target Effects of RNA Interference. *J. Neurosci.* **26**:7820–7825. [10]

Amat, J., M. V. Baratta, E. Paul, et al. 2005. Medial Prefrontal Cortex Determines How Stressor Controllability Affects Behavior and Dorsal Raphé Nucleus. *Nature Neurosci.* **8(3)**:365–371. [12]

Amir, R. E., I. B. Van den Veyver, M. Wan, et al. 1999. Rett Syndrome Is Caused by Mutations in X-Linked MECP2, Encoding Methyl-CpG-Binding Protein 2. *Nat. Genet.* **23(2)**:185–188. [2]

Amminger, G. P., M. R. Schafer, K. Papageorgiou, et al. 2010. Long-Chain Omega-3 Fatty Acids for Indicated Prevention of Psychotic Disorders: A Randomized, Placebo-Controlled Trial. *Arch. Gen. Psychiatry* **67(2)**:146–154. [4]

Amoroso, M. W., G. F. Croft, D. J. Williams, et al. 2013. Accelerated High-Yield Generation of Limb-Innervating Motor Neurons from Human Stem Cells. *J. Neurosci.* **33**:574–586. [11]

Anderson, M. H. C., M. R. Munafo, and E. S. J. Robinson. 2013. Investigating the Psychopharmacology of Cognitive Affective Bias in Rats Using an Affective Tone Discrimination Task. *Psychopharmacology* **226**:601–613. [12]

Anderson, S., and P. Vanderhaeghen. 2014. Cortical Neurogenesis from Pluripotent Stem Cells: Complexity Emerging from Simplicity. *Curr. Opin. Neurobiol.* **27C**:151–157. [10]

Andreasen, N. C. 1989. The American Concept of Schizophrenia. *Schizophr. Bull.* **15(4)**:519–531. [4]

———. 2007. DSM and the Death of Phenomenology in America: An Example of Unintended Consequences. *Schizophr. Bull.* **33(1)**:108–112. [4]

Anney, R., L. Klei, D. Pinto, et al. 2012. Individual Common Variants Exert Weak Effects on the Risk for Autism Spectrum Disorders. *Hum. Mol. Genet.* **21(21)**:4781–4492. [2]

Antal, A., L. Chaieb, V. Moliadze, et al. 2010. Brain-Derived Neurotrophic Factor (BDNF) Gene Polymorphisms Shape Cortical Plasticity in Humans. *Brain Stimul.* **3**:230–237. [10]

Araki, T., Y. Sasaki, and J. Milbrandt. 2004. Increased Nuclear NAD Biosynthesis and SIRT1 Activation Prevent Axonal Degeneration. *Science* **305(5686)**:1010–1013. [7]

Arguello, P. A., and J. Gogos. 2006. Modeling Madness in Mice: One Piece at a Time. *Neuron* **52**:179–196. [12]

Arrowsmith, J. 2011. Trial Watch: Phase III and Submission Failures: 2007–2010. *Nat. Rev. Drug Discov.* **10(2)**:87. [13]

Asanuma, K., C. Tang, Y. Ma, et al. 2006. Network Modulation in the Treatment of Parkinson's Disease. *Brain* **12**:2667–2678. [9]

Asrican, B., G. J. Augustine, K. Berglund, et al. 2013. Next-Generation Transgenic Mice for Optogenetic Analysis of Neural Circuits. *Front. Neural Circuits* **7**:160. [14]

Atasoy, D., J. N. Betley, H. H. Su, and S. M. Sternson. 2012. Deconstruction of a Neural Circuit for Hunger. *Nature* **488**:172–177. [10]

Atwal, J. K., Y. Chen, C. Chiu, et al. 2011. A Therapeutic Antibody Targeting BACE1 Inhibits Amyloid-Beta Production *in Vivo*. *Sci. Transl. Med.* **3(84)**:84ra43. [7]

Autism Genome Project Consortium, P. Szatmari, A. D. Paterson, et al. 2007. Mapping Autism Risk Loci Using Genetic Linkage and Chromosomal Rearrangements. *Nat. Genet.* **39(3)**:319–328. [4]

Ayutyanont, N., J. B. Langbaum, S. B. Hendrix, et al. 2014. The Alzheimer's Prevention Initiative Composite Cognitive Test Score: Sample Size Estimates for the Evaluation of Preclinical Alzheimer's Disease Treatments in Presenilin 1 E280a Mutation Carriers. *J. Clin. Psychiatry* **75(6)**:652–660. [7]

Ba, W., J. van der Raadt, and N. Nadif Kasri. 2013. Rho GTPase Signaling at the Synapse: Implications for Intellectual Disability. *Exp. Cell Res.* **319(15)**:2368–2374. [2]

Baddeley, A. 1986. Working Memory. Oxford: Clarendon Press. [12]

Balleine, B. W., and J. P. O'Doherty. 2010. Human and Rodent Homologies in Action Control: Corticostriatal Determinants of Goal-Directed and Habitual Action. *Neuropharmacology* **35**:48–69. [12]

Barch, D. M., J. Bustillo, W. Gaebel, et al. 2013. Logic and Justification for Dimensional Assessment of Symptoms and Related Clinical Phenomena in Psychosis: Relevance to DSM-5. *Schizophr. Res.* **150(1)**:15–20. [4]

Bari, A., D. M. Eagle, A. C. Mar, E. S. J. Robinson, and T.W.Robbins. 2009. Dissociable Effects of Noradrenaline, Dopamine, and Serotonin Uptake Blockade on Stop Task Performance in Rats. *Psychopharmacology* **205**:273–283. [12]

Bari, A., D. E. Theobald, D. Caprioli, et al. 2010. Serotonin Modulates Sensitivity to Reward and Negative Feedback in a Probabilistic Reversal Learning Task in Rats. *Neuropsychopharmacology* **35(6)**:1290–1301. [12]

Barrett, C. E., A. C. Keebaugh, T. H. Ahern, et al. 2013. Variation in Vasopressin Receptor (Avpr1a) Expression Creates Diversity in Behaviors Related to Monogamy in Prairie Voles. *Horm. Behav.* **63**:518–526. [10]

Bartus, R. T., M. S. Weinberg, and R. J. Samulski. 2014. Parkinson's Disease Gene Therapy: Success by Design Meets Failure by Efficacy. *Mol. Ther.* **22(3)**:487–497. [7]

Bateman, R. J., C. Xiong, T. L. Benzinger, et al. 2012. Clinical and Biomarker Changes in Dominantly Inherited Alzheimer's Disease. *N. Engl. J. Med.* **367(9)**:795–804. [6, 7, 9]

Baudouin, S. J., J. Gaudias, S. Gerharz, et al. 2012. Shared Synaptic Pathophysiology in Syndromic and Nonsyndromic Rodent Models of Autism. *Science* **338(6103)**:128–132. [2]

Bear, M. F., G. Dölen, E. Osterweil, and N. Nagarajan. 2008. Fragile X: Translation in Action. *Neuropsychopharmacology* **33**:84–87. [10, 13]

Bear, M. F., K. M. Huber, and S. T. Warren. 2004. The mGluR Theory of Fragile X Mental Retardation. *Trends Neurosci.* **27(7)**:370–377. [13]

Beck, H., I. V. Goussakov, A. Lie, C. Helmstaedter, and C. E. Elger. 2000. Synaptic Plasticity in the Human Dentate Gyrus. *J. Neurosci.* **20**:7080–7086. [10]

Belin, D., A. C. Mar, J. W. J.W. Dalley, T. W. Robbins, and B. J. Everitt. 2008. High Impulsivity Predicts the Switch to Compulsive Cocaine-Taking. *Science* **320**:1352–1355. [12]

Benabid, A. L., S. Chabardes, J. Mitrofanis, and P. Pollak. 2009. Deep Brain Stimulation of the Subthalamic Nucleus for the Treatment of Parkinson's Disease. *Lancet Neurol.* **8(1)**:67–81. [14]

Benilova, I., E. Karran, and B. De Strooper. 2012. The Toxic Abeta Oligomer and Alzheimer's Disease: An Emperor in Need of Clothes. *Nat. Neurosci.* **15(3)**:349–357. [7]

Benzinger, T. L., T. Blazey, C. R. Jack, Jr., et al. 2013. Regional Variability of Imaging Biomarkers in Autosomal Dominant Alzheimer's Disease. *PNAS* **110**:E4502–E4509. [9]

Bergman, H., T. Wichmann, B. Karmon, and M. R. DeLong. 1994. The Primate Subthalamic Nucleus. II. Neuronal Activity in the MPTP Model of Parkinsonism. *J. Neurophysiol.* **72**:507–521. [9]

Bergmann, C., K. Zerres, J. Senderek, et al. 2003. Oligophrenin 1 (OPHN1) Gene Mutation Causes Syndromic X-Linked Mental Retardation with Epilepsy, Rostral Ventricular Enlargement and Cerebellar Hypoplasia. *Brain* **126(Pt 7)**:1537–1544. [2]

Berk, J. L., O. B. Suhr, L. Obici, et al. 2013. Repurposing Diflunisal for Familial Amyloid Polyneuropathy: A Randomized Clinical Trial. *JAMA* **310(24)**:2658–2667. [7]

Berkel, S., C. R. Marshall, B. Weiss, et al. 2010. Mutations in the SHANK2 Synaptic Scaffolding Gene in Autism Spectrum Disorder and Mental Retardation. *Nat. Genet.* **42(6)**:489–491. [4]

Berndt, A., S. Y. Lee, C. Ramakrishnan, and K. Deisseroth. 2014. Structure-Guided Transformation of Channelrhodopsin into a Light-Activated Chloride Channel. *Science* **344(6182)**:420–424. [8]

Bero, A. W., A. Q. Bauer, F. R. Stewart, et al. 2012. Bidirectional Relationship between Functional Connectivity and Amyloid-Beta Deposition in Mouse Brain. *J. Neurosci.* **32**:4334–4340. [9]

Bero, A. W., P. Yan, J. H. Roh, et al. 2011. Neuronal Activity Regulates the Regional Vulnerability to Amyloid-Beta Deposition. *Nat. Neurosci.* **14**:750–756. [9]

Berry-Kravis, E., D. Hessl, L. Abbeduto, et al. 2013. Outcome Measures for Clinical Trials in Fragile X Syndrome. *J. Dev. Behav. Pediatr.* **34**:508–522. [10]

Bersani, F. S., A. Minichino, P. G. Enticott, et al. 2013. Deep Transcranial Magnetic Stimulation as a Treatment for Psychiatric Disorders: A Comprehensive Review. *Eur. J. Psychiatry* **28**:30–39. [10]

Bertram, L., C. Lange, K. Mullin, et al. 2008. Genome-Wide Association Analysis Reveals Putative Alzheimer's Disease Susceptibility Loci in Addition to APOE. *Am. J. Hum. Genet.* **83(5)**:623–632. [7]

Betancur, C., and J. D. Buxbaum. 2013. SHANK3 Haploinsufficiency: A "Common" but Underdiagnosed Highly Penetrant Monogenic Cause of Autism Spectrum Disorders. *Mol. Autism* **4(1)**:17. [4]

Bezard, E., and S. Przedborski. 2011. A Tale on Animal Models of Parkinson's Disease. *Mov. Disord.* **26(6)**:993–1002. [13]

Bezzina, G., S. Body, T. H. Cheung, et al. 2008. Effect of Disconnecting the Orbital Prefrontal Cortex from the Nucleus Accumbens Core on Inter-Temporal Choice Behavior: A Quantitative Analysis. *Behav. Brain Res.* **191**:272–279. [12]

Bien-Ly, N., A. K. Gillespie, D. Walker, S. Y. Yoon, and Y. Huang. 2012. Reducing Human Apolipoprotein E Levels Attenuates Age-Dependent Abeta Accumulation in Mutant Human Amyloid Precursor Protein Transgenic Mice. *J. Neurosci.* **32(14)**:4803–4811. [7]

Biglan, K. M., Y. Zhang, J. D. Long, et al. 2013. Refining the Diagnosis of Huntington Disease: The PREDICT-HD Study. *Front. Aging Neurosci.* **5**:12. [7]

Bilican, B., A. Serio, S. J. Barmada, et al. 2012. Mutant Induced Pluripotent Stem Cell Lines Recapitulate Aspects of TDP-43 Proteinopathies and Reveal Cell-Specific Vulnerability. *PNAS* **198(15)**:5803–5808. [5]

———. 2013. Comment on "Drug Screening for ALS Using Patient-Specific Induced Pluripotent Stem Cells." *Sci. Transl. Med.* **5(188)**:188le182. [5]

Birks, J. 2006. Cholinesterase Inhibitors for Alzheimer's Disease. *Cochrane Database Syst. Rev.***(1)**:CD005593. [7]

Birman, S. 2008. Neurodegeneration: RNA Turns Number One Suspect in Polyglutamine Diseases. *Curr. Biol.* **18(15)**:R659–R661. [5]

Birrell, J. M., and V. J. Brown. 2000. Medial Frontal Cortex Mediates Perceptual Attentional Set-Shifting in the Rat. *J. Neurosci.* **20**:4320–4324. [12]

Bishop, D. V. 2010. Which Neurodevelopmental Disorders Get Researched and Why? *PLoS One* **5(11)**:e15112. [2]

Blennow, K., H. Zetterberg, and A. M. Fagan. 2012. Fluid Biomarkers in Alzheimer Disease. *Cold Spring Harb. Perspect. Med.* **2(9)**:a006221. [13]

Blennow, K., H. Zetterberg, C. Haass, and T. Finucane. 2013. Semagacestat's Fall: Where Next for AD Therapies? *Nat. Med.* **19(10)**:1214–1215. [6]

Bliss, T. V. P., and S. F. Cooke. 2011. Long-Term Potentiation and Long-Term Depression: A Clinical Perspective. *Clinics (Sao Paulo)* **66(Suppl. 1)**:3–17. [10]

Bloom, D. E., E. T. Cafiero, E. Jané-Llopis, S. Abrahams-Gessel, and L. R. e. a. Bloom. 2011. The Global Economic Burden of Noncommunicable Diseases. Geneva: World Economic Forum. [1]

Blumensath, T., S. Jbabdi, M. F. Glasser, et al. 2013. Spatially Constrained Hierarchical Parcellation of the Brain with Resting-State fMRI. *Neuroimage* **76**:313–324. [9]

Boch, J., H. Scholze, S. Schornack, et al. 2009. Breaking the Code of DNA Binding Specificity of TAL-Type III Effectors. *Science* **326**:1509–1512. [8]

Boissart, C., A. Poulet, P. Georges, et al. 2013. Differentiation from Human Pluripotent Stem Cells of Cortical Neurons of the Superficial Layers Amenable to Psychiatric Disease Modeling and High-Throughput Drug Screening. *Transl. Psychiatry* **3**:e294. [2, 4]

Bonaglia, M. C., R. Giorda, S. Beri, et al. 2011. Molecular Mechanisms Generating and Stabilizing Terminal 22q13 Deletions in 44 Subjects with Phelan/Mcdermid Syndrome. *PLoS Genet.* **7(7)**:e1002173. [4]

Bourgeron, T. 2009. A Synaptic Trek to Autism. *Curr. Opin. Neurobiol.* **19(2)**:231–234. [2]

Boutajangout, A., J. Ingadottir, P. Davies, and E. M. Sigurdsson. 2011. Passive Immunization Targeting Pathological Phospho-Tau Protein in a Mouse Model Reduces Functional Decline and Clears Tau Aggregates from the Brain. *J. Neurochem.* **118(4)**:658–667. [7]

Boutros, N., G. Zouridakis, and J. Overall. 1991. Replication and Extension of P50 Findings in Schizophrenia. *Clin. Electroencephalogr.* **22**:40–45. [10]

Boyden, E. S., F. Zhang, E. Bamberg, G. Nagel, and K. Deisseroth. 2005. Millisecond-Timescale, Genetically Targeted Optical Control of Neural Activity. *Nat. Neurosci.* **8(9)**:1263–1268. [8, 10, 14]

Bozdagi, O., T. Sakurai, D. Papapetrou, et al. 2010. Haploinsufficiency of the Autism-Associated Shank3 Gene Leads to Deficits in Synaptic Function, Social Interaction, and Social Communication. *Mol. Autism* **1(1)**:15. [2]

Brecht, M., R. Naumann, F. Anjum, et al. 2011. The Neurobiology of Etruscan Shrew Active Touch. *Philos. Trans. R. Soc. Lond. B* **366**:3026–3036. [10]

Brennand, K. J., A. Simone, J. Jou, et al. 2011. Modelling Schizophrenia Using Human Induced Pluripotent Stem Cells. *Nature* **473**:221–225. [11]

Brier, M. R., J. B. Thomas, A. Z. Snyder, et al. 2012. Loss of Intranetwork and Internetwork Resting State Functional Connections with Alzheimer's Disease Progression. *J. Neurosci.* **32**:8890–8899. [9]

Briggman, K. L., M. Helmstaedter, and W. Denk. 2011. Wiring Specificity in the Direction-Selectivity Circuit of the Retina. *Nature* **471(7337)**:183–188. [8]

Brockhaus-Dumke, A., R. Mueller, U. Faigle, and J. Klosterkoetter. 2008. Sensory Gating Revisited: Relation between Brain Oscillations and Auditory Evoked Potentials in Schizophrenia. *Schizophr. Res.* **99**:238–249. [10]

Brown, C. A., M. C. Campbell, M. Karimi, et al. 2012. Dopamine Pathway Loss in Nucleus Accumbens and Ventral Tegmental Area Predicts Apathetic Behavior in MPTP-Lesioned Monkeys. *Exp. Neurol.* **236**:190–197. [9]

Brown, C. A., M. K. Karimi, L. Tian, et al. 2013. Validation of Midbrain Positron Emission Tomography Measures for Nigrostriatal Neurons in Macaques. *Ann. Neurol.* **74**:602–610. [9, 10]

Burchell, V. S., D. E. Nelson, A. Sanchez-Martinez, et al. 2013. The Parkinson's Disease-Linked Proteins Fbxo7 and Parkin Interact to Mediate Mitophagy. *Nat. Neurosci.* **16(9)**:1257–1265. [5]

Bussey, T. J., A. Holmes, L. Lyon, et al. 2012. New Translational Assays for Preclinical Modeling of Cognition in Schizophrenia; the Touch-Screen Testing Method for Mice and Rats. *Neuropharmacology* **62**:1191–1203. [12]

Bussey, T. J., J. L. Muir, and T. W. Robbins. 1994. A Novel Automated Touchscreen Procedure for Assessing Learning in the Rat Using Computer Graphic Stimuli. *Neurosci. Res. Comm.* **15**:103–110. [12]

Bütefisch, C. M., V. Khurana, L. Kopylev, and L. G. Cohen. 2004. Enhancing Encoding of a Motor Memory in the Primary Motor Cortex by Cortical Stimulation. *J. Neurophysiol.* **91**:2110–2116. [10]

Button, K. S., J. P. Ioannidis, C. Mokrysz, et al. 2013. Power Failure: Why Small Sample Size Undermines the Reliability of Neuroscience. *Nat. Rev. Neurosci.* **14(5)**:365–376. [3]

Cahan, P., and G. Q. Daley. 2013. Origins and Implications of Pluripotent Stem Cell Variability and Heterogeneity. *Nat. Rev. Mol. Cell. Biol.* **14**:357–368. [11]

Callaway, E. M. 2008. Transneuronal Circuit Tracing with Neurotropic Viruses. *Curr. Opin. Neurobiol.* **18**:617–623. [10]

Campbell, C. D., and E. E. Eichler. 2013. Properties and Rates of Germline Mutations in Humans. *Trends Genet.* **29(10)**:575–584. [2]

Campbell, M. C., J. Markham, H. Flores, et al. 2013. Principal Component Analysis of PiB Distribution in Parkinson and Alzheimer Diseases. *Neurology* **81**:520–527. [10]

Cannon, T. D. 2014. Neurocognitive Growth Charts and Psychosis: Is the Relationship Predictive? *JAMA Psychiatry* **71(4)**:361–362. [4]

Cannon, T. D., K. Cadenhead, B. Cornblatt, et al. 2008. Prediction of Psychosis in Youth at High Clinical Risk: A Multisite Longitudinal Study in North America. *Arch. Gen. Psychiatry* **65(1)**:28–37. [4]

Cao, G., J. Platisa, V. A. Pieribone, et al. 2013. Genetically Targeted Optical Electrophysiology in Intact Neural Circuits. *Cell* **154**:904–913. [10]

Capitanio, J. P., and M. E. Emborg. 2008. Contributions of Non-Human Primates to Neuroscience Research. *Lancet* **371(9618)**:1126–1135. [13]

Cardinal, R. N., D. R. Pennicott, C. L. Sugathapala, T. W. Robbins, and B. J. Everitt. 2001. Impulsive Choice Induced in Rats by Lesions of the Nucleus Accumbens Core. *Science* **292**:2499–2501. [12]

Cardno, A. G., E. J. Marshall, B. Coid, et al. 1999. Heritability Estimates for Psychotic Disorders: The Maudsley Twin Psychosis Series. *Arch. Gen. Psychiatry* **56**:162–168. [8]

Cardno, A. G., and M. J. Owen. 2014. Genetic Relationships between Schizophrenia, Bipolar Disorder and Schizoaffective Disorder. *Schizophr. Bull.* **40(3)**:504–515. [3]

Carey, J. R., F. Fregni, and A. Pascual-Leone. 2006. rTMS Combined with Motor Learning Training in Healthy Subjects. *Restor. Neurol. Neurosci.* **24**:191–199. [10]

Carli, M., J. L. Evenden, and T. W. Robbins. 1985. Depletion of Unilateral Striatal Dopamine Impairs Initiation of Contralateral Actions and Not Sensory Attention. *Nature* **313**:679–682. [12]

Carlsson, A. 2002. Treatment of Parkinson's with L-DOPA. The Early Discovery Phase, and a Comment on Current Problems. *J. Neural Transm.* **109(5-6)**:777–787. [13]

Carrion, R. J., and J. L. Patterson. 2012. An Animal Model That Reflects Human Disease: The Common Marmoset (*Callithrix jacchus*). *Curr. Opin. Virol.* **2**:357–362. [8]

Castner, S. A., P. S. Goldman-Rakic, and G. V. Williams. 2004. Animal Models of Working Memory: Insights for Targeting Cognitive Dysfunction in Schizophrenia. *Psychopharmacology* **174(1)**:111–125. [13]

Castner, S. A., G. V. Williams, and P. S. Goldman-Rakic. 2000. Reversal of Antipsychotic-Induced Working Memory Deficits by Short-Term Dopamine D1 Receptor Stimulation. *Science* **287(5460)**:2020–2022. [13]

Caughey, B., and P. T. Lansbury. 2003. Protofibrils, Pores, Fibrils, and Neurodegeneration: Separating the Responsible Protein Aggregates from the Innocent Bystanders. *Annu. Rev. Neurosci.* **26**:267–298. [7]

Ceaser, A. E., T. E. Goldberg, M. F. Egan, et al. 2008. Set-Shifting Ability and Schizophrenia: A Marker of Clinical Illness or an Intermediate Phenotype? *Biol. Psychiatry* **64**:782–788. [12]

Cenik, B., C. F. Sephton, B. Kutluk Cenik, J. Herz, and G. Yu. 2012. Progranulin: A Proteolytically Processed Protein at the Crossroads of Inflammation and Neurodegeneration. *J. Biol. Chem.* **287(39)**:32298–32306. [7]

Cepeda, C., L. Galvan, S. M. Holley, et al. 2013. Multiple Sources of Striatal Inhibition Are Differentially Affected in Huntington's Disease Mouse Models. *J. Neurosci.* **33**:7393–7406. [9]

Chamberlain, S. R., L. Menzies, A. Hampshire, et al. 2008. Orbitofrontal Dysfunction in Patients with Obsessive-Compulsive Disorder and Their Unaffected Relatives. *Science* **321**:421–422. [12]

Chamberlain, S. R., U. Müller, A. D. Blackwell, et al. 2006. Neurochemical Modulation of Response Inhibition and Probabilistic Learning in Humans. *Science* **311**:861–863. [12]

Chanda, S., S. Marro, M. Wernig, and T. C. Südhof. 2013. Neurons Generated by Direct Conversion of Fibroblasts Reproduce Synaptic Phenotype Caused by Autism-Associated Neuroligin-3 Mutation. *PNAS* **110**:16622–16627. [11]

Chari, S., and I. Dworkin. 2013. The Conditional Nature of Genetic Interactions: The Consequences of Wild-Type Backgrounds on Mutational Interactions in a Genome-Wide Modifier Screen. *PLoS Genet.* **9(8)**:e1003661. [2]

Chaudhuri, K. R., D. G. Healy, and A. H. Schapira. 2006. Non-Motor Symptoms of Parkinson's Disease: Diagnosis and Management. *Lancet Neurol.* **5(3)**:235–245. [7]

Cheeran, B., P. Talelli, F. Mori, et al. 2008. A Common Polymorphism in the Brain-Derived Neurotrophic Factor Gene (BDNF) Modulates Human Cortical Plasticity and the Response to rTMS. *J. Physiol.* **586**:5717–5725. [10]

Chen, H. K., Z. Liu, A. Meyer-Franke, et al. 2012. Small Molecule Structure Correctors Abolish Detrimental Effects of Apolipoprotein E4 in Cultured Neurons. *J. Biol. Chem.* **287(8)**:5253–5266. [7]

Chen, P. C., B. J. Bhattacharyya, J. Hanna, et al. 2011. Ubiquitin Homeostasis Is Critical for Synaptic Development and Function. *J. Neurosci.* **31(48)**:17505–17513. [5]

Chen, T., K. Koga, G. Descalzi, et al. 2014a. Postsynaptic Potentiation of Corticospinal Projecting Neurons in the Anterior Cingulate Cortex after Nerve Injury. *Mol. Pain* **10**:33. [10]

Chen, T., J. S. Lu, Q. Song, et al. 2014b. Pharmacological Rescue of Cortical Synaptic and Network Potentiation in a Mouse Model for Fragile X Syndrome. *Neuropsychopharmacology* **39(8)**:1955–1967. [13]

Chen, T. W., T. J. Wardill, Y. Sun, et al. 2013. Ultrasensitive Fluorescent Proteins for Imaging Neuronal Activity. *Nature* **499(7458)**:295–300. [8]

Chen, W. R., S. Lee, K. Kato, et al. 1996. Long-Term Modifications of Synaptic Efficacy in the Human Inferior and Middle Temporal Cortex. *PNAS* **93**:8011–8015. [10]

Chiba, A. A., R. P. Kesner, and C. J. Gibson. 1997. Memory for Temporal Order of New and Familiar Spatial Location Sequences: Role of the Medial Prefrontal Cortex. *Learn. Mem.* **4**:311–317. [12]

Chilian, B., H. Abdollahpour, T. Bierhals, et al. 2013. Dysfunction of SHANK2 and CHRNA7 in a Patient with Intellectual Disability and Language Impairment Supports Genetic Epistasis of the Two Loci. *Clin. Genet.* **84(6)**:560–565. [2]

Chu, C.-P., G.-Y. Zhao, R. Jin, et al. 2014. Properties of 4 Hz Stimulation-Induced Parallel Fiber-Purkinje Cell Presynaptic Long-Term Plasticity in Mouse Cerebellar Cortex *in Vivo*. *Eur. J. Neurosci.* **39**:1624–1631. [10]

Chung, K., J. Wallace, S. Y. Kim, et al. 2013. Structural and Molecular Interrogation of Intact Biological Systems. *Nature* **497(7449)**:332–337. [8, 10]

Church, D. M., L. Goodstadt, L. W. Hillier, et al. 2009. Lineage-Specific Biology Revealed by a Finished Genome Assembly of the Mouse. *PLoS Biol.* **7**:e1000112. [10]

Churchill, G. A., D. C. Airey, H. Allayee, et al. 2004. The Collaborative Cross, a Community Resource for the Genetic Analysis of Complex Traits. *Nat. Genet.* **36**:1133–1137. [8]

Clapp, W. C., I. J. Kirk, J. P. Hamm, D. Shepherd, and T. J. Teyler. 2005. Induction of LTP in the Human Auditory Cortex by Sensory Stimulation. *Eur. J. Neurosci.* **22**:1135–1140. [10]

Clarke, H. F., J. W. Dalley, H. S. Crofts, T. W. Robbins, and A. C. Roberts. 2004. Cognitive Inflexibility after Prefrontal Serotonin Depletion. *Science* **304**:878–880. [12]

Clavaguera, F., H. Akatsu, G. Fraser, et al. 2013. Brain Homogenates from Human Tauopathies Induce Tau Inclusions in Mouse Brain. *PNAS* **110(23)**:9535–9540. [5]

Clavaguera, F., F. Grueninger, and M. Tolnay. 2014a. Intercellular Transfer of Tau Aggregates and Spreading of Tau Pathology: Implications for Therapeutic Strategies. *Neuropharmacology* **76(Pt A)**:9–15. [5]

Clavaguera, F., J. Hench, I. Lavenir, et al. 2014b. Peripheral Administration of Tau Aggregates Triggers Intracerebral Tauopathy in Transgenic Mice. *Acta Neuropathol.* **1(1)**:82. [5]

Clayton, N. S., and A. Dickinson. 1998. Episodic-Like Memory During Cache Recovery by Scrub-Jays. *Nature* **395**:272–274. [12]

Coelho, T., L. F. Maia, A. Martins da Silva, et al. 2012. Tafamidis for Transthyretin Familial Amyloid Polyneuropathy: A Randomized, Controlled Trial. *Neurology* **79(8)**:785–792. [7]

Cong, L., F. A. Ran, D. Cox, et al. 2013. Multiplex Genome Engineering Using CRISPR/Cas Systems. *Science* **339**:819–823. [8]

Cooke, S. F., and M. F. Bear. 2012. Stimulus-Selective Response Plasticity in the Visual Cortex: An Assay for the Assessment of Pathophysiology and Treatment of Cognitive Impairment Associated with Psychiatric Disorders. *Biol. Psychiatry* **71**:487–495. [10]

Cooper, J. J., and F. Ovsiew. 2013. The Relationship between Schizophrenia and Frontotemporal Dementia. *J. Geriatr. Psychiatry Neurol.* **26(3)**:131–137. [13]

Corlett, P. R., G. K. Murray, G. D. Honey, et al. 2007. Disrupted Prediction-Error Signal in Psychosis: Evidence for an Associative Account of Delusions. *Brain* **130(9)**:2387–2400. [12]

Corti, S., M. Nizzardo, C. Simone, et al. 2012. Genetic Correction of Human Induced Pluripotent Stem Cells from Patients with Spinal Muscular Atrophy. *Sci. Transl. Med.* **4**:165ra162. [11]

Cosentino, G., B. Fierro, P. Paladino, et al. 2012. Transcranial Direct Current Stimulation Preconditioning Modulates the Effect of High-Frequency Repetitive Transcranial Magnetic Stimulation in the Human Motor Cortex. *Eur. J. Neurosci.* **35**:119–124. [10]

Costa-Mattioli, M., and L. M. Monteggia. 2013. mTOR Complexes in Neurodevelopmental and Neuropsychiatric Disorders. *Nat. Neurosci.* **16(11)**:1537–1543. [2]

Courtine, G., M. B. Bunge, J. W. Fawcett, et al. 2007. Can Experiments in Nonhuman Primates Expedite the Translation of Treatments for Spinal Cord Injury in Humans? *Nat. Med.* **13(5)**:561–566. [13]

Craddock, N., M. C. O'Donovan, and M. J. Owen. 2007. Phenotypic and Genetic Complexity of Psychosis. Invited Commentary on Schizophrenia: A Common Disease Caused by Multiple Rare Alleles. *Br. J. Psychiatry* **190**:200–203. [3]

Craddock, N., and M. J. Owen. 2005. The Beginning of the End for the Kraepelinian Dichotomy. *Br. J. Psychiatry* **186**:364–366. [3]

———. 2010. The Kraepelinian Dichotomy: Going, Going...But Still Not Gone. *Br. J. Psychiatry* **196**:92–95. [3]

Craddock, N., and P. Sklar. 2013. Genetics of Bipolar Disorder. *Lancet* **381**:1654–1662. [8]

Cramer, P. E., J. R. Cirrito, D. W. Wesson, et al. 2012. Apoe-Directed Therapeutics Rapidly Clear Beta-Amyloid and Reverse Deficits in AD Mouse Models. *Science* **335(6075)**:1503–1506. [7]

Criswell, S. R., J. S. Perlmutter, T. O. Videen, et al. 2011. Reduced Uptake of [^{18}F] FDOPA PET in Asymptomatic Welders with Occupational Manganese Exposure. *Neurology* **76**:1296–1301. [10]

Cross-Disorder Group of the Psychiatric Genomics, C., S. H. Lee, S. Ripke, et al. 2013. Genetic Relationship between Five Psychiatric Disorders Estimated from Genome-Wide SNPs. *Nat. Genet.* **45(9)**:984–994. [2, 3]

Cross-Disorder Group of the Psychiatric Genomics Consortium, and Genetic Risk Outcome of Psychosis Consortium. 2013. Identification of Risk Loci with Shared Effects on Five Major Psychiatric Disorders: A Genome-Wide Analysis. *Lancet* **381(9875)**:1371–1379. [4]

Cuervo, A. M., and E. Wong. 2013. Chaperone-Mediated Autophagy: Roles in Disease and Aging. *Cell Res.* **24**:92–104. [5]

Cuthbert, B. N. 2014. The RDoC Framework: Facilitating Transition from ICD/DSM to Dimensional Approaches That Integrate Neuroscience and Psychopathology. *World Psychiatry* **13**:28–35. [12]

Cuthbert, B. N., and T. R. Insel. 2013. Toward the Future of Psychiatric Diagnosis: The Seven Pillars of RDoC. *BMC Med.* **11**:126. [3]

Czupryn, A., Y. D. Zhou, X. Chen, et al. 2011. Transplanted Hypothalamic Neurons Restore Leptin Signaling and Ameliorate Obesity in db/db Mice. *Science* **334**:1133–1137. [11]

Dalley, J. W., T. D. Fryer, L. Brichard, et al. 2007. Nucleus Accumbens D2/3 Receptors Predict Trait Impulsivity and Cocaine Reinforcement. *Science* **315**:1267–1270. [12]

Daniels, A. M., A. K. Halladay, A. Shih, L. M. Elder, and G. Dawson. 2014. Approaches to Enhancing the Early Detection of Autism Spectrum Disorders: A Systematic Review of the Literature. *J. Am. Acad. Child Adolesc. Psychiatry* **53(2)**:141–152. [4]

Davies, G., A. Tenesa, A. Payton, et al. 2011. Genome-Wide Association Studies Establish That Human Intelligence Is Highly Heritable and Polygenic. *Mol. Psychiatry* **16(10)**:996–1005. [2]

Davis, M., K. Ressler, B. O. Rothbaum, and R. Richardson. 2006. Effects of D-Cycloserine on Extinction: Translation from Preclinical to Clinical Work. *Biol. Psychiatry* **60**:369–375. [12]

Deary, I. J., J. Yang, G. Davies, et al. 2012. Genetic Contributions to Stability and Change in Intelligence from Childhood to Old Age. *Nature* **482(7384)**:212–215. [2]

de Calignon, A., M. Polydoro, M. Suarez-Calvet, et al. 2012. Propagation of Tau Pathology in a Model of Early Alzheimer's Disease. *Neuron* **73(4)**:685–697. [5]

Deisseroth, K. 2014. Circuit Dynamics of Adaptive and Maladaptive Behaviour. *Nature* **505(7483)**:309–317. [13, 14]

Deisseroth, K., and M. J. Schnitzer. 2013. Engineering Approaches to Illuminating Brain Structure and Dynamics. *Neuron* **80**:568–577. [10]

Dell'Mour, V., F. Range, and L. Huber. 2009. Social Learning and Mother's Behavior in Manipulative Tasks in Infant Marmosets. *Am. J. Primatol.* **71**:503–509. [8]

DeLong, G. R., C. R. Ritch, and S. Burch. 2002. Fluoxetine Response in Children with Autistic Spectrum Disorders: Correlation with Familial Major Affective Disorder and Intellectual Achievement. *Dev. Med. Child Neurol.* **44(10)**:652–659. [4]

DeLong, G. R., L. A. Teague, and M. McSwain Kamran. 1998. Effects of Fluoxetine Treatment in Young Children with Idiopathic Autism. *Dev. Med. Child Neurol.* **40(8)**:551–562. [4]

Delorme, R., E. Ey, R. Toro, et al. 2013. Progress toward Treatments for Synaptic Defects in Autism. *Nat. Med.* **19(6)**:685–694. [2]

Demeule, M., J. C. Currie, Y. Bertrand, et al. 2008. Involvement of the Low-Density Lipoprotein Receptor-Related Protein in the Transcytosis of the Brain Delivery Vector Angiopep-2. *J. Neurochem.* **106(4)**:1534–1544. [7]

Denny, J. C., L. Nastarache, M. S. Ritchie, et al. 2013. Systematic Comparison of Phenome-Wide Association Study of Electronic Medical Record Data and Genome-Wide Association Study Data. *Nat. Biotechnol.* **31**:1102–1111. [10]

Dere, E., J. P. Huston, and M. A. D. S. Silva. 2007. The Pharmacology, Neuroanatomy and Neurogenetics of One-Trial Object Recognition in Rodents. *Neurosci. Biobehav. Res.* **31**:673–704. [12]

De Rubeis, S., X. He, A. P. Goldberg, et al. 2014. Synaptic, Transcriptional and Chromatin Genes Disrupted in Autism. *Nat. Chem. Biol.* **515**:209–215. [1]

Developmental Disabilities Monitoring Network Surveillance Year Principal Investigators. 2014. Prevalence of Autism Spectrum Disorder among Children Aged 8 Years: Autism and Developmental Disabilities Monitoring Network, 11 Sites, United States, 2010. *MMWR Surveill. Summ.* **63(Suppl 2)**:1–21. [2]

Dias, R., T. W. Robbins, and A. C. Roberts. 1996. Dissociation in Prefrontal Cortex of Affective and Attentional Shifts. *Nature* **380**:69–72. [12]

Dickerson, F., C. Stallings, A. Origoni, J. Boronow, and R. Yolken. 2007. Elevated Serum Levels of C-Reactive Protein Are Associated with Mania Symptoms in Outpatients with Bipolar Disorder. *Prog. Neuropsychopharmacol. Biol. Psychiatry* **31(4)**:952–955. [13]

Diester, I., M. T. Kaufman, M. Mogri, et al. 2011. An Optogenetic Toolbox Designed for Primates. *Nat. Neurosci.* **14**:387–397. [10, 13]

Di Giorgio, F. P., M. A. Carrasco, M. C. Siao, T. Maniatis, and K. Eggan. 2007. Non-Cell Autonomous Effect of Glia on Motor Neurons in an Embryonic Stem Cell-Based ALS Model. *Nature Neurosci.* **10**:608–614. [11]

Dimitrov, D., Y. He, H. Mutoh, et al. 2007. Engineering and Characterization of an Enhanced Fluorescent Protein Voltage Sensor. *PLoS One* **2**:e440. [10]

Dodart, J. C., R. A. Marr, M. Koistinaho, et al. 2005. Gene Delivery of Human Apolipoprotein E Alters Brain Abeta Burden in a Mouse Model of Alzheimer's Disease. *PNAS* **102(4)**:1211–1216. [7]

Dölen, G., and M. F. Bear. 2009. Fragile X Syndrome and Autism: From Disease Model to Therapeutic Targets. *J. Neurodev. Disord.* **1**:133–140. [10, 13]

Dölen, G., A. Darvishzadeh, K. W. Huang, and R. C. Malenka. 2013. Social Reward Requires Coordinated Activity of Nucleus Accumbens Oxytocin and Serotonin. *Nature* **501**:179–184. [10]

Dölen, G., E. Osterweil, B. S. S. Rao, et al. 2007. Correction of Fragile X Syndrome in Mice. *Neuron* **56**:955–962. [10, 13]

Dong, S., S. C. Rogan, and B. L. Roth. 2010. Directed Molecular Evolution of DREADDs: A Generic Approach to Creating Next-Generation Rassls. *Nat. Protoc.* **5**:561–573. [10]

Doody, R. S., R. Raman, M. Farlow, et al. 2013. A Phase 3 Trial of Semagacestat for Treatment of Alzheimer's Disease. *N. Engl. J. Med.* **369(4)**:341–350. [6]

Dubois, B., H. H. Feldman, C. Jacova, et al. 2010. Revising the Definition of Alzheimer's Disease: A New Lexicon. *Lancet Neurol.* **9(11)**:1118–1127. [6]

Dunnett, S. B. 1985. Comparative Effects of Cholinergic Drugs and Lesions of Nucleus Basalis or Fimbria-Fornix on Delayed Matching in Rats. *Psychopharmacology* **87**:357–363. [12]

Durand, C. M., C. Betancur, T. M. Boeckers, et al. 2007. Mutations in the Gene Encoding the Synaptic Scaffolding Protein SHANK3 Are Associated with Autism Spectrum Disorders. *Nat. Genet.* **39(1)**:25–27. [2, 4]

Dutch-Belgian Fragile X Consortium. 1994. Fmr1 Knockout Mice: A Model to Study Fragile X Mental Retardation. *Cell* **78(1)**:23–33. [13]

Eacott, M. J., A. Easton, and A. Zinkivskay. 2005. Recollection in an Episodic-Like Memory Task in the Rat. *Memory* **12**:221–223. [12]

Eagle, D. M., and T. W. Robbins. 2003. Lesions of the Medial Prefrontal Cortex or Nucleus Accumbens Core Do Not Impair Inhibitory Control in Rats Performing a Stop-Signal Reaction Time Task. *Behav. Brain Res.* **146**:131–144. [12]

Ebert, A. D., J. Yu, F. F. Rose, Jr., et al. 2009. Induced Pluripotent Stem Cells from a Spinal Muscular Atrophy Patient. *Nature* **457**:277–280. [11]

Eiraku, M., K. Watanabe, M. Matsuo-Takasaki, et al. 2008. Self-Organized Formation of Polarized Cortical Tissues from ESCs and Its Active Manipulation by Extrinsic Signals. *Cell Stem Cell* **3**:519–532. [8]

Elliott, R., B. J. Sahakian, A. P. McKay, et al. 1996. Neuropsychological Impairments in Unipolar Depression: The Influence of Perceived Failure on Subsequent Performance. *Psychol. Med.* **26**:975–989. [12]

Emre, M., D. Aarsland, A. Albanese, et al. 2004. Rivastigmine for Dementia Associated with Parkinson's Disease. *N. Engl. J. Med.* **351(24)**:2509–2518. [13]

Engle, S. J., and D. Puppala. 2013. Integrating Human Pluripotent Stem Cells into Drug Development. *Cell Stem Cell* **12**:669–677. [11]

Enkel, T., D. Gholizadeh, U. von Bohlen, et al. 2010. Ambiguous Cue Interpretation Is Biased under Stress- and Depression-Like States in Rats. *Neuropsychopharmacology* **35**:1008–1015 [12]

Ennaceur, A., and J. Delacour. 1988. A New One-Trial Test for Neurobiological Studies of Memory in Rats. 1: Behavioral Data. *Behav. Brain Res.* **31**:47–59. [12]

Ennis, L., and T. Wykes. 2013. Impact of Patient Involvement in Mental Health Research: Longitudinal Study. *Br. J. Psychiatry* **203(5)**:381–386. [4]

Ersche, K. D., P. S. Jones, G. B. Williams, et al. 2012. Abnormal Brain Structure Implicated in Stimulant Drug Addiction. *Science* **335**:601–604. [12]

Espuny-Camacho, I., K. A. Michelsen, D. Gall, et al. 2013. Pyramidal Neurons Derived from Human Pluripotent Stem Cells Integrate Efficiently into Mouse Brain Circuits *in Vivo*. *Neuron* **77**:440–456. [11]

Evenden, J. L. 1999. Varieties of Impulsivity. *Psychopharmacology* **146**:348–361. [12]

Everitt, B. J., and T. W. Robbins. 1997. Central Cholinergic Systems and Cognition. *Annu. Rev. Psychol.* **48**:649–684. [13]

Everitt, B. J., and T. W. Robbins. 2005. Neural Systems of Reinforcement for Drug Addiction: From Actions to Habits to Compulsion. *Nature Neurosci.* **8**:1481–1489. [12]

Ey, E., C. S. Leblond, and T. Bourgeron. 2011. Behavioral Profiles of Mouse Models for Autism Spectrum Disorders. *Autism Res.* **4(1)**:5–16. [2]

Fagan, A. M., M. A. Mintun, R. H. Mach, et al. 2006. Inverse Relation between *in Vivo* Amyloid Imaging Load and Cerebrospinal Fluid Abeta42 in Humans. *Ann. Neurol.* **59(3)**:512–519. [7]

Fernandez, F., W. Morishita, E. Zuniga, et al. 2007. Pharmacotherapy for Cognitive Impairment in a Mouse Model of Down Syndrome. *Nat. Neurosci.* **10(4)**:411–413. [7]

Fernando, A. B., and T. W. Robbins. 2011. Animal Models of Neuropsychiatric Disorders. *Ann. Rev. Clin. Psychol.* **7**:39–61. [12]

Figurov, A., L. D. Pozz-Miller, P. Olafsson, T. Wang, and B. Lu. 1996. Regulation of Synaptic Responses to High-Frequency Stimulation and LTP by Neurotrophins in the Hippocampus. *Nature* **381**:706–709. [10]

Flannick, J., G. Thorleifsson, N. L. Beer, et al. 2014. Loss-of-Function Mutations in Slc30a8 Protect against Type 2 Diabetes. *Nat. Genet.* **46(4)**:357–363. [2, 4]

Fontes, A., and U. Lakshmipathy. 2013. Advances in Genetic Modification of Pluripotent Stem Cells. *Biotechnol. Adv.* **31**:994–1001. [8]

Foust, K. D., E. Nurre, C. L. Montgomery, et al. 2009. Intravascular AAV9 Preferentially Targets Neonatal-Neurons and Adult-Astrocytes in CNS. *Nat. Biotechnol.* **27**:59–65. [10]

Fox, M. D., M. A. Halko, M. C. Eldaief, and A. Pascual-Leone. 2012. Measuring and Manipulating Brain Connectivity with Resting State Functional Connectivity Magnetic Resonance Imaging (fcMRI) and Transcranial Magnetic Stimulation (TMS). *Neuroimage* **62(4)**:2232–2243. [13]

Frank, S. A. 2014. Somatic Mosaicism and Disease. *Curr. Biol.* **24(12)**:R577–R581. [2]

Freitas, C., F. Farzan, and A. Pascual-Leone. 2013. Assessing Brain Plasticity across the Lifespan with Transcranial Magnetic Stimulation: Why, How, and What Is the Ultimate Goal? *Front. Neurosci.* **7**:42. [13]

Freund, P., E. Schmidlin, T. Wannier, et al. 2006. NoGo: A-Specific Antibody Treatment Enhances Sprouting and Functional Recovery after Cervical Lesion in Adult Primates. *Nat. Med.* **12(7)**:790–792. [13]

Fritsch, B., J. Reis, K. Martinowich, et al. 2010. Direct Current Stimulation Promotes BDNF-Dependent Synaptic Plasticity: Potential Implications for Motor Learning. *Neuron* **66**:198–204. [10]

Frodl, T., J. Scheuerecker, V. Schoepf, et al. 2011. Different Effects of Mirtazapine and Venlafaxine on Brain Activation: An Open Randomized Controlled fMRI Study. *J. Clin. Psychiatry* **72**:448–457. [10]

Fromer, M., A. J. Pocklington, D. H. Kavanagh, et al. 2014. *De Novo* Mutations in Schizophrenia Implicate Synaptic Networks. *Nature* **506(7487)**:179–184. [3, 4]

Frost, B., and M. I. Diamond. 2010. Prion-Like Mechanisms in Neurodegenerative Diseases. *Nat. Rev. Neurosci.* **11(3)**:155–159. [7]

Fusar-Poli, P., S. Borgwardt, A. Bechdolf, et al. 2013. The Psychosis High-Risk State: A Comprehensive State-of-the-Art Review. *JAMA Psychiatry* **70(1)**:107–120. [4]

Gaffan, D., R. C. Saunders, E. A. Gaffan, et al. 1984. Effects of Fornix Transection Upon Associative Memory in Monkeys: Role of the Hippocampus in Learned Action. *Q. J. Exp. Psychol. B* **36**:173–227. [12]

Gafni, O., L. Weinberger, A. A. Mansour, et al. 2013. Derivation of Novel Human Ground State Naive Pluripotent Stem Cells. *Nature* **504(7479)**:282–286. [13]

Galvan, A., X. Hu, Y. Smith, and T. Wichmann. 2012. *In Vivo* Optogenetic Control of Striatal and Thalamic Neurons in Non-Human Primates. *PLoS One* **7**:e50808. [10]

Gantois, I., A. S. Pop, C. E. de Esch, et al. 2013. Chronic Administration of AFQ056/Mavoglurant Restores Social Behaviour in Fmr1 Knockout Mice. *Behav. Brain Res.* **239**:72–79. [13]

Gaskin, P. L., S. P. Alexander, and K. C. Fone. 2014. Neonatal Phencyclidine Administration and Post-Weaning Social Isolation as a Dual-Hit Model of "Schizophrenia-Like" Behaviour in the Rat. *Psychopharmacology* **231(12)**:2533–2545. [13]

Gastambide, F., M. C. Cotel, G. Gilmour, et al. 2012. Selective Remediation of Reversal Learning Deficits in the Neurodevelopmental MAM Model of Schizophrenia by a Novel Mglu5 Positive Allosteric Modulator. *Neuropsychopharmacology* **37(4)**:1057–1066. [13]

Gatev, P., and T. Wichmann. 2009. Interactions between Cortical Rhythms and Spiking Activity of Single Basal Ganglia Neurons in the Normal and Parkinsonian State. *Cereb. Cortex* **19**:1330–1344. [9]

Geyer, M. A., K. Krebs-Thomson, D. L. Braff, and N. R. Swerdlow. 2001. Pharmacological Studies of Prepulse Inhibition Models of Sensorimotor Gating Deficits in Schizophrenia: A Decade in Review. *Psychopharmacology* **156**:117–154. [12]

Gilissen, C., J. Y. Hehir-Kwa, D. T. Thung, et al. 2014. Genome Sequencing Identifies Major Causes of Severe Intellectual Disability. *Nature* **511(7509)**:344–347. [2]

Gillberg, C. 2010. The ESSENCE in Child Psychiatry: Early Symptomatic Syndromes Eliciting Neurodevelopmental Clinical Examinations. *Res. Dev. Disabil.* **31(6)**:1543–1551. [2]

Giovanoli, S., H. Engler, A. Engler, et al. 2013. Stress in Puberty Unmasks Latent Neuropathological Consequences of Prenatal Immune Activation in Mice. *Science* **339(6123)**:1095–1099. [13]

Girirajan, S., J. A. Rosenfeld, B. P. Coe, et al. 2012. Phenotypic Heterogeneity of Genomic Disorders and Rare Copy-Number Variants. *N. Engl. J. Med.* **367(14)**:1321–1331. [2]

Girirajan, S., J. A. Rosenfeld, G. M. Cooper, et al. 2010. A Recurrent 16p12.1 Microdeletion Supports a Two-Hit Model for Severe Developmental Delay. *Nat. Genet.* **42(3)**:203–209. [2]

Giusti-Rodríguez, P., and P. F. Sullivan. 2013. The Genomics of Schizophrenia: Update and Implications. *J. Clin. Invest.* **123**:4557–4563. [8]

Glass, C. K., K. Saijo, B. Winner, M. C. Marchetto, and F. H. Gage. 2010. Mechanisms Underlying Inflammation in Neurodegeneration. *Cell* **140(6)**:918–934. [5]

Goetghebeur, P., and R. Dias. 2008. Comparison of Haloperidol, Risperidone, Sertindole, and Modafinil to Reverse an Attentional Set-Shifting Impairment Following Subchronic Pcp Administration in the Rat: A Back Translational Study. *Psychopharmacology* **202**:202. [12]

Gokhale, A., J. Larimore, E. Werner, et al. 2012. Quantitative Proteomic and Genetic Analyses of the Schizophrenia Susceptibility Factor Dysbindin Identify Novel Roles of the Biogenesis of Lysosome-Related Organelles Complex 1. *J. Neurosci.* **32(11)**:3697–3711. [2]

Golde, T. E., L. S. Schneider, and E. H. Koo. 2011. Anti-Abeta Therapeutics in Alzheimer's Disease: The Need for a Paradigm Shift. *Neuron* **69(2)**:203–213. [6]

Goldman-Rakic, P. S. 1996. The Prefrontal Landscape: Implications of Functional Architecture for Understanding Human Mentation and the Central Executive. In: The Prefrontal Cortex, ed. A. C. Roberts et al., pp. 87–102. Oxford: Oxford Univ. Press. [12]

Goldman-Rakic, P. S., S. A. Castner, T. H. Svensson, L. J. Siever, and G. V. Williams. 2004. Targeting the Dopamine D1 Receptor in Schizophrenia: Insights for Cognitive Dysfunction. *Psychopharmacology* **174(1)**:3–16. [13]

Goldsworthy, M. R., J. B. Pitcher, and M. C. Ridding. 2012. The Application of Spaced Theta Burst Protocols Induces Long-Lasting Neuroplastic Changes in the Human Motor Cortex. *Eur. J. Neurosci.* **35**:125–134. [10]

Goodin, D. S., and D. Bates. 2009. Treatment of Early Multiple Sclerosis: The Value of Treatment Initiation after a First Clinical Episode. *Mult. Scler.* **15(10)**:1175–1182. [6]

Gottesman, I. I., and T. D. Gould. 2003. The Endophenotype Concept in Psychiatry: Etymology and Strategic Intentions. *Am. J. Psychiatry* **160(4)**:636–645. [12]

Grabrucker, A. M., M. J. Schmeisser, M. Schoen, and T. M. Boeckers. 2011. Postsynaptic ProSAP/Shank Scaffolds in the Cross-Hair of Synaptopathies. *Trends Cell. Biol.* **21(10)**:594–603. [4, 10]

Gradinaru, V., M. Mogri, K. R. Thompson, J. M. Henderson, and K. Deisseroth. 2009. Optical Deconstruction of Parkinsonian Neural Circuitry. *Sci. Transl. Med.* **324**:354–359. [9]

Graf, T. 2011. Historical Origins of Transdifferentiation and Reprogramming. *Cell Stem Cell* **9(6)**:504–516. [13]

Granger, A. J., and R. A. Nicoll. 2014. Expression Mechanisms Underlying Long-Term Potentiation: A Postsynaptic View, 10 Years on. *Philos. Trans. R. Soc. Lond. B* **369(1633)**:20130136. [10]

Grant, S. G., M. C. Marshall, K. L. Page, M. A. Cumiskey, and J. D. Armstrong. 2005. Synapse Proteomics of Multiprotein Complexes: en Route from Genes to Nervous System Diseases. *Hum. Mol. Genet.* **14(2)**:R225–R234. [3, 4]

Gratten, J., N. R. Wray, M. C. Keller, and P. M. Visscher. 2014. Large-Scale Genomics Unveils the Genetic Architecture of Psychiatric Disorders. *Nat. Neurosci.* **17(6)**:782–790. [2]

Graves, A. R., S. J. Moore, E. B. Bloss, et al. 2012. Hippocampal Pyramidal Neurons Comprise Two Distinct Cell Types That Are Countermodulated by Metabotropic Receptors. *Neuron* **76**:776–789. [10]

Graybiel, A. M. 2008. Habits, Rituals, and the Evaluative Brain. *Annu. Rev. Neurosci.* **31**:359–387. [8]

Green, E. K., D. Grozeva, I. Jones, et al. 2010. The Bipolar Disorder Risk Allele at CACNA1C Also Confers Risk of Recurrent Major Depression and of Schizophrenia. *Mol. Psychiatry* **15(10)**:1016–1022. [3]

Green, M. F., P. D. Butler, Y. Chen, et al. 2009. Perception Measurement in Clinical Trials of Schizophrenia: Promising Paradigms from CNTRICS. *Schizophr. Bull.* **35**:163–181. [12]

Greicius, M. D. 2008. Resting-State Functional Connectivity in Neuropsychiatric Disorders. *Curr. Opin. Neurol.* **21**:424–430. [10]

Greicius, M. D., G. Srivastava, A. L. Reiss, and V. Menon. 2004. Default-Mode Network Activity Distinguishes Alzheimer's Disease from Healthy Aging: Evidence from Functional MRI. *PNAS* **101(13)**:4637–4642. [7]

Grimm, D., J. S. Lee, L. Wang, et al. 2008. *In Vitro* and *in Vivo* Gene Therapy Vector Evolution via Multispecies Interbreeding and Retargeting of Adeno-Associated Viruses. *J. Virol.* **82**:5887–5911. [10]

Grozeva, D., G. Kirov, D. F. Conrad, et al. 2013. Reduced Burden of Very Large and Rare CNVs in Bipolar Affective Disorder. *Bipolar Dis.* **15(8)**:893–898. [3]

Grozeva, D., G. Kirov, D. Ivanov, et al. 2010. Rare Copy Number Variants: A Point of Rarity in Genetic Risk for Bipolar Disorder and Schizophrenia. *Arch. Gen. Psychiatry* **67**:318–327. [3]

Guerreiro, R., A. Wojtas, J. Bras, et al. 2013. TREM2 Variants in Alzheimer's Disease. *N. Engl. J. Med.* **368(2)**:117–127. [7]

Guilfoyle, D. N., S. V. Gerum, J. L. Sanchez, et al. 2013. Functional Connectivity fMRI in Mouse Brain at 7t Using Isoflurane. *J. Neurosci. Meth.* **214**:144–148. [9]

Guilmatre, A., G. Huguet, R. Delorme, and T. Bourgeron. 2014. The Emerging Role of Shank Genes in Neuropsychiatric Disorders. *Dev. Neurobiol.* **74(2)**:113–122. [4]

Gulsuner, S., T. Walsh, A. C. Watts, et al. 2013. Spatial and Temporal Mapping of *de Novo* Mutations in Schizophrenia to a Fetal Prefrontal Cortical Network. *Cell* **154**:518–529. [10]

Guo, Z., L. Zhang, Z. Wu, et al. 2014. *In Vivo* Direct Reprogramming of Reactive Glial Cells into Functional Neurons after Brain Injury in an Alzheimer's Disease Model. *Cell Stem Cell* **14**:188–202. [11]

Gur, R. C., M. E. Calkins, T. D. Satterthwaite, et al. 2014. Neurocognitive Growth Charting in Psychosis Spectrum Youths. *JAMA Psychiatry* **71(4)**:366–374. [4]

Haber, S. N., and B. Knutson. 2010. The Reward Circuit: Linking Primate Anatomy to Human Imaging. *Neuropsychopharmacol. Rev.*:4–26. [12]

Hacker, C. D., J. S. Perlmutter, S. R. Criswell, B. M. Ances, and A. Z. Snyder. 2012. Resting State Functional Connectivity of the Striatum in Parkinson's Disease. *Brain* **135**:3699–3711. [9]

Hall, E. L., S. E. Robson, P. G. Morris, and M. J. Brookes. 2013. The Relationship between MEG and fMRI. *Neuroimage* **102(Pt 1)**:80–91. [9]

Hall, F. S., L. S. Wilkinson, T. Humby, et al. 1998. Isolation Rearing in Rats: Pre- and Postsynaptic Changes in Striatal Dopaminergic Systems. *Pharmacol. Biochem. Behav.* **59(4)**:859–872. [13]

Hall, M.-H., G. Taylor, P. Sham, et al. 2011. The Early Auditory Gamma-Band Response Is Heritable and a Putative Endophenotype of Schizophrenia. *Schizophr. Bull.* **37**:778–787. [10]

Hamasaki, M., N. Furuta, A. Matsuda, et al. 2013. Autophagosomes Form at ER-Mitochondria Contact Sites. *Nature* **495(7441)**:389–393. [5]

Han, K., J. L. Holder, Jr., C. P. Schaaf, et al. 2013a. SHANK3 Overexpression Causes Manic-Like Behaviour with Unique Pharmacogenetic Properties. *Nature* **503(7474)**:72–77. [2, 4]

Han, X., M. Chen, F. Wang, et al. 2013b. Forebrain Engraftment by Human Glial Progenitor Cells Enhances Synaptic Plasticity and Learning in Adult Mice. *Cell Stem Cell* **12(3)**:342–353. [13]

Hanlon, F. M., M. P. Weisend, and D. A. Hamilton. 2006. Impairment on the Hippocampal-Dependent Virtual Morris Water Task in Schizophrenia. *Schizophr. Res.* **87**:67–80. [12]

Haqqani, A. S., N. Caram-Salas, W. Ding, et al. 2013. Multiplexed Evaluation of Serum and CSF Pharmacokinetics of Brain-Targeting Single-Domain Antibodies Using a NanoLC-SRM-ILIS Method. *Molec. Pharm.* **10(5)**:1542–1556. [7]

Hara, T., K. Nakamura, M. Matsui, et al. 2006. Suppression of Basal Autophagy in Neural Cells Causes Neurodegenerative Disease in Mice. *Nature* **441(7095)**:885–889. [5]

Harding, E. J., E. S. Paul, and M. Mendl. 2004. Animal Behavior: Cognitive Bias and Affective State. *Nature* **427**:312–312. [12]

Hayden, E. Y., and D. B. Teplow. 2013. Amyloid Beta-Protein Oligomers and Alzheimer's Disease. *Alzheimers Res. Ther.* **5(6)**:60. [7]

Hayhurst, M., A. Wagner, M. Cerletti, A. J. Wagers, and L. L. Rubin. 2012. A Cell-Autonomous Defect in Skeletal Muscle Satellite Cells Expressing Low Levels of Survival of Motor Neuron Protein. *Devel. Biol.* **368**:323–334. [11]

Hayworth, K. J., N. Kasthuri, R. Schalek, and J. W. Lichtman. 2006. Automating the Collection of Ultrathin Serial Sections for Large Volume TEM Reconstructions. *Microsc. Microanal.* **12(S02)**:86–87. [8]

Heckers, S. 2008. Making Progress in Schizophrenia Research. *Schizophr. Bull.* **34(4)**:591–594. [4]

Heiman, M., A. Schaefer, S. Gong, et al. 2008. A Translational Profiling Approach for the Molecular Characterization of CNS Cell Types. *Cell* **135(4)**:738–748. [7]

Heiman-Patterson, T. D., R. B. Sher, E. A. Blankenhorn, et al. 2011. Effect of Genetic Background on Phenotype Variability in Transgenic Mouse Models of Amyotrophic Lateral Sclerosis: A Window of Opportunity in the Search for Genetic Modifiers. *Amyotroph. Lateral Scler.* **12**:79–86. [9]

Helmich, R. C., L. C. Derikx, M. Bakker, et al. 2009. Spatial Remapping of Cortico-Striatal Connectivity in Parkinson's Disease. *Cereb. Cortex* **20(5)**:1175–1186. [9]

Heneka, M. T., M. P. Kummer, A. Stutz, et al. 2013. NLRP3 Is Activated in Alzheimer's Disease and Contributes to Pathology in APP/PS1 Mice. *Nature* **493(7434)**:674–678. [7]

Hershey, T., K. J. Black, J. L. Carl, and J. S. Perlmutter. 2000. Dopa-Induced Blood Flow Responses in Non-Human Primates. *Exp. Neurol.* **166**:342–349. [9]

Hertz, N. T., A. Berthet, M. L. Sos, et al. 2013. A Neo-Substrate That Amplifies Catalytic Activity of Parkinson's-Disease-Related Kinase PINK1. *Cell* **154(4)**:737–747. [5]

Hnasko, T. S., F. A. Perez, A. D. Scouras, et al. 2006. Cre Recombinase-Mediated Restoration of Nigrostriatal Dopamine in Dopamine-Deficient Mice Reverses Hypophagia and Bradykinesia. *PNAS* **103**:8858–8863. [10]

Ho, B. C., M. Flaum, W. Hubbard, S. Arndt, and N. C. Andreasen. 2004. Validity of Symptom Assessment in Psychotic Disorders: Information Variance across Different Sources of History. *Schizophr. Res.* **68(2-3)**:299–307. [4]

Hoff, P. 1985. Zum Krankheitsbegriff Bei Emil Kraepelin (Concept of Disease of Emil Kraepelin). *Nervenarzt* **56(9)**:510–513. [4]

Hoffman, S. G., J. Q. Wu, and H. Boettcher. 2013. D-Cycloserine as an Augmentation Strategy for Cognitive Behavioral Therapy of Anxiety Disorders. *Biol. Mood Anxiety Disord.* **3**:11. [12]

Hoischen, A., N. Krumm, and E. E. Eichler. 2014. Prioritization of Neurodevelopmental Disease Genes by Discovery of New Mutations. *Nat. Neurosci.* **17(6)**:764–772. [2]

Holland, D., L. K. McEvoy, and A. M. Dale. 2012. Unbiased Comparison of Sample Size Estimates from Longitudinal Structural Measures in ADNI. *Hum. Brain Mapping* **33(11)**:2586–2602. [6]

Hollingworth, P., D. Harold, R. Sims, et al. 2011. Common Variants at ABCA7, MS4A6A/MS4A4E, EPHA1, CD33 and CD2AP Are Associated with Alzheimer's Disease. *Nat. Genet.* **43(5)**:429–435. [7]

Holmes, C., D. Boche, D. Wilkinson, et al. 2008. Long-Term Effects of Abeta42 Immunisation in Alzheimer's Disease: Follow-up of a Randomised, Placebo-Controlled Phase I Trial. *Lancet* **3372(9634)**:216–223. [6]

Holtzman, D. M., J. Herz, and G. Bu. 2012. Apolipoprotein E and Apolipoprotein E Receptors: Normal Biology and Roles in Alzheimer Disease. *Cold Spring Harb. Perspect. Med.* **2(3)**:a006312. [7]

Holtzman, D. M., J. C. Morris, and A. M. Goate. 2011. Alzheimer's Disease: The Challenge of the Second Century. *Sci. Transl. Med.* **3(77)**:77sr71. [6, 7]

Hong, L. E., A. Summerfelt, B. D. Mitchell, et al. 2008. Sensory Gating Endophenotype Based on Its Neural Oscillatory Pattern and Heritability Estimate. *Arch. Gen. Psychiatry* **65**:1008–1016. [10]

Hood, L., and S. H. Friend. 2011. Predictive, Personalized, Preventive, Participatory (P4) Cancer Medicine. *Nat. Rev. Clin. Oncol.* **8(3)**:184–187. [2]

Hoon, M., T. Soykan, B. Falkenburger, et al. 2011. Neuroligin-4 Is Localized to Glycinergic Postsynapses and Regulates Inhibition in the Retina. *PNAS* **108**:3053–3058. [10]

Hornykiewicz, O., and W. Birkmayer. 1961. Biochemisch-Pharmakologische Grundlagen für Die Klinische Anwendung Von L-Dioxyphenylalanin Beim Parkinson-Syndrom. *Wien Klein. Wochenschr.* **73**:839–840. [9]

Howes, O. D., A. J. Montgomery, M. C. Asselin, et al. 2009. Elevated Striatal Dopamine Function Linked to Prodromal Signs of Schizophrenia. *Arch. Gen. Psychiatry* **66(1)**:13–20. [13]

Howes, O. D., and R. M. Murray. 2014 Schizophrenia: An Integrated Sociodevelopmental-Cognitive Model. *Lancet* **383(9929)**:1677–1687. [3]

Hsu, P. D., E. S. Lander, and F. Zhang. 2014. Development and Applications of CRISPR-Cas9 for Genome Engineering. *Cell* **157**:1262–1278. [8]

Huang, Y., and L. Mucke. 2012. Alzheimer Mechanisms and Therapeutic Strategies. *Cell* **148**:1204–1222. [5]

Huber, K. M., S. M. Gallagher, S. T. Warren, and M. F. Bear. 2002. Altered Synaptic Plasticity in a Mouse Model of Fragile X Mental Retardation. *PNAS* **99(11)**:7746–7750. [13]

Hudry, E., J. Dashkoff, A. D. Roe, et al. 2013. Gene Transfer of Human *Apoe* Isoforms Results in Differential Modulation of Amyloid Deposition and Neurotoxicity in Mouse Brain. *Sci. Transl. Med.* **5(212)**:212ra161. [7]

Huguet, G., E. Ey, and T. Bourgeron. 2013. The Genetic Landscapes of Autism Spectrum Disorders. *Annu. Rev. Hum. Genet.* **14**:191–213. [2, 4]

Hulme, S. R., O. D. Jones, and W. C. Abraham. 2013. Emerging Roles of Metaplasticity in Behaviour and Disease. *Trends Neurosci.* **36**:353–362. [10]

Hung, A. Y., K. Futai, C. Sala, et al. 2008. Smaller Dendritic Spines, Weaker Synaptic Transmission, but Enhanced Spatial Learning in Mice Lacking Shank1. *J. Neurosci.* **28(7)**:1697–1708. [4]

Hurd, M. D., P. Martorell, A. Delavande, K. J. Mullen, and K. M. Langa. 2013. Monetary Costs of Dementia in the United States. *N. Engl. J. Med.* **368**:1326–1334. [7]

Hutchison, R. M., and S. Everling. 2012. Monkey in the Middle: Why Non-Human Primates Are Needed to Bridge the Gap in Resting-State Investigations. *Front. Neuroanat.* **6**:29. [9]

Hutchison, R. M., T. Womelsdorf, E. A. Allen, et al. 2013. Dynamic Functional Connectivity: Promise, Issues, and Interpretations. *Neuroimage* **80**:360–378. [9]

Hutton, M., C. L. Lendon, P. Rizzu, et al. 1998. Association of Missense and 5'-Splice-Site Mutations in Tau with the Inherited Dementia FTDP-17. *Nature* **393(6686)**:702–705. [13]

Hyman, S. E. 2010. The Diagnosis of Mental Disorders: The Problem of Reification. *Annu. Rev. Clin. Psychol.* **6**:155–179. [3, 14]

———. 2012. Revolution Stalled. *Transl. Med.* **4(155)**:155cm111. [14]

———. 2014. Revitalizing Psychiatric Therapeutics. *Neuropsychopharmacology* **39**: 220–229. [1, 11]

Iadecola, C. 2004. Neurovascular Regulation in the Normal Brain and in Alzheimer's Disease. *Nature Neurosci.* **5**:347–360. [7]

Iliff, J. J., M. Wang, Y. Liao, et al. 2012. A Paravascular Pathway Facilitates CSF Flow through the Brain Parenchyma and the Clearance of Interstitial Solutes, Including Amyloid ß. *Sci. Transl. Med.* **4(147)**:147ra111. [5, 7]

Imaizumi, Y., and H. Okano. 2014. Modeling Human Neurological Disorders with Induced Pluripotent Stem Cells. *J. Neurochem.* **129**:388–399. [8]

Institute of Medicine. 2013. Neurodegeneration: Exploring Commonalities across Diseases. Workshop Summary. Washington, D.C.: National Academies Press. [5]

International Schizophrenia Consortium, S. M. Purcell, N. R. Wray, et al. 2009. Common Polygenic Variation Contributes to Risk of Schizophrenia and Bipolar Disorder. *Nature* **460(7256)**:748–752. [3]

Ioannidis, J. P. 2005. Why Most Published Research Findings Are False. *PloS Med.* **2(8)**:e124. [6]

Iossifov, I., M. Ronemus, D. Levy, et al. 2012. *De Novo* Gene Disruptions in Children on the Autistic Spectrum. *Neuron* **74(2)**:285–299. [2]

Irwin, I., L. E. DeLanney, L. S. Forno, et al. 1990. The Evolution of Nigrostriatal Neurochemical Changes in the MPTP-Treated Squirrel Monkey. *Brain Res.* **531**:242–252. [9]

Irwin, S. 1968. Comprehensive Observational Assessment: Ia. A Systematic, Quantitative Procedure for Assessing the Behavioral and Physiologic State of the Mouse. *Psychopharmacology* **13**:222–257. [12]

Jack, C. R., Jr., and D. M. Holtzman. 2013. Biomarker Modeling of Alzheimer's Disease. *Neuron* **80(6)**:1347–1358. [7, 14]

Jamain, S., H. Quach, C. Betancur, et al. 2003. Mutations of the X-Linked Genes Encoding Neuroligins NLGN3 and NLGN4 Are Associated with Autism. *Nat. Genet.* **34(1)**:27–29. [2, 4]

Jamain, S., K. Radyushkin, K. Hammerschmidt, et al. 2008. Reduced Social Interaction and Ultrasonic Communication in a Mouse Model of Monogenic Heritable Autism. *PNAS* **105(5)**:1710–1715. [2]

Javitt, D. C., A. Shelley, and W. Ritter. 2000. Associated Deficits in Mismatch Negativity Generation and Tone Matching in Schizophrenia. *Clin. Neurophysiol.* **111**:1733–1737. [10]

Jenner, P. 2003. The Contribution of the MPTP-Treated Primate Model to the Development of New Treatment Strategies for Parkinson's Disease. *Parkinsonism Relat. Disord.* **9(3)**:131–137. [13]

Jensen, K., J. Call, and M. Tomasello. 2007. Chimpanzees Are Rational Maximizers in an Ultimatum Game. *Science* **318**:108–109. [12]

Jiang, Y. H., and M. D. Ehlers. 2013. Modeling Autism by Shank Gene Mutations in Mice. *Neuron* **78(1)**:8–27. [4]

Johnson, K. A., R. A. Sperling, C. M. Gidicsin, et al. 2013. Florbetapir (F18-AV-45) PET to Assess Amyloid Burden in Alzheimer's Disease Dementia, Mild Cognitive Impairment, and Normal Aging. *Alzheimers Dement.* **9(Suppl. 5)**:S72–S83. [6]

Jones, W., and A. Klin. 2013. Attention to Eyes Is Present but in Decline in 2-6-Month-Old Infants Later Diagnosed with Autism. *Nature* **504(7480)**:427–431. [4]

Jonsson, T., J. K. Atwal, S. Steinberg, et al. 2012. A Mutation in APP Protects against Alzheimer's Disease and Age-Related Cognitive Decline. *Nature* **488(7409)**:96–99. [7]

Jonsson, T., H. Stefansson, S. Steinberg, et al. 2013. Variant of TREM2 Associated with the Risk of Alzheimer's Disease. *N. Engl. J. Med.* **368(2)**:107–116. [7]

Joober, R., N. Schmitz, L. Annable, and P. Boksa. 2012. Publication Bias: What Are the Challenges and Can They Be Overcome? *J. Psychiatry Neurosci.* **37(3)**:149–152. [2, 4]

Ju, Y. E., B. P. Lucey, and D. M. Holtzman. 2014. Sleep and Alzheimer Disease Pathology: A Bidirectional Relationship. *Nat. Rev. Neurol.* **10(2)**:115–119. [7]

Julkunen, P., L. Säisänen, N. Danner, et al. 2009. Comparison of Navigated and Non-Navigated Transcranial Magnetic Stimulation for Motor Cortex Mapping, Motor Threshold and Motor Evoked Potentials. *NeuroImage* **44**:790–795. [10]

Jurado, S., D. Goswami, Y. Zhang, et al. 2014. LTP Requires a Unique Postsynaptic Snare Fusion Machinery. *Neuron* **77**:542–558. [10]

Kaale, A., M. W. Fagerland, E. W. Martinsen, and L. Smith. 2014. Preschool-Based Social Communication Treatment for Children with Autism: 12-Month Follow-up of a Randomized Trial. *J. Am. Acad. Child Adolesc. Psychiatry* **53(2)**:188–198. [4]

Kadoshima, T., H. Sakaguchi, T. Nakano, et al. 2013. Self-Organization of Axial Polarity, inside-out Layer Pattern, and Species-Specific Progenitor Dynamics in Human ES Cell-Derived Neocortex. *PNAS* **110(50)**:20284–20289. [13]

Kalthoff, D., C. Po, D. Wiedermann, and M. Hoehn. 2013. Reliability and Spatial Specificity of Rat Brain Sensorimotor Functional Connectivity Networks Are Superior under Sedation Compared with General Anesthesia. *NMR Biomed.* **26**:638–650. [9]

Karimi, M., L. Tian, C. A. Brown, et al. 2013. Validation of Nigrostriatal Positron Emission Tomography Measures: Critical Limits. *Ann. Neurol.* **73**:390–396. [9, 10]

Kawakubo, Y., S. Kamio, T. Nose, et al. 2007. Phonetic Mismatch Negativity Predicts Social Skills Acquisition in Schizophrenia. *Psychiatry Res.* **152**:261–265. [10]

Kawakubo, Y., and K. Kasai. 2006. Support for an Association between Mismatch Negativity and Social Functioning in Schizophrenia. *Prog. Neuropsychopharmacol. Biol. Psychiatry* **30**:1367–1368. [10]

Kawakubo, Y., K. Kasai, N. Kudo, et al. 2006. Phonetic Mismatch Negativity Predicts Verbal Memory Deficits in Schizophrenia. *Neuroreport* **17**:1043–1046. [10]

Keeler, J. F., and T. W. Robbins. 2011. Translating Cognition from Animals to Humans. *Biochem. Pharmacol.* **81**:1356–1366. [12]

Kelleher, R. J., III, and M. F. Bear. 2008. The Autistic Neuron: Troubled Translation? *Cell* **135(3)**:401–406. [2]

Kendell, R. E., J. E. Cooper, A. J. Gourlay, et al. 1971. Diagnostic Criteria of American and British Psychiatrists. *Arch. Gen. Psychiatry* **25(2)**:123–130. [4]

Kendler, K. S. 2006. Reflections on the Relationship between Psychiatric Genetics and Psychiatric Nosology. *Am. J. Psychiatry* **163(7)**:1138–1146. [3]

———. 2013. What Psychiatric Genetics Has Taught Us About the Nature of Psychiatric Illness and What Is Left to Learn. *Mol. Psychiatry* **18**:1058–1066. [8]

Kettleborough, R. N., E. M. Busch-Nentwich, S. A. Harvey, et al. 2013. A Systematic Genome-Wide Analysis of Zebrafish Protein-Coding Gene Function. *Nature* **496**:494–497. [8]

Kim, E. J., G. D. Rabinovici, W. W. Seeley, et al. 2007. Patterns of MRI Atrophy in Tau Positive and Ubiquitin Positive Frontotemporal Lobar Degeneration. *J. Neurol. Neurosurg. Psychiatry* **78**:1375–1378. [9]

Kim, J., J. M. Basak, and D. M. Holtzman. 2009. The Role of Apolipoprotein E in Alzheimer's Disease. *Neuron* **63(3)**:287–303. [7]

Kim, J., A. E. Eltorai, H. Jiang, et al. 2012. Anti-apoE Immunotherapy Inhibits Amyloid Accumulation in a Transgenic Mouse Model of Aß Amyloidosis. *J. Exp. Med.* **209(12)**:2149–2156. [7]

Kim, J., H. Jiang, S. Park, et al. 2011. Haploinsufficiency of Human APOE Reduces Amyloid Deposition in a Mouse Model of Amyloid-Beta Amyloidosis. *J. Neurosci.* **31(49)**:18007–18012. [7]

Kim, Y. S., and M. W. State. 2014. Recent Challenges to the Psychiatric Diagnostic Nosology: A Focus on the Genetics and Genomics of Neurodevelopmental Disorders. *Int. J. Epidemiol.* **43(2)**:465–475. [2]

Kirov, G., A. J. Pocklington, P. Holmans, et al. 2012. *De Novo* CNV Analysis Implicates Specific Abnormalities of Postsynaptic Signalling Complexes in the Pathogenesis of Schizophrenia. *Mol. Psychiatry* **17(2)**:142–153. [3, 4]

Kirov, G., E. Rees, J. T. Walters, et al. 2014. The Penetrance of Copy Number Variations for Schizophrenia and Developmental Delay. *Biol. Psychiatry* **75(5)**:378–385. [3, 13]

Kirov, G., D. Rujescu, A. Ingason, et al. 2009. Neurexin 1 (NRXN1) Deletions in Schizophrenia. *Schizophr. Bull.* **35(5)**:851–854. [3]

Klapoetke, N. C., Y. Murata, S. S. Kim, et al. 2014. Independent Optical Excitation of Distinct Neural Populations. *Nat. Methods* **11(3)**:338–346. [8, 10]

Klei, L., S. J. Sanders, M. T. Murtha, et al. 2012. Common Genetic Variants, Acting Additively, Are a Major Source of Risk for Autism. *Mol. Autism* **3(1)**:9. [2]

Kleim, J. A., S. Chan, E. Pringle, et al. 2006. BDNF Val66met Polymorphism Is Associated with Modified Experience-Dependent Plasticity in Human Motor Cortex. *Nat. Neurosci.* **9**:735–737. [10]

Klumpp, H., D. A. Fitzgerald, and K. L. Phan. 2013. Neural Predictors and Mechanisms of Cognitive Behavioral Therapy on Threat Processing in Social Anxiety Disorder. *Prog. Neuropsychopharmacol. Biol. Psychiatry* **45**:83–91. [10]

Klunk, W. E., H. Engler, A. Nordberg, et al. 2004. Imaging Brain Amyloid in Alzheimer's Disease with Pittsburgh Compound-B. *Ann. Neurol.* **55(3)**:306–319. [7]

Kneeland, R. E., and S. H. Fatemi. 2013. Viral Infection, Inflammation and Schizophrenia. *Prog. Neuropsychopharmacol. Biol. Psychiatry* **42**:35–48. [13]

Knox, A., A. Schneider, F. Abucayan, et al. 2012. Feasibility, Reliability, and Clinical Validity of the Test of Attentional Performance for Children (KiTAP) in Fragile X Syndrome (FXS). *J. Neurodev. Disord.* **4**:2. [10]

Koga, H., M. Martinez-Vicente, E. Arias, et al. 2011. Constitutive Upregulation of Chaperone-Mediated Autophagy in Huntington's Disease. *J. Neurosci.* **31(50)**:18492–18505. [5]

Koldamova, R., N. F. Fitz, and I. Lefterov. 2010. The Role of ATP-Binding Cassette Transporter A1 in Alzheimer's Disease and Neurodegeneration. *Biochim. Biophys. Acta* **1801(8)**:824–830. [7]

Komatsu, M., S. Waguri, T. Chiba, et al. 2006. Loss of Autophagy in the Central Nervous System Causes Neurodegeneration in Mice. *Nature* **441(7095)**:880–884. [5]

Kong, A., M. L. Frigge, G. Masson, et al. 2012. Rate of *de Novo* Mutations and the Importance of Father's Age to Disease Risk. *Nature* **488(7412)**:471–475. [2]

Kordower, J. H., C. W. Olanow, H. B. Dodiya, et al. 2013. Disease Duration and the Integrity of the Nigrostriatal System in Parkinson's Disease. *Brain* **136**:2419–2431. [10]

Krabbe, G., A. Halle, V. Matyash, et al. 2013. Functional Impairment of Microglia Coincides with Beta-Amyloid Deposition in Mice with Alzheimer-Like Pathology. *PLoS One* **8(4)**:e60921. [7]

Kravitz, A. V., B. S. Freeze, P. R. Parker, et al. 2010. Regulation of Parkinsonian Motor Behaviours by Optogenetic Control of Basal Ganglia Circuitry. *Nature* **466**:622–626. [9, 10]

Krezowski, J., D. Knudson, C. Ebeling, et al. 2004. Identification of Loci Determining Susceptibility to the Lethal Effects of Amyloid Precursor Protein Transgene Overexpression. *Hum. Mol. Genet.* **13(18)**:1989–1997. [6]

LaFerla, F. M., and K. N. Green. 2012. Animal Models of Alzheimer Disease. *Cold Spring Harb. Perspect. Med.* **2(11)** [13]

Lambert, J. C., C. A. Ibrahim-Verbaas, D. Harold, et al. 2013. Meta-Analysis of 74,046 Individuals Identifies 11 New Susceptibility Loci for Alzheimer's Disease. *Nat. Genet.* **45**:1452–1458. [1, 14]

Lancaster, M. A., M. Renner, C. A. Martin, et al. 2013. Cerebral Organoids Model Human Brain Development and Microcephaly. *Nature* **501(7467)**:373–379. [8, 11, 13, 14]

Lang, A. E., S. Gill, N. K. Patel, et al. 2006. Randomized Controlled Trial of Intraputamenal Glial Cell Line-Derived Neurotrophic Factor Infusion in Parkinson Disease. *Ann. Neurol.* **59(3)**:459–466. [7]

Lang, N., H. R. Siebner, D. Ernst, et al. 2004. Preconditioning with Transcranial Direct Current Stimulation Sensitizes the Motor Cortex to Rapid-Rate Transcranial Magnetic Stimulation and Controls the Direction of after-Effects. *Biol. Psychiatry* **56**:634–639. [10]

Laumonnier, F., F. Bonnet-Brilhault, M. Gomot, et al. 2004. X-Linked Mental Retardation and Autism Are Associated with a Mutation in the NLGN4 Gene, a Member of the Neuroligin Family. *Am. J. Hum. Genet.* **74(3)**:552–557. [4]

Lawrie, S. M., B. Olabi, J. Hall, and A. M. McIntosh. 2011. Do We Have Any Solid Evidence of Clinical Utility About the Pathophysiology of Schizophrenia? *World Psychiatry* **10(1)**:19–31. [3, 10]

Leblond, C. S., J. Heinrich, R. Delorme, et al. 2012. Genetic and Functional Analyses of SHANK2 Mutations Suggest a Multiple Hit Model of Autism Spectrum Disorders. *PLoS Genet.* **8(2)**:e1002521. [2, 4]

LeDoux, J. E. 2000. Emotion Circuits in the Brain. *Annu. Rev. Neurosci.* **23(1)**:155–184. [12]

LeDoux, J. E. 2014. Coming to Terms with Fear. *PNAS* **111**:2871–2878. [8]

Lee, B. H., M. J. Lee, S. Park, et al. 2010. Enhancement of Proteasome Activity by a Small-Molecule Inhibitor of USP14. *Nat. Chem. Biol.* **467(7312)**:179–184. [5]

Lee, H.-J., E.-J. Bae, and S.-J. Lee. 2014. Extracellular [Alpha]-Synuclein: A Novel and Crucial Factor in Lewy Body Diseases. *Nat. Rev. Neurol.* **10(2)**:92–98. [7]

Lee, H.-M., P. M. Giguere, and B. L. Roth. 2013. DREADDs: Novel Tools for Drug Discovery and Development. *Drug Discov. Today* **19**:469–473. [10]

Lee, K.-H., L. M. Williams, M. Breakspear, and E. Gordon. 2003. Synchronous Gamma Activity: A Review and Contribution to an Integrative Neuroscience Model of Schizophrenia. *Brain Res. Rev.* **41**:57–78. [10]

Lee, S. H., T. R. DeCandia, S. Ripke, et al. 2012. Estimating the Proportion of Variation in Susceptibility to Schizophrenia Captured by Common SNPs. *Nat. Genet.* **44(3)**:247–250. [3]

Lemon, R. N. 2008. Descending Pathways in Motor Control. *Annu. Rev. Neurosci.* **31**:195–218. [13]

Lenin, V. I. 1917/1990. Што Делать [What Is to Be Done?], translated by J. Fineberg and G. Hanna. London: Penguin Classics. [1]

Leonhard, K., H. Beckmann, and C. H. Cahn, eds. 1999. Classification of Endogenous Psychoses and Their Differentiated Etiology. Heidelberg: Springer. [3]

Liang, G., and Y. Zhang. 2013. Genetic and Epigenetic Variations in iPSCs: Potential Causes and Implications for Application. *Cell Stem Cell* **13**:149–159. [11]

Lichtenstein, P., B. H. Yip, C. Björk, et al. 2009. Common Genetic Determinants of Schizophrenia and Bipolar Disorder in Swedish Families: A Population-Based Study. *Lancet* **373(9659)**:234–239. [3]

Light, G. A., and D. L. Braff. 2005. Stability of Mismatch Negativity Deficits and Their Relationship to Functional Impairments in Chronic Schizophrenia. *Am. J. Psychiatry* **162**:1741–1743. [10]

Light, G. A., and R. Näätänen. 2013. Mismatch Negativity Is a Breakthrough Biomarker for Understanding and Treating Psychotic Disorders. *PNAS* **110**:15175–15176. [10]

Lim, A. L., D. A. Taylor, and D. T. Malone. 2012. A Two-Hit Model: Behavioural Investigation of the Effect of Combined Neonatal MK-801 Administration and Isolation Rearing in the Rat. *J. Psychopharmacol.* **26(9)**:1252–1264. [13]

Lim, E. T., S. Raychaudhuri, S. J. Sanders, et al. 2013. Rare Complete Knockouts in Humans: Population Distribution and Significant Role in Autism Spectrum Disorders. *Neuron* **77(2)**:235–242. [2]

Lima, S. Q., and G. Miesenböck. 2005. Remote Control of Behavior through Genetically Targeted Photostimulation of Neurons. *Cell* **121**:141–152. [10]

Lin, J. Y., M. Z. Lin, P. Steinbach, and R. Y. Tsien. 2009. Characterization of Engineered Channelrhodopsin Variants with Improved Properties and Kinetics. *Biophys. J.* **96**:1803–1814. [10]

Ling, S. C., M. Polymenidou, and D. W. Cleveland. 2013. Converging Mechanisms in ALS and FTD: Disrupted RNA and Protein Homeostasis. *Neuron* **79(3)**:416–438. [5, 6]

Lipska, B. K., and D. R. Weinberger. 2000. To Model a Psychiatric Disorder in Animals: Schizophrenia as a Reality Test. *Neuropsychopharmacology* **23(3)**:223–239. [13]

Lista, S., F. G. Garaci, M. Ewers, et al. 2014. CSF Abeta1-42 Combined with Neuroimaging Biomarkers in the Early Detection, Diagnosis and Prediction of Alzheimer's Disease. *Alzheimers Dement.* **10(3)**:381–392. [13]

Liu, G. H., J. Qu, K. Suzuki, et al. 2012. Progressive Degeneration of Human Neural Stem Cells Caused by Pathogenic LRRK2. *Nature* **491**:603–607. [11]

Liu, H., Y. Chen, Y. Niu, et al. 2014. TALEN-Mediated Gene Mutagenesis in Rhesus and Cynomolgus Monkeys. *Cell Stem Cell* **14**:323–328. [8]

Lobb, C. J., A. K. Zaheer, Y. Smith, and D. Jaeger. 2013. *In Vivo* Electrophysiology of Nigral and Thalamic Neurons in Alpha-Synuclein-Overexpressing Mice Highlights Differences from Toxin-Based Models of Parkinsonism. *J. Neurophysiol.* **110**:2792–2805. [9]

Luo, L., E. Callaway, and K. Svoboda. 2008. Genetic Dissection of Neural Circuits. *Neuron* **57**:634–660. [10]

Lüscher, C., and R. C. Malenka. 2011. Drug-Evoked Synaptic Plasticity in Addiction: From Molecular Changes to Circuit Remodeling. *Neuron* **69**:650–663. [10]

Lustig, C., R. Kozak, M. Sarter, J. W. Young, and T. W. Robbins. 2013. CNTRICS Final Animal Model Task Selection: Control of Attention. *Neurosci. Biobehav. Res.* **37**:2099–2110. [12]

Ma, Y., S. Peng, P. G. Spetsieris, et al. 2012. Abnormal Metabolic Brain Networks in a Nonhuman Primate Model of Parkinsonism. *J. Cereb. Blood Flow Metab.* **32**:633–642. [9]

Mackintosh, N. J. 1983. Conditioning and Associative Learning. Oxford: Oxford Univ. Press. [12]

Mahley, R. W., and Y. Huang. 2012. Apolipoprotein E Sets the Stage: Response to Injury Triggers Neuropathology. *Neuron* **76(5)**:871–885. [7]

Malaspina, D., M. J. Owen, S. Heckers, et al. 2013. Schizoaffective Disorder in the DSM-5. *Schizophr. Res.* **150(1)**:21–25. [3]

Maldonado, T. A., R. E. Jones, and D. O. Norris. 2002. Intraneuronal Amyloid Precursor Protein (APP) and Appearance of Extracellular Beta-Amyloid Peptide (Abeta) in the Brain of Aging Kokanee Salmon. *J. Neurobiol.* **53(1)**:11–20. [6]

Malhotra, D., and J. Sebat. 2012. CNVs: Harbingers of a Rare Variant Revolution in Psychiatric Genetics. *Cell* **148(6)**:1223–1241. [3]

Mali, P., K. M. Esvelt, and G. M. Church. 2013a. Cas9 as a Versatile Tool for Engineering Biology. *Nat. Methods* **19**:957–963. [8]

Mali, P., L. Yang, K. M. Esvelt, et al. 2013b. RNA-Guided Human Genome Engineering via Cas9. *Science* **339**:823–826. [8]

Manabe, T., D. J. Wyllie, D. J. Perkel, and R. A. Nicoll. 1993. Modulation of Synaptic Transmission and Long-Term Potentiation: Effects on Paired Pulse Facilitation and EPSC Variance in the CA1 Region of the Hippocampus. *J. Neurophysiol.* **70**:1451–1459. [10]

Mandelkow, E. M., and E. Mandelkow. 2012. Biochemistry and Cell Biology of Tau Protein in Neurofibrillary Degeneration. *Cold Spring Harb. Perspect. Med.* **2(7)**:a006247. [7]

Manly, J. J. 2008. Critical Issues in Cultural Neuropsychology: Profit from Diversity. *Neuropsychol. Rev.* **18(3)**:179–183. [2]

Mantini, D., A. Gerits, K. Nelissen, et al. 2011. Default Mode of Brain Function in Monkeys. *J. Neurosci.* **31**:12954–12962. [9]

Marchetto, M. C., C. Carromeu, A. Acab, et al. 2010. A Model for Neural Development and Treatment of Rett Syndrome Using Human Induced Pluripotent Stem Cells. *Cell* **143**:527–539. [11]

Marie, H., and R. C. Malenka. 2006. Acute *in Vivo* Expression of Recombinant Proteins in Rat Brain Using Sindbis Virus. In: The Dynamic Synapse: Molecular Methods in Ionotropic Receptor Biology, ed. J. T. Kittler and S. J. Moss. Boca Raton: CRC Press. [10]

Marshall, M., and J. Rathbone. 2011. Early Intervention for Psychosis. *Cochrane Database Syst. Rev.* **6**:CD004718. [4]

Martin-Ordas, G., D. Haun, and J. Call. 2010. Keeping Track of Time: Evidence for Episodic-Like Memory in Great Apes. *Anim. Cogn.* **13**:331–340. [12]

Maruyama, M., H. Shimada, T. Suhara, et al. 2013. Imaging of Tau Pathology in a Tauopathy Mouse Model and in Alzheimer Patients Compared to Normal Controls. *Neuron* **79(6)**:1094–1108. [7]

Masliah, E., E. Rockenstein, I. Veinbergs, et al. 2001. Beta-Amyloid Peptides Enhance Alpha-Synuclein Accumulation and Neuronal Deficits in a Transgenic Mouse Model Linking Alzheimer's Disease and Parkinson's Disease. *PNAS* **98(21)**:12245–12250. [7]

Maxwell, C. R., Y. Liang, B. D. Weightman, et al. 2004. Effects of Chronic Olanzapine and Haloperidol Differ on the Mouse N1 Auditory Evoked Potential. *Neuropsychopharmacology* **29**:739–746. [10]

Mazur, J. E., and R. J. Herrnstein. 1988. On the Functions Relating Delay, Reinforcer Value, and Behavior. *Behav. Brain Sci.* **11**:690–691. [12]

McCarroll, S. A., G. Feng, and S. E. Hyman. 2014. Genome-Scale Neurogenetics: Methodology and Meaning. *Nat. Neurosci.* **17(6)**:756–763. [2, 8]

McCarroll, S. A., and S. E. Hyman. 2013. Progress in Genetics of Polygenic Brain Disorders: Significant New Challenges for Neurobiology. *Neuron* **80**:578–587. [8, 14]

McClintock, S. M., C. Freitas, L. Oberman, S. H. Lisanby, and A. Pascual-Leone. 2011. Transcranial Magnetic Stimulation: A Neuroscientific Probe of Cortical Function in Schizophrenia. *Biol. Psychiatry* **70(1)**:19–27. [13]

McClure, S. M., D. I. Laibson, G. Loewenstein, and J. D. Cohen. 2004. Separate Neural Systems Value Immediate and Delayed Monetary Rewards. *Science* **306**:503–507. [12]

McConnell, E. R., M. A. McClain, J. Ross, W. R. Lefew, and T. J. Shafer. 2012. Evaluation of Multi-Well Microelectrode Arrays for Neurotoxicity Screening Using a Chemical Training Set. *Neurotoxicology* **33**:1048–1057. [8]

McFarland, H. F., F. Barkhof, J. Antel, and D. H. Miller. 2002. The Role of MRI as a Surrogate Outcome Measure in Multiple Sclerosis. *Mult. Scler.* **8(1)**:40–51. [6]

McGaugh, J. L., and B. Roozendaal. 2008. Drug Enhancement of Memory Consolidation: Historical Perspective and Neurobiological Implications. *Psychopharmacology* **202**:3–14. [12]

McGoldrick, P., P. I. Joyce, E. M. Fisher, and L. Greensmith. 2013. Rodent Models of Amyotrophic Lateral Sclerosis. *Biochim. Biophys. Acta* **1832(9)**:1421–1436. [6]

McKhann, G. M., D. S. Knopman, H. Chertkow, et al. 2011. The Diagnosis of Dementia Due to Alzheimer's Disease: Recommendations from the National Institute on Aging-Alzheimer's Association Workgroups on Diagnostic Guidelines for Alzheimer's Disease. *Alzheimers Dement.* **7(3)**:263–269. [6]

McTeague, L. M., and P. J. Lang. 2012. The Anxiety Spectrum and the Reflex Physiology of Defense: From Circumscribed Fear to Broad Distress. *Depress. Anxiety* **29(4)**:264–281. [4]

Medland, S. E., N. Jahanshad, B. M. Neale, and P. M. Thompson. 2014. Whole-Genome Analyses of Whole-Brain Data: Working within an Expanded Search Space. *Nat. Neurosci.* **17(6)**:791–800. [2]

Mehta, D., T. Klengel, K. N. Conneely, et al. 2013. Childhood Maltreatment Is Associated with Distinct Genomic and Epigenetic Profiles in Posttraumatic Stress Disorder. *PNAS* **110(20)**:8302–8307. [4]

Mendell, J. T., and E. N. Olson. 2012. Micrornas in Stress Signaling and Human Disease. *Cell* **148(6)**:1172–1187. [7]

Merkle, F. T., and K. Eggan. 2013. Modeling Human Disease with Pluripotent Stem Cells: From Genome Association to Function. *Cell Stem Cell* **12**:656–668. [11]

Meyer-Luehmann, M., T. L. Spires-Jones, C. Prada, et al. 2008. Rapid Appearance and Local Toxicity of Amyloid-Beta Plaques in a Mouse Model of Alzheimer's Disease. *Nature* **451(7179)**:720–724. [7]

Michalon, A., A. Bruns, C. Risterucci, et al. 2014. Chronic Metabotropic Glutamate Receptor 5 Inhibition Corrects Local Alterations of Brain Activity and Improves Cognitive Performance in Fragile X Mice. *Biol. Psychiatry* **75(3)**:189–197. [13]

Michie, P. T. 2001. What Has MMN Revealed About the Auditory System in Schizophrenia? *Int. J. Psychophysiol.* **42**:177–194. [10]

Michie, P. T., T. W. Budd, J. Todd, et al. 2000. Duration and Frequency Mismatch Negativity in Schizophrenia. *Clin. Neurophysiol.* **111**:1054–1065. [10]

Miller, C. T., K. Mandel, and X. Wang. 2010. The Communicative Content of the Common Marmoset Phee Call During Antiphonal Calling. *Am. J. Primatol.* **72**:974–980. [8]

Miller, J. D., Y. M. Ganat, S. Kishinevsky, et al. 2013a. Human iPSC-Based Modeling of Late-Onset Disease via Progerin-Induced Aging. *Cell Stem Cell* **13**:691–705. [11, 13]

Miller, T. M., A. Pestronk, W. David, et al. 2013b. An Antisense Oligonucleotide against SOD1 Delivered Intrathecally for Patients with SOD1 Familial Amyotrophic Lateral Sclerosis: A Phase 1, Randomised, First-in-Man Study. *Lancet Neurol.* **12(5)**:435–442. [7]

Milton, A. L., and B. J. Everitt. 2012. The Psychological and Neurochemical Mechanisms of Drug Memory Consolidation: Implications for the Treatment of Addiction. *Eur. J. Neurosci.* **31**:2308–2319. [12]

Ming, G. L., O. Brustle, A. Muotri, et al. 2011. Cellular Reprogramming: Recent Advances in Modeling Neurological Diseases. *J. Neurosci.* **31(45)**:16070–16075. [13]

Mingozzi, F., and K. A. High. 2011. Therapeutic *in Vivo* Gene Transfer for Genetic Disease Using AAV: Progress and Challenges. *Nat. Rev. Genet.* **12**:341–355. [10]

Mink, J. W. 1996. The Basal Ganglia: Focused Selection and Inhibition of Competing Motor Programs. *Prog. Neurobiol.* **50**:381–425. [9]

Miocinovic, S., S. Somayajula, S. Chitnis, and J. L. Vitek. 2013. History, Applications, and Mechanisms of Deep Brain Stimulation. *JAMA Neurol.* **70**:163–171. [9]

Mitsui, K., Y. Tokuzawa, H. Itoh, et al. 2003. The Homeoprotein Nanog Is Required for Maintenance of Pluripotency in Mouse Epiblast and ES Cells. *Cell* **113(5)**:631–642. [13]

Moghaddam, B. 2013. A Mechanistic Approach to Preventing Schizophrenia in At-Risk Individuals. *Neuron* **78(1)**:1–3. [13]

Moore, H., J. D. Jentsch, M. Ghajarnia, M. A. Geyer, and A. A. Grace. 2006. A Neurobehavioral Systems Analysis of Adult Rats Exposed to Methylazoxymethanol Acetate on E17: Implications for the Neuropathology of Schizophrenia. *Biol. Psychiatry* **60(3)**:253–264. [13]

Moreno-De-Luca, A., S. M. Myers, T. D. Challman, et al. 2013. Developmental Brain Dysfunction: Revival and Expansion of Old Concepts Based on New Genetic Evidence. *Lancet Neurol.* **12(4)**:406–414. [2, 4]

Morfini, G. A., M. Burns, L. I. Binder, et al. 2009. Axonal Transport Defects in Neurodegenerative Diseases. *J. Neurosci.* **29(41)**:12776–12786. [5]

Morris, K. V., and J. S. Mattick. 2014. The Rise of Regulatory RNA. *Nat. Rev. Genet.* **15(6)**:423–437. [10]

Morris, R. G. M., P. Garrud, J. N. P. Rawlins, and J. O'Keefe. 1982. Place Navigation Impaired in Rats with Hippocampal Lesions. *Nature* **297**:681–683. [12]

Morrow, E. M., S. Y. Yoo, S. W. Flavell, et al. 2008. Identifying Autism Loci and Genes by Tracing Recent Shared Ancestry. *Science* **321(5886)**:218–223. [2]

Moskvina, V., N. Craddock, P. Holmans, et al. 2009. Gene-Wide Analyses of Genome-Wide Association Data Sets: Evidence for Multiple Common Risk Alleles for Schizophrenia and Bipolar Disorder and for Overlap in Genetic Risk. *Mol. Psychiatry* **14(3)**:252–260. [3]

Mottron, L. 2011. Changing Perceptions: The Power of Autism. *Nature* **479(7371)**:33–35. [2]

Moulder, K. L., B. J. Snider, S. L. Mills, et al. 2013. Dominantly Inherited Alzheimer Network: Facilitating Research and Clinical Trials. *Alzheimers Res. Ther.* **5(5)**:48. [7]

Moussavi, S., S. Chatterji, E. Verdes, et al. 2007. Depression, Chronic Diseases, and Decrements in Health: Results from the World Health Surveys. *Lancet* **370**:851–858. [1]

Moy, S. S., J. J. Nadler, A. Perez, et al. 2004. Sociability and Preference for Social Novelty in Five Inbred Strains: An Approach to Assess Autistic-Like Behavior in Mice. *Genes Brain Behav.* **3**:287–302. [12]

Murphy, F. C., B. J. Sahakian, J. S. Rubinsztein, et al. 1999. Emotional Bias and Inhibitory Control Processes in Mania and Depression. *Psychol. Med.* **29**:1307–1321. [12]

Murray, C. J., T. Vos, R. Lozano, et al. 2012. Disability-Adjusted Life Years (Dalys) for 291 Diseases and Injuries in 21 Regions, 1990–2010: A Systematic Analysis for the Global Burden of Disease Study 2010. *Lancet* **380**:2163–2196. [1]

Murray, E. A., T. J. Bussey, and L. M. Saksida. 2007. Visual Perception and Memory: A New View of Medial Temporal Lobe Function in Primates and Rodents. *Annu. Rev. Neurosci.* **30**:99–122. [12]

Murray, R. M., and S. W. Lewis. 1987. Is Schizophrenia a Neurodevelopmental Disorder? *Br. Med. J. (Clin. Res. Ed.)* **295(6600)**:681–682. [3, 13]

Musiek, E. S., and D. M. Holtzman. 2012. Origins of Alzheimer's Disease: Reconciling Cerebrospinal Fluid Biomarker and Neuropathology Data Regarding the Temporal Sequence of Amyloid-Beta and Tau Involvement. *Curr. Opin. Neurobiol.* **25(6)**:715–720. [7]

Näätänen, R. 2008. Mismatch Negativity (MMN) as an Index of Central Auditory System Plasticity. *Int. J. Audiol.* **47(Suppl 2)**:S16–20. [10]

Näätänen, R., and S. Kähkönen. 2009. Central Auditory Dysfunction in Schizophrenia as Revealed by the Mismatch Negativity (MMN) and Its Magnetic Equivalent Mmnm: A Review. *Int. J. Neuropsychopharmacol.* **12**:125–135. [10]

Nader, K., and O. Hardt. 2009. A Single Standard for Memory: The Case for Reconsolidation. *Nat. Rev. Neurosci.* **10(3)**:224–234. [12]

Nader, K., G. E. Schafe, and J. E. L. Doux. 2000. Fear Memories Require Protein Synthesis in the Amygdala for Reconsolidation after Retrieval. *Nature* **406**:722–726. [12]

Nagai, M., D. B. Re, T. Nagata, et al. 2007. Astrocytes Expressing ALS-Linked Mutated SOD1 Release Factors Selectively Toxic to Motor Neurons. *Nat. Neurosci.* **10**:615–622. [11]

Nagai, T., M. Tada, K. Kirihara, et al. 2013. Mismatch Negativity as a "Translatable" Brain Marker toward Early Intervention for Psychosis: A Review. *Front. Psychiatry* **4**:115. [10]

Naj, A. C., G. Jun, G. W. Beecham, et al. 2011. Common Variants at MS4A4/MS4A6E, CD2AP, CD33 and EPHA1 Are Associated with Late-Onset Alzheimer's Disease. *Nat. Genet.* **43(5)**:436–441. [7]

Nakano, T., S. Ando, N. Takata, et al. 2012. Self-Formation of Optic Cups and Storable Stratified Neural Retina from Human ESCs. *Cell Stem Cell* **10(6)**:771–785. [13]

Neale, B. M., Y. Kou, L. Liu, et al. 2012. Patterns and Rates of Exonic *de Novo* Mutations in Autism Spectrum Disorders. *Nature* **485(7397)**:242–245. [2]

Nelson, E. E., and J. T. Winslow. 2009. Non-Human Primates: Model Animals for Developmental Psychopathology. *Neuropsychopharmacology* **34(1)**:90–105. [13]

Nestler, E. J., E. Gould, H. Manji, et al. 2002. Preclinical Models: Status of Basic Research in Depression. *Biol. Psychiatry* **52**:503–528. [12]

Neve, R. L., K. A. Neve, E. J. Nestler, and W. A. Carlezon. 2005. Use of Herpes Virus Amplicon Vectors to Study Brain Disorders. *Biotechniques* **39**:381–391. [10]

Nguyen, H. N., B. Byers, B. Cord, et al. 2011. LRRK2 Mutant iPSC-Derived DA Neurons Demonstrate Increased Susceptibility to Oxidative Stress. *Cell Stem Cell* **8**:267–280. [11]

Nielsen, E. B., M. Lyon, and G. Ellison. 1983. Apparent Hallucinations in Monkeys During around-the Clock Amphetamine for Seven to Fourteen Days: Possible Relevance to Amphetamine Psychosis. *J. Nerv. Ment. Disord.* **171(4)**:222–233. [12]

Niewoehner, J., B. Bohrmann, L. Collin, et al. 2014. Increased Brain Penetration and Potency of a Therapeutic Antibody Using a Monovalent Molecular Shuttle. *Neuron* **81(1)**:49–60. [7]

Nishida, Y., S. Arakawa, K. Fujitani, et al. 2009. Discovery of Atg5/Atg7-Independent Alternative Macroautophagy. *Nat. Chem. Biol.* **461(7264)**:654–658. [5]

Nithianantharajah, J., N. H. Komiyama, A. McKechanie, et al. 2013. Synaptic Scaffold Evolution Generated Components of Vertebrate Cognitive Complexity. *Nat. Neurosci.* **16**:16–24. [12]

Nitsche, M. A., F. Müller-Dahlhaus, W. Paulus, and U. Ziemann. 2012. The Pharmacology of Neuroplasticity Induced by Non-Invasive Brain Stimulation: Building Models for the Clinical Use of CNS Active Drugs. *J. Physiol.* **590**:4641–4662. [10]

Nitsche, M. A., A. Roth, M.-F. Kuo, et al. 2007. Timing-Dependent Modulation of Associative Plasticity by General Network Excitability in the Human Motor Cortex. *J. Neurosci.* **27**:3807–3812. [10]

Niu, Y., B. Shen, Y. Cui, et al. 2014. Generation of Gene-Modified Cynomolgus Monkey via Cas9/RNA-Mediated Gene Targeting in One-Cell Embryos. *Cell* **156(4)**:836–843. [6, 8, 13]

Novarino, G., P. El-Fishawy, H. Kayserili, et al. 2012. Mutations in BCKD-Kinase Lead to a Potentially Treatable Form of Autism with Epilepsy. *Science* **338(6105)**:394–397. [2]

Obianyo, O., and K. Ye. 2013. Novel Small Molecule Activators of the Trk Family of Receptor Tyrosine Kinases. *Biochim. Biophys. Acta* **1834(10)**:2213–2218. [7]

O'Donovan, M. C., N. Craddock, N. Norton, et al. 2008. Identification of Loci Associated with Schizophrenia by Genome-Wide Association and Follow-Up. *Nat. Genet.* **40(9)**:1053–1055. [3]

Okano, H., K. Hikishima, A. Iriki, and E. Sasaki. 2012. The Common Marmoset as a Novel Animal Model System for Biomedical and Neuroscience Research Applications. *Semin. Fetal Neonatal Med.* **17**:336–340. [8]

Olanow, C. W., O. Rascol, R. Hauser, et al. 2009. A Double-Blind, Delayed-Start Trial of Rasagiline in Parkinson's Disease. *N. Engl. J. Med.* **361(13)**:1268–1278. [6]

Olton, D. S., and B. C. Paras. 1979. Spatial Memory and Hippocampal Function. *Neuropsychologia* **17**:669–682. [12]

Orenstein, S. J., S. H. Kuo, I. Tasset, et al. 2013. Interplay of LRRK2 with Chaperone-Mediated Autophagy. *Nat. Neurosci.* **16(4)**:394–406. [5]

O'Roak, B. J., P. Deriziotis, C. Lee, et al. 2011. Exome Sequencing in Sporadic Autism Spectrum Disorders Identifies Severe *de Novo* Mutations. *Nat. Genet.* **43(6)**:585–589. [2]

O'Roak, B. J., L. Vives, W. Fu, et al. 2012a. Multiplex Targeted Sequencing Identifies Recurrently Mutated Genes in Autism Spectrum Disorders. *Science* **338(6114)**:1619–1622. [2]

O'Roak, B. J., L. Vives, S. Girirajan, et al. 2012b. Sporadic Autism Exomes Reveal a Highly Interconnected Protein Network of *de Novo* Mutations. *Nature* **485(7397)**:246–250. [2]

Ostergard, T., and J. P. Miller. 2014. Deep Brain Stimulation: New Directions. *J. Neurosurg. Sci.* **58(4)**:191–198. [14]

Oswal, A., P. Brown, and V. Litvak. 2013. Synchronized Neural Oscillations and the Pathophysiology of Parkinson's Disease. *Curr. Opin. Neurol.* **26**:662–670. [9]

Owen, M. J. 2011. Is There a Schizophrenia to Diagnose? *World Psychiatry* **10(1)**:34–35. [3]

———. 2012a. Implications of Genetic Findings for Understanding Schizophrenia. *Schizophr. Bull.* **38**:904–907. [3, 8]

———. 2012b. Intellectual Disability and Major Psychiatric Disorders: A Continuum of Neurodevelopmental Causality. *Br. J. Psychiatry* **200**:268–269. [3]

Owen, M. J., M. C. O'Donovan, A. Thapar, and N. Craddock. 2011. Neurodevelopmental Hypothesis of Schizophrenia. *Br. J. Psychiatry* **198(3)**:173–175. [3]

Ozden, I., J. Wang, Y. Lu, et al. 2013. A Coaxial Optrode as Multifunction Write-Read Probe for Optogenetic Studies in Non-Human Primates. *J. Neurosci. Meth.*:142–154. [9]

Ozerdema, A., B. Güntekind, M. I. Atagüne, and E. Başar. 2013. Brain Oscillations in Bipolar Disorder in Search of New Biomarkers. *Suppl. Clin. Neurophysiol.* **62**:207–221. [10]

Pankevich, D. E., B. M. Altevogt, J. Dunlop, F. H. Gage, and S. E. Hyman. 2014. Improving and Accelerating Drug Development for Nervous System Disorders. *Neuron* **84**:546–553. [1, 14]

Paoletti, P., C. Bellone, and Q. Zhou. 2013. NMDA Receptor Subunit Diversity: Impact on Receptor Properties, Synaptic Plasticity and Disease. *Nat. Rev. Neurosci.* **14**:383–400. [10]

Parikshak, N. N., R. Luo, A. Zhang, et al. 2013. Integrative Functional Genomic Analyses Implicate Specific Molecular Pathways and Circuits in Autism. *Cell* **155(5)**:1008–1021. [2, 10]

Park, C., and A. M. Cuervo. 2013. Selective Autophagy: Talking with the UPS. *Cell Biochem. Biophys.* **67(1)**:3–13. [5]

Park, H. J., C. H. Kim, E. S. Park, et al. 2013. Increased GABA-a Receptor Binding and Reduced Connectivity at the Motor Cortex in Children with Hemiplegic Cerebral Palsy: A Multimodal Investigation Using 18F-Fluoroflumazenil PET, Immunohistochemistry, and MR Imaging. *J. Nucl. Med.* **54**:1263–1269. [9]

Parkinson Study Group. 1989. Effect of Deprenyl on the Progression of Disability in Early Parkinson's Disease. *N. Engl. J. Med.* **321(20)**:1364–1371. [6]

———. 1993. Effects of Tocopherol and Deprenyl on the Progression of Disability in Early Parkinson's Disease. *N. Engl. J. Med.* **328(3)**:176–183. [6]

Pasamanick, B., M. E. Rogers, and A. M. Lilienfeld. 1956. Pregnancy Experience and the Development of Behavior Disorders in Children. *Am. J. Psychiatry* **112**:613–618. [3]

Paterson, N. E., and A. Markou. 2007. Animal Models and Treatments for Addiction and Deprssion Co-Morbidity. *Neurotox. Res.* **11**:1–32. [12]

Patil, S. T., L. Zhang, F. Martenyi, et al. 2007. Activation of mGlu2/3 Receptors as a New Approach to Treat Schizophrenia: A Randomized Phase 2 Clinical Trial. *Nat. Med.* **13(9)**:1102–1107. [13]

Paul, B. D., J. I. Sbodio, R. Xu, et al. 2014. Cystathionine γ-Lyase Deficiency Mediates Neurodegeneration in Huntington's Disease. *Nature* **508**:96–100. [10]

Paulsen, J. S., J. D. Long, C. A. Ross, et al. 2014. Prediction of Manifest Huntington's Disease with Clinical and Imaging Measures: A Prospective Observational Study. *Lancet Neurol.* **13(12)**:1193–1201. [5]

Peca, J., C. Feliciano, J. T. Ting, et al. 2011. Shank3 Mutant Mice Display Autistic-Like Behaviours and Striatal Dysfunction. *Nature* **472(7344)**:437–442. [2, 4]

Penn, D. C., and D. J. Povinelli. 2007. On the Lack of Evidence That Non-Human Animals Possess Anything Remotely Resembling a "Theory of Mind." *Philos. Trans. R. Soc. Lond.* **362**:731–744. [12]

Perkins, D. O. 2006. Review: Longer Duration of Untreated Psychosis Is Associated with Worse Outcome in People with First Episode Psychosis. *Evid. Based Ment. Health* **9(2)**:36. [4]

Perou, R., R. H. Bitsko, S. J. Blumberg, et al. 2013. Mental Health Surveillance among Children: United States, 2005–2011. *MMWR Surveill. Summ.* **62(Suppl 2)**:1–35. [2]

Perrat, P. N., S. DasGupta, J. Wang, et al. 2013. Transposition-Driven Genomic Heterogeneity in the *Drosophila* Brain. *Science* **340(6128)**:91–95. [5]

Perry, D. C., and B. L. Miller. 2013. Frontotemporal Dementia. *Semin. Neurol.* **33(4)**:336–341. [7]

Petersen, R. C., R. G. Thomas, M. Grundman, et al. 2005. Vitamin E and Donepezil for the Treatment of Mild Cognitive Impairment. *N. Engl. J. Med.* **352(23)**::2379–2388. [6]

Pinto, D., A. T. Pagnamenta, L. Klei, et al. 2010. Functional Impact of Global Rare Copy Number Variation in Autism Spectrum Disorders. *Nature* **466(7304)**:368–372. [2]

Poduri, A., G. D. Evrony, X. Cai, and C. A. Walsh. 2013. Somatic Mutation, Genomic Variation, and Neurological Disease. *Science* **341(6141)**:1237758. [2]

Poline, J. B., J. L. Breeze, S. Ghosh, et al. 2012. Data Sharing in Neuroimaging Research. *Front. Neuroinform.* **6**:9. [2]

Polymeropoulos, M. H. 2000. Genetics of Parkinson's Disease. *Ann. NY Acad. Sci.* **920**:28–32. [13]

Pop, A. S., J. Levenga, C. E. de Esch, et al. 2014. Rescue of Dendritic Spine Phenotype in Fmr1 KO Mice with the mGluR5 Antagonist AFQ056/Mavoglurant. *Psychopharmacology* **231(6)**:1227–1235. [13]

Posner, M. I., and S. E. Petersen. 1990. The Attention System of the Human Brain. *Annu. Rev. Neurosci.* **13**:25–42. [12]

Posporelis, S., A. Sawa, G. S. Smith, et al. 2014. Promoting Careers in Academic Research to Psychiatry Residents. *Acad. Psychiatry* **38(2)**:185–190. [13]

Pouladi, M. A., A. J. Morton, and M. R. Hayden. 2013. Choosing an Animal Model for the Study of Huntington's Disease. *Nat. Rev. Neurosci.* **14(10)**:708–721. [6]

Pouladi, M. A., L. M. Stanek, Y. Xie, et al. 2012. Marked Differences in Neurochemistry and Aggregates Despite Similar Behavioural and Neuropathological Features of Huntington Disease in the Full-Length Bachd and Yac128 Mice. *Hum. Mol. Genet.* **21**:2219–2232. [9]

Preuss, T. M. 1995. Do Rats Have Prefrontal Cortex? The Rose-Woolsey-Akert Program Reconsidered. *J. Cogn. Neurosci.* **7**:1–24. [8]

Pringle, A., M. Browning, P. J. Cowen, and C. J. Harmer. 2011. A Cognitive Neuropsychological Model of Antidepressant Drug Action. *Prog. Neuropsychopharmacol. Biol. Psychiatry* **35**:1586–1592. [12]

Prinz, F., T. Schlange, and K. Asadullah. 2011a. Believe It or Not: How Much Can We Rely on Published Data on Potential Drug Targets? *Nat. Rev. Drug Discov.* **10(9)**:712. [6]

Prinz, M., J. Priller, S. S. Sisodia, and R. M. Ransohoff. 2011b. Heterogeneity of CNS Myeloid Cells and Their Roles in Neurodegeneration. *Nat. Neurosci.* **14(10)**:1227–1235. [7]

Purcell, S. M., J. L. Moran, M. Fromer, et al. 2014. A Polygenic Burden of Rare Disruptive Mutations in Schizophrenia. *Nature* **506(7487)**:185–190. [3, 4]

Raby, C. R., D. M. Alexis, A. Dickinson, and N. S. Clayton. 2007. Planning for the Future by Western Scrub-Jays. *Nature* **445**:919–921. [10]

Racette, B. A., L. Good, J. A. Antenor, et al. 2006. [18F]FDOPA PET as an Endophenotype for Parkinson's Disease Linkage Studies. *Am. J. Med. Genet. B Neuropsychiatr. Genet.* **141**:245–249. [9]

Rafii, M. S., T. L. Baumann, R. A. Bakay, et al. 2014. A Phase1 Study of Stereotactic Gene Delivery of AAV2-NGF for Alzheimer's Disease. *Alzheimers Dement.* [7]

Ramaswami, M., J. P. Taylor, and R. Parker. 2013. Altered Ribostasis: RNA-Protein Granules in Degenerative Disorders. *Cell* **154(4)**:727–736. [5]

Ramocki, M. B., and H. Y. Zoghbi. 2008. Failure of Neuronal Homeostasis Results in Common Neuropsychiatric Phenotypes. *Nature* **455(7215)**:912–918. [2]

Ran, F. A., P. D. Hsu, J. Wright, et al. 2013. Genome Engineering Using the CRISPR-Cas9 System. *Nat. Protoc.* **8**:2281–2308. [10]

Ravina, B., D. Eidelberg, J. E. Ahlskog, et al. 2005. The Role of Radiotracer Imaging in Parkinson Disease. *Neurology* **64**:208–215. [9]

Reardon, S. 2014. NIH Rethinks Psychiatry Trials. *Nature* **507(7492)**:288. [4]
Rees, E., J. T. Walters, L. Georgieva, et al. 2014. Analysis of Copy Number Variations at 15 Schizophrenia-Associated Loci. *Br. J. Psychiatry* **204(2)**:108–114. [3]
Regier, D. A., W. E. Narrow, D. E. Clarke, et al. 2013. DSM-5 Field Trials in the United States and Canada, Part II: Test-Retest Reliability of Selected Categorical Diagnoses. *Am. J. Psychiatry* **170(1)**:59–70. [4]
Reilly, M. T., G. J. Faulkner, J. Dubnau, I. Ponomarev, and F. H. Gage. 2013. The Role of Transposable Elements in Health and Diseases of the Central Nervous System. *J. Neurosci.* **33(45)**:17577–17586. [5]
Reiman, E. M., J. B. Langbaum, A. S. Fleisher, et al. 2011. Alzheimer's Prevention Initiative: A Plan to Accelerate the Evaluation of Presymptomatic Treatments. *J. Alzheimer. Dis.* **26 (Suppl. 3)**:321–329. [6]
Riley, B. E., S. J. Gardai, D. Emig-Agius, et al. 2014. Systems-Based Analyses of Brain Regions Functionally Impacted in Parkinson's Disease Reveals Underlying Causal Mechanisms. *PLoS One* **9(8)**:e102909. [14]
Ripke, S., C. O'Dushlaine, K. Chambert, et al. 2013. Genome-Wide Association Analysis Identifies 13 New Risk Loci for Schizophrenia. *Nat. Genet.* **45(10, 13)**:1150–1159. [3, 4, 13]
Rissling, A. J., S. Makeig, D. L. Braff, and G. A. Light. 2010. Neurophysiologic Markers of Abnormal Brain Activity in Schizophrenia. *Curr. Psychiatry Rep.* **12**:572–578. [10]
Rissling, A. J., S.-H. Park, J. W. Young, et al. 2013. Demand and Modality of Directed Attention Modulate "Pre-Attentive" Sensory Processes in Schizophrenia Patients and Nonpsychiatric Controls. *Schizophr. Res.* **146**:326–335. [10]
Robbins, T. W. 2002. The 5-Choice Serial Reaction Time Task: Behavioral Pharmacology and Functional Neurochemistry. *Psychopharmacology* **163**:362–380. [12]
———. 2014. The Neuropsychopharmacology of Attention. In: The Handbook of Attention, ed. K. Nobre and S. Kastner, pp. 509–541. Oxford: Oxford Univ. Press. [12]
Robbins, T. W., and A. F. Arnsten. 2009. The Neuropsychopharmacology of Fronto-Executive Function: Monoaminergic Modulation. *Annu. Rev. Neurosci.* **32**:267–287. [13]
Robins, E., and S. B. Guze. 1970. Establishment of Diagnostic Validity in Psychiatric Illness: Its Application to Schizophrenia. *Am. J. Psychiatry* **126**:983–987. [3]
Robinson, E. S. J., D. M. Eagle, A. C. Mar, et al. 2008. Similar Effects of the Selective Noradrenaline Reuptake Inhibitor Atomoxetine on Three Distinct Forms of Impulsivity in the Rat. *Neuropsychopharmacology* **33**:1028–1037. [12]
Robinson, R. G., K. L. Kubos, L. B. Starr, K. Rao, and T. R. Price. 1984. Mood Disorders in Stroke Patients. Importance of Location of Lesion. *Brain* **107 (Pt 1)**:81–93. [13]
Rodriguez-Navarro, J. A., S. Kaushik, H. Koga, et al. 2012. Inhibitory Effect of Dietary Lipids on Chaperone-Mediated Autophag. *PNAS* **109(12)**:E705–714. [5]
Romberg, C., M. P. Mattson, M. R. Mughal, T. J. Bussey, and L. M. Saksida. 2011. Impaired Attention in the 3xTgAD Mouse Model of Alzheimer's Disease: Rescue by Donepezil (Aricept). *J. Neurosci.* **31(9)**:3500–3507. [13]
Rosen, R. F., A. S. Farberg, M. Gearing, et al. 2008. Tauopathy with Paired Helical Filaments in an Aged Chimpanzee. *J. Comp. Neurol.* **509(3)**:259–270. [6]
Rosin, B., M. Slovik, R. Mitelman, et al. 2011. Closed-Loop Deep Brain Stimulation Is Superior in Ameliorating Parkinsonism. *Neuron* **72(2)**:370–384. [13]
Ross, C. A., E. H. Aylward, E. J. Wild, et al. 2014. Huntington Disease: Natural History, Biomarkers and Prospects for Therapeutics. *Nat. Rev. Neurol.* **10(4)**:204–216. [13]

Ross, R. M., N. A. McNair, S. L. Fairhall, et al. 2008. Induction of Orientation-Specific LTP-Like Changes in Human Visual Evoked Potentials by Rapid Sensory Stimulation. *Brain Res. Bull.* **76**:97–101. [10]

Rossi, S., M. Hallett, P. M. Rossini, and A. Pascual-Leone. 2009. Safety, Ethical Considerations, and Application Guidelines for the Use of Transcranial Magnetic Stimulation in Clinical Practice and Research. *Clin. Neurophysiol.* **120**:2008–2039. [10]

Rossini, P. M., S. Rossi, C. Babiloni, and J. Polich. 2007. Clinical Neurophysiology of Aging Brain: From Normal Aging to Neurodegeneration. *Prog. Neurobiol.* **83**:375–400. [9]

Rosti, R. O., A. A. Sadek, K. K. Vaux, and J. G. Gleeson. 2014. The Genetic Landscape of Autism Spectrum Disorders. *Dev. Med. Child Neurol.* **56**:12–18. [8]

Roth, G. S., J. A. Mattison, M. A. Ottinger, et al. 2004. Aging in Rhesus Monkeys: Relevance to Human Health Interventions. *Science* **305(5689)**:1423–1426. [13]

Rothkegel, H., M. Sommer, and W. Paulus. 2010. Breaks During 5 Hz rTMS Are Essential for Facilitatory after Effects. *Clin. Neurophysiol.* **121**:426–430. [10]

Rubin, L. L. 2008. Stem Cells and Drug Discovery: The Beginning of a New Era? *Cell* **132**:549–552. [11]

Ruderfer, D. M., A. H. Fanous, S. Ripke, et al. 2013. Polygenic Dissection of Diagnosis and Clinical Dimensions of Bipolar Disorder and Schizophrenia. *Mol. Psychiatry* **19(9)**:1017–1024. [3]

Rutter, M., J. Kim-Cohen, and B. Maughan. 2006. Continuities and Discontinuities in Psychopathology between Childhood and Adult Life. *J. Child Psychol. Psychiatry* **47(3-4)**: 276–295. [3]

Ryan, S. D., N. Dolatabadi, S. F. Chan, et al. 2013. Isogenic Human iPSC Parkinson's Model Shows Nitrosative Stress-Induced Dysfunction in MEF2-PGC1alpha Transcription. *Cell* **155(6)**:1351–1364. [5]

Sahakian, B. J., A. M. Owen, N. J. Morant, et al. 1993. Further Analysis of the Cognitive Effects of Tetrahydroaminoacridine (THA) in Alzheimer's Disease: Assessment of Attentional and Mnemonic Function Using CANTAB. *Psychopharmacology* **110**:395–401. [12, 13]

Sahakian, K., and R. Jaenisch. 2009. Technical Challenges in Using Human Induced Pluripotent Stem Cells to Model Disease. *Cell Stem Cell* **5(6)**:584–595. [13]

Salinas, S., G. Schiavo, and E. J. Kremer. 2010. A Hitchhiker's Guide to the Nervous System: The Complex Journey of Viruses and Toxins. *Nat. Rev. Microbiol.* **8**:645–655. [10]

Salloway, S., R. Sperling, and H. R. Brashear. 2014a. Phase 3 Trials of Solanezumab and Bapineuzumab for Alzheimer's Disease. *N. Engl. J. Med.* **370(15)**:1460. [7]

Salloway, S., R. Sperling, N. C. Fox, et al. 2014b. Two Phase 3 Trials of Bapineuzumab in Mild-to-Moderate Alzheimer's Disease. *N. Engl. J. Med.* **370(4)**:322–333. [6]

Sanchez, P. E., L. Zhu, L. Verret, et al. 2012. Levetiracetam Suppresses Neuronal Network Dysfunction and Reverses Synaptic and Cognitive Deficits in an Alzheimer's Disease Model. *PNAS* **109(42)**:E2895–2903. [7]

Sanders, S. J., A. G. Ercan-Sencicek, V. Hus, et al. 2011. Multiple Recurrent *de Novo* CNVs, Including Duplications of the 7q11.23 Williams Syndrome Region, Are Strongly Associated with Autism. *Neuron* **70(5)**:863–885. [2]

Sanders, S. J., M. T. Murtha, A. R. Gupta, et al. 2012. *De Novo* Mutations Revealed by Whole-Exome Sequencing Are Strongly Associated with Autism. *Nature* **485(7397)**:237–241. [2]

SantaCruz, K., J. Lewis, T. Spires, et al. 2005. Tau Suppression in a Neurodegenerative Mouse Model Improves Memory Function. *Science* **309(5733)**:476–481. [6]

Sasai, Y. 2013. Next-Generation Regenerative Medicine: Organogenesis from Stem Cells in 3D Culture. *Cell Stem Cell* **12**:520–530. [11]

Sasaki, E., H. Suemizu, A. Shimada, et al. 2009. Generation of Transgenic Non-Human Primates with Germline Transmission. *Nature* **459(7246)**:523–527. [6, 13]

Sato, D., A. C. Lionel, C. S. Leblond, et al. 2012. Shank1 Deletions in Males with Autism Spectrum Disorder. *Am. J. Hum. Genet.* **90(5)**:879–887. [4]

Schafer, D. P., E. K. Lehrman, A. G. Kautzman, et al. 2012. Microglia Sculpt Postnatal Neural Circuits in an Activity and Complement-Dependent Manner. *Neuron* **74(4)**:691–705. [7]

Schiel, N., and L. Huber. 2006. Social Influences on the Development of Foraging Behavior in Free-Living Common Marmosets (*Callithrix jacchus*). *Am. J. Primatol.* **68**:1150–1160. [8]

Schizophrenia Working Group of the Psychiatric Genomics Consortium. 2014. Biological Insights from 108 Schizophrenia-Associated Genetic Loci. *Nature* **511**:421–427. [1, 3, 10]

Schmeisser, M. J., E. Ey, S. Wegener, et al. 2012. Autistic-Like Behaviours and Hyperactivity in Mice Lacking ProSAP1/Shank2. *Nature* **486(7402)**:256–260. [2, 4]

Schmiedt, C., A. Brand, H. Hildebrandt, and C. Basar-Eroglu. 2005. Event-Related Theta Oscillations During Working Memory Tasks in Patients with Schizophrenia and Healthy Controls. *Cogn. Brain Res.* **25**:936–947. [10]

Schneider, J. S., E. Y. Pioli, Y. Jianzhong, Q. Li, and E. Bezard. 2013. Levodopa Improves Motor Deficits but Can Further Disrupt Cognition in a Macaque Parkinson Model. *Mov. Disord.* **28**:663–667. [9]

Schneider, J. S., G. Unguez, A. Yuwiler, S. C. Berg, and C. H. Markham. 1988. Deficits in Operant Behaviour in Monkeys Treated with N-Methyl-4-Phenyl-1,2,3,6-Tetrahydropyridine (MPTP). *Brain* **111(Pt 6)**:1265–1285. [9]

Schrijvers, E. M., B. F. Verhaaren, P. J. Koudstaal, et al. 2012. Is Dementia Incidence Declining? Trends in Dementia Incidence since 1990 in the Rotterdam Study. *Neurology* **78(19)**:1456–1463. [7]

Schultz, W., and A. Dickinson. 2000. Neuronal Coding of Prediction Errors. *Annu. Rev. Neurosci.* **23**:473–500. [12]

Schwab, M. E. 2004. Nogo and Axon Regeneration. *Curr. Opin. Neurobiol.* **14(1)**:118–124. [13]

Schwarz, S., L. Froelich, and A. Burns. 2012. Pharmacological Treatment of Dementia. *Curr. Opin. Psychiatry* **25(6)**:542–550. [13]

Sekar, S., E. Jonckers, M. Verhoye, et al. 2013. Subchronic Memantine Induced Concurrent Functional Disconnectivity and Altered Ultra-Structural Tissue Integrity in the Rodent Brain: Revealed by Multimodal MRI. *Psychopharmacology* **227**:479–491. [9]

Seminowicz, D. A., H. S. Mayberg, A. R. McIntosh, et al. 2004. Limbic-Frontal Circuitry in Major Depression: A Path Modeling Metanalysis. *NeuroImage* **22**:409–418. [10]

Seok, J., H. S. Warren, A. G. Cuenca, et al. 2013. Inflammation and Host Response to Injury, Large Scale Collaborative Research Program. Genomic Responses in Mouse Models Poorly Mimic Human Inflammatory Diseases. *PNAS* **110**:3507–3512. [11]

Serio, A., B. Bilican, S. J. Barmada, et al. 2013. Astrocyte Pathology and the Absence of Non-Cell Autonomy in an Induced Pluripotent Stem Cell Model of TDP-43 Proteinopathy. *PNAS* **110(12)**:4697–4702. [5]

Shafi, M. M., M. B. Westover, M. D. Fox, and A. Pascual-Leone. 2012. Exploration and Modulation of Brain Network Interactions with Noninvasive Brain Stimulation in Combination with Neuroimaging. *Eur. J. Neurosci.* **35(6)**:805–825. [13]

Shao, K., R. Huang, J. Li, et al. 2010. Angiopep-2 Modified PE-PEG Based Polymeric Micelles for Amphotericin B Delivery Targeted to the Brain. *J. Control Release* **147(1)**:118–126. [7]

Sharma, K., S. Y. Choi, Y. Zhang, et al. 2013. High-Throughput Genetic Screen for Synaptogenic Factors: Identification of LRP6 as Critical for Excitatory Synapse Development. *Cell Rep.* **5**:1330–1341. [8]

Shay, T., V. Jojic, O. Zuk, et al. 2013. Conservation and Divergence in the Transcriptional Programs of the Human and Mouse Immune Systems. *PNAS* **110(8)**:2946–2951. [6]

Shcheglovitov, A., O. Shcheglovitova, M. Yazawa, et al. 2013. SHANK3 and IGF1 Restore Synaptic Deficits in Neurons from 22q13 Deletion Syndrome Patients. *Nature* **503(7475)**:267–271. [2, 4, 11]

Sheline, Y. I., M. E. Raichle, A. Z. Snyder, et al. 2010. Amyloid Plaques Disrupt Resting State Default Mode Network Connectivity in Cognitively Normal Elderly. *Biol. Psychiatry* **67**:584–587. [9]

Sidorov, M. S., B. D. Auerbach, and M. F. Bear. 2013. Fragile X Mental Retardation Protein and Synaptic Plasticity. *Mol. Brain* **6**:15. [10]

Siebner, H. R., T. O. Bergmann, S. Bestmann, et al. 2009. Consensus Paper: Combining Transcranial Stimulation with Neuroimaging. *Brain Stimul.* **2**:58–80. [10]

Siegel, M. S., and E. Y. Isacoff. 1997. A Genetically Encoded Optical Probe of Membrane Voltage. *Neuron* **19**:735–741. [10]

Silverman, J. L., M. Yang, C. Lord, and J. N. Crawley. 2010. Behavioural Phenotyping Assays for Mouse Models of Autism. *Nat. Rev. Neurosci.* **11(7)**:490–502. [2]

Sliwkowski, M. X., and I. Mellman. 2013. Antibody Therapeutics in Cancer. *Science* **341(6151)**:1192–1198. [13]

Smith, S. M., C. F. Beckmann, J. Andersson, et al. 2013. Resting-State fMRI in the Human Connectome Project. *NeuroImage* **80**:144–168. [14]

Solanto, M. V., H. Abikoff, E. Sonuga-Barke, et al. 2001. The Ecological Validity of Delay Aversion and Response Inhibition as Measures of Impulsivity in AD/HD: A Supplement to the NIMH Multimodal Treatment Study of AD/HD. *J. Abnorm. Child Psychol.* **29(3)**:215–228. [12]

Sowman, P. F., S. S. Dueholm, J. H. Rasmussen, and N. Mrachacz-Kersting. 2014. Induction of Plasticity in the Human Motor Cortex by Pairing an Auditory Stimulus with TMS. *Front. Hum. Neurosci.* **8**:398. [10]

Spencer, B., R. A. Marr, R. Gindi, et al. 2011. Peripheral Delivery of a CNS Targeted, Metalo-Protease Reduces Abeta Toxicity in a Mouse Model of Alzheimer's Disease. *PLoS One* **6(1)**:e16575. [7]

Sperling, R. A., P. S. Aisen, L. A. Beckett, et al. 2011a. Toward Defining the Preclinical Stages of Alzheimer's Disease: Recommendations from the National Institute on Aging-Alzheimer's Association Workgroups on Diagnostic Guidelines for Alzheimer's Disease. *Alzheimers Dement.* **7(3)**:280–292. [7]

Sperling, R. A., C. R. Jack, Jr., and P. S. Aisen. 2011b. Testing the Right Target and Right Drug at the Right Stage. *Sci. Transl. Med.* **3(111)**:111cm133. [6]

Spiegel, R., M. Berres, A. R. Miserez, and A. U. Monsch. 2011. For Debate: Substituting Placebo Controls in Long-Term Alzheimer's Prevention Trials. *Alzheimers Res. Ther.* **3(2)**:9. [6]

Spooren, W., L. Lindemann, A. Ghosh, and L. Santarelli. 2012. Synapse Dysfunction in Autism: A Molecular Medicine Approach to Drug Discovery in Neurodevelopmental Disorders. *Trends Pharmacol. Sci.* **33(12)**:669–684. [2]

Squire, L. R. 1992. Memory and the Hippocampus: A Synthesis from Findings with Rats, Monkeys, and Humans. *Psychol. Rev.* **99**:195–231. [12]

Stein, T. D., V. E. Alvarez, and A. C. McKee. 2014. Chronic Traumatic Encephalopathy: A Spectrum of Neuropathological Changes Following Repetitive Brain Trauma in Athletes and Military Personnel. *Alzheimers Res. Ther.* **6(1)**:4. [7]

Steinberg, E. E., J. R. Boivin, B. T. Saunders, et al. 2014. Positive Reinforcement Mediated by Midbrain Dopamine Neurons Requires D1 and D2 Receptor Activation in the Nucleus Accumbens. *PLoS One* **14(9)**:e94771. [14]

Steinberg, J., and C. Webber. 2013. The Roles of FMRP-Regulated Genes in Autism Spectrum Disorder: Single- and Multiple-Hit Genetic Etiologies. *Am. J. Hum. Genet.* **93(5)**:825–839. [4]

Stella, S. L., A. Vila, A. Y. Hung, et al. 2012. Association of Shank 1a Scaffolding Protein with Cone Photoreceptor Terminals in the Mammalian Retina. *PLoS One* **7**:e43463. [10]

Stephan, A. H., D. V. Madison, J. M. Mateos, et al. 2013. A Dramatic Increase of C1q Protein in the CNS During Normal Aging. *J. Neurosci.* **33(33)**:13460–13474. [5]

Stewart, A. M., O. Braubach, J. Spitsbergen, R. Gerlai, and A. V. Kalueff. 2014. Zebrafish Models for Translational Neuroscience Research: From Tank to Bedside. *Trends Neurosci.* **37**:264–278. [8]

St. Johnston, D. 2002. The Art and Design of Genetic Screens: *Drosophila Melanogaster*. *Nat. Rev. Genet.* **3**:176–188. [8]

Stoner, R., M. L. Chow, M. P. Boyle, et al. 2014. Patches of Disorganization in the Neocortex of Children with Autism. *N. Engl. J. Med.* **370(13)**:1209–1219. [2]

St-Pierre, F., J. D. Marshall, Y. Yang, et al. 2014. High-Fidelity Optical Reporting of Neuronal Electrical Activity with an Ultrafast Fluorescent Voltage Sensor. *Nature Neurosci.* **17(6)**:884–889. [10]

Strittmatter, W. J., and A. D. Roses. 1996. Apolipoprotein E and Alzheimer's Disease. *Annu. Rev. Neurosci.* **19**:53–77. [7]

Stuart, S. A., P. Butler, M. Munafo, D. J. Nutt, and E. S. J. Robinson. 2013. A Translational Rodent Assay of Affective Biases in Depression and Antidepressant Therapy. *Neuropsychopharmacology* **38**:1625–1635. [12]

Sullivan, P. F., M. J. Daly, and M. C. O'Donovan. 2012. Genetic Architectures of Psychiatric Disorders: The Emerging Picture and Its Implications. *Nat. Rev. Genet.* **13(8)**:537–551. [3]

Sullivan, P. F., K. S. Kendler, and M. C. Neale. 2003. Schizophrenia as a Complex Trait: Evidence from a Meta-Analysis of Twin Studies. *Arch. Gen. Psychiatry* **60(12)**:1187–1192. [13]

Swainson, R., J. R. Hodges, C. J. Galton, et al. 2001. Early Detection and Differential Diagnosis of Alzheimer's Disease and Depression with Neuropsychological Tasks. *Dement. Geri. Cogn. Dis.* **12**:265–280. [12]

Szatmari, P., A. D. Paterson, L. Zwaigenbaum, et al. 2007. Mapping Autism Risk Loci Using Genetic Linkage and Chromosomal Rearrangements. *Nat. Genet.* **39(3)**:319–328. [2]

Tabbal, S. D., L. Tian, M. Karimi, et al. 2012. Low Nigrostriatal Reserve for Motor Parkinsonism in Nonhuman Primates. *Exp. Neurol.* **237**:355–362. [9]

Tabuchi, K., J. Blundell, M. R. Etherton, et al. 2007. A Neuroligin-3 Mutation Implicated in Autism Increases Inhibitory Synaptic Transmission in Mice. *Science* **318(5847)**:71–76. [2]

Takahashi, D. Y., D. Z. Narayanan, and A. A. Ghazanfar. 2013. Coupled Oscillator Dynamics of Vocal Turn-Taking in Monkeys. *Curr. Biol.* **23**:2162–2168. [8]

Takahashi, K., K. Tanabe, M. Ohnuki, et al. 2007. Induction of Pluripotent Stem Cells from Adult Human Fibroblasts by Defined Factors. *Cell* **131**:861–872. [8, 11]

Takahashi, K., and S. Yamanaka. 2006. Induction of Pluripotent Stem Cells from Mouse Embryonic and Adult Fibroblast Cultures by Defined Factors. *Cell* **126**:663–676. [8, 11]

Talpos, J. C., B. D. Winters, R. Dias, L. M. Saksida, and T. J. Bussey. 2009. A Novel Touchscreen Automated Paired-Associate Learning (PAL) Task Sensitive to Pharmacological Manipulation of the Hippocampus: A Translational Rodent Model of Cognitive Impairments in Neurodegenerative Disease. *Psychopharmacology* **205**:157–168. [12]

Taylor Tavares, J. V., L. Clark, M. L. Furey, et al. 2008. Neural Basis of Abnormal Response to Negative Feedback in Unmedicated Mood Disorders. *NeuroImage* **42(3)**:1118–1126. [12]

Testa-Silva, G., M. B. Verhoog, N. A. Goriounova, et al. 2010. Human Synapses Show a Wide Temporal Window for Spike-Timing-Dependent Plasticity. *Front. Synaptic Neurosci.* **2**:12. [10]

Teyler, T. J., J. P. Hamm, W. C. Clapp, et al. 2005. Long-Term Potentiation of Human Visual Evoked Responses. *Eur. J. Neurosci.* **21**:2045–2050. [10]

Thut, G., and A. Pascual-Leone. 2010. A Review of Combined TMS-EEG Studies to Characterize Lasting Effect of Repetitive TMS and Assess Their Usefuleness in Cognitive and Clinical Neuroscience. *Brain Topogr.* **22**:219–232. [10]

Tian, G., and D. Finley. 2012. Cell Biology: Destruction Deconstructed. *Nat. Chem. Biol.* **482(7384)**:170–171. [5]

Toro, R., M. Konyukh, R. Delorme, et al. 2010. Key Role for Gene Dosage and Synaptic Homeostasis in Autism Spectrum Disorders. *Trends Genet.* **26(8)**:363–372. [2]

Toro, R., J. B. Poline, G. Huguet, et al. 2014. Genomic Architecture of Human Neuroanatomical Diversity. *Mol. Psychiatry*: doi: 10.1038/mp.2014.1099. [2]

Toyomaki, A., I. Kusumi, T. Matsuyama, et al. 2008. Tone Duration Mismatch Negativity Deficits Predict Impairment of Executive Function in Schizophrenia. *Prog. Neuropsychopharmacol. Biol. Psychiatry* **32**:95–99. [10]

Tremblay, M. E., and A. K. Majewska. 2011. A Role for Microglia in Synaptic Plasticity? *Commun. Integr. Biol.* **4(2)**:220–222. [7]

Tritsch, N. X., J. B. Ding, and B. L. Sabatini. 2012. Dopaminergic Neurons Inhibit Striatal Output through Non-Canonical Release of GABA. *Nature* **490**:262–266. [10]

Tropea, D., E. Giacometti, N. R. Wilson, et al. 2009. Partial Reversal of Rett Syndrome-Like Symptoms in MeCP2 Mutant Mice. *PNAS* **106**:2029–2034. [10]

Tulving, E. 1983. Elements of Episodic Memory. Oxford Oxford Univ. Press. [12]

Turetsky, B. I., M. E. Calkins, G. A. Light, et al. 2007a. Neurophysiological Endophenotypes of Schizophrenia: The Viability of Selected Candidate Measures. *Schizophr. Bull.* **33**:69–94. [10]

Turetsky, B. I., C. G. Kohler, T. Indersmitten, et al. 2007b. Facial Emotion Recognition in Schizophrenia: When and Why Does It Go Awry? *Schizophr. Res.* **94**:253–263. [10]

Turner, D. C., L. Clark, and E. Pomarol-Clotet. 2004. Modafinil Improves Cognition and Attentional Set Shifting in Patients with Chronic Schizophrenia. *Neuropsychopharmacology* **29**:1363–1373. [12]

Tye, K. M., and K. Deisseroth. 2012. Optogenetic Investigation of Neural Circuits Underlying Brain Disease in Animal Models. *Nat. Rev. Neurosci.* **13(4)**:251–266. [8]

Tye, K. M., J. J. Mirzabekov, M. R. Warden, et al. 2013. Dopamine Neurons Modulate Neural Encoding and Expression of Depression-Related Behaviour. *Nature* **493**:537–541. [9]

Tye, K. M., R. Prakash, S.-Y. Kim, et al. 2011. Amygdala Circuitry Mediating Reversible and Bidirectional Control of Anxiety. *Nature* **471(7338)**:358–362. [8]

Ugurbil, K., J. Xu, E. J. Auerbach, et al. 2013. Pushing Spatial and Temporal Resolution for Functional and Diffusion MRI in the Human Connectome Project. *Neuroimage* **80**:80–104. [9]

Uhlhaas, P. J., and W. Singer. 2006. Neural Synchrony in Brain Disorders: Relevance for Cognitive Dysfunctions and Pathophysiology. *Neuron* **52**:155–168. [10]

Umbricht, D., and S. Krljes. 2005. Mismatch Negativity in Schizophrenia: A Meta-Analysis. *Schizophr. Res.* **76**:1–23. [10]

van de Leemput, I. A., M. Wichers, A. O. Cramer, et al. 2014. Critical Slowing Down as Early Warning for the Onset and Termination of Depression. *PNAS* **111(1)**:87–92. [4]

van den Broeke, E. N., C. M. van Rijn, J. A. Biurrun Manresa, et al. 2010. Neurophysiological Correlates of Nociceptive Heterosynaptic Long-Term Potentiation in Humans. *J. Neurophysiol.* **103**:2107–2113. [10]

Van Essen, D. C., S. M. Smith, D. M. Barch, et al. 2013. The WU-Minn Human Connectome Project: An Overview. *Neuroimage* **80**:62–79. [7]

van Os, J., P. Delespaul, J. Wigman, I. Myin-Germeys, and M. Wichers. 2013. Beyond DSM and ICD: Introducing “Precision Diagnosis” for Psychiatry Using Momentary Assessment Technology. *World Psychiatry* **12(2)**:113–117. [4]

Varga, V., A. Losonczy, B. V. Zemelman, et al. 2009. Fast Synaptic Subcortical Control of Hippocampal Circuits. *Science* **326**:449–453. [10]

Varoqueaux, F., G. Aramuni, R. L. Rawson, et al. 2006. Neuroligins Determine Synapse Maturation and Function. *Neuron* **51(6)**:741–754. [2]

Varrone, A., and C. Halldin. 2012. New Developments of Dopaminergic Imaging in Parkinson’s Disease. *Q. J. Nucl. Med. Mol. Imaging* [7]

Vellucci, S. V., P. J. Martin, and B. J. Everitt. 1988. The Discriminative Stimulus Produced by Penntylenetrazole: Effects of Systemic Anxiolytics and Anxiogenics. Aggressive Defeat and Midazolam or Muscimol Infused into the Amygdala. *J. Psychopharmacol.* **2**:809–893. [12]

Verkerk, A. J., M. Pieretti, J. S. Sutcliffe, et al. 1991. Identification of a Gene (FMR-1) Containing a CGG Repeat Coincident with a Breakpoint Cluster Region Exhibiting Length Variation in Fragile X Syndrome. *Cell* **65(5)**:905–914. [13]

Verret, L., E. O. Mann, G. B. Hang, et al. 2012. Inhibitory Interneuron Deficit Links Altered Network Activity and Cognitive Dysfunction in Alzheimer Model. *Cell* **149(3)**:708–721. [13]

Vierbuchen, T., A. Ostermeier, Z. P. Pang, et al. 2010. Direct Conversion of Fibroblasts to Functional Neurons by Defined Factors. *Nature* **463(7284)**:1035–1041. [13]

Vierbuchen, T., and M. Wernig. 2011. Direct Lineage Conversions: Unnatural but Useful? *Nat. Biotechnol.* **29(10)**:892–907. [13]

Vincent, J. L., G. H. Patel, M. D. Fox, et al. 2007. Intrinsic Functional Architecture in the Anaesthetized Monkey Brain. *Nature* **447**:83–86. [9]
Voineagu, I., X. Wang, P. Johnston, et al. 2011. Transcriptomic Analysis of Autistic Brain Reveals Convergent Molecular Pathology. *Nature* **474(7351)**:380–384. [2]
Voisine, C., J. S. Pedersen, and R. I. Morimoto. 2010. Chaperone Networks: Tipping the Balance in Protein Folding Diseases. *Neurobiol. Dis.* **40(1)**:12–20. [5]
Volk, D. W., and D. A. Lewis. 2010. Prefrontal Cortical Circuits in Schizophrenia. *Curr. Top. Behav. Neurosci.* **4**:485–508. [8]
Voon, V., M. A. Irvine, K. Derbyshire, et al. 2014. Measuring "Waiting" Impulsivity in Substance Addictions and Binge Eating Disorder in a Novel Analogue of Rodent Serial Reaction Time Task. *Biol. Psychiatry* **75**:148–155. [12]
Voronin, L. L. 1994. Quantal Analysis of Hippocampal Long-Term Potentiation. *Rev. Neurosci.* **5**:141–170. [10]
Vos, T., A. D. Flaxman, M. Naghavi, et al. 2012. Years Lived with Disability (YLDs) for 1160 Sequelae of 289 Diseases and Injuries 1990–2010: A Systematic Analysis for the Global Burden of Disease Study 2010. *Lancet* **380**:2163–2196. [1]
Vossel, K. A., A. J. Beagle, G. D. Rabinovici, et al. 2013. Seizures and Epileptiform Activity in the Early Stages of Alzheimer Disease. *JAMA Neurol.* **70(9)**:1158–1166. [6, 13]
Walker, L. C., M. I. Diamond, K. E. Duff, and B. T. Hyman. 2013. Mechanisms of Protein Seeding in Neurodegenerative Diseases. *JAMA Neurol.* **70**:304–310. [9]
Wang, D., S. S. El-Amouri, M. Dai, et al. 2013a. Engineering a Lysosomal Enzyme with a Derivative of Receptor-Binding Domain of apoE Enables Delivery across the Blood-Brain Barrier. *PNAS* **110(8)**:2999–3004. [7]
Wang, H., H. Yang, C. S. Shivalila, et al. 2013b. One-Step Generation of Mice Carrying Mutations in Multiple Genes by CRISPR/Cas-Mediated Genome Engineering. *Cell* **153**:910–918. [9]
Wang, J., F. Wagner, D. A. Borton, et al. 2012. Integrated Device for Combined Optical Neuromodulation and Electrical Recording for Chronic *in Vivo* Applications. *J. Neural Eng.* **9(1)**:016001. [13]
Wang, J., X. Zuo, and Y. He. 2010a. Graph-Based Network Analysis of Resting-State Functional MRI. *Front. Syst. Neurosci.* **4**:15. [9]
Wang, P., Y. Xue, X. Shang, and Y. Liu. 2010b. Diphtheria Toxin Mutant CRM197-Mediated Transcytosis across Blood-Brain Barrier *in vitro*. *Cell. Mol. Neurobiol.* **30(5)**:717–725. [7]
Wang, W. Y., B. J. Barratt, D. G. Clayton, and J. A. Todd. 2005. Genome-Wide Association Studies: Theoretical and Practical Concerns. *Nat. Rev. Genet.* **6**:109–118. [3]
Wang, X., P. A. McCoy, R. M. Rodriguiz, et al. 2011. Synaptic Dysfunction and Abnormal Behaviors in Mice Lacking Major Isoforms of Shank3. *Hum. Mol. Genet.* **20(15)**:3093–3108. [4]
Wang, Y., H.-F. Guo, T. A. Pologruto, et al. 2004. Stereotyped Odor-Evoked Activity in the Mushroom Body of *Drosophila* Revealed by Green Fluorescent Protein-Based Ca^{2+} Imaging. *J. Neurosci.* **24**:6507–6514. [10]
Ward, N. M., and V. J. Brown. 1996. Covert Orienting of Attention in the Rat and the Role of Striatal Dopamine. *J. Neurosci.* **16**:3082–3088. [12]
Watson, K. K., and M. L. Platt. 2012. Of Mice and Monkeys: Using Non-Human Primate Models to Bridge Mouse- and Human-Based Investigations of Autism Spectrum Disorders. *J. Neurodev. Disord.* **4(1)**:21. [13]

Watts, J. C., K. Giles, A. Oehler, et al. 2013. Transmission of Multiple System Atrophy Prions to Transgenic Mice. *PNAS* **110(48)**:19555–19560. [5]

Wei, F., and M. Zhuo. 2001. Potentiation of Sensory Responses in the Anterior Cingulate Cortex Following Digit Amputation in the Anaesthetised Rat. *J. Physiol.* **532**:823–833. [10]

Weinberger, D. R. 1987. Implications of Normal Brain Development for the Pathogenesis of Schizophrenia. *Arch. Gen. Psychiatry* **44(7)**:660–669. [3, 13]

Weiner, I. 2003. The "Two-Headed" Latent Inhibition Model of Schizophrenia: Modeling Positive and Negative Symptoms and Their Treatment. *Psychopharmacology* **169**:257–297. [12]

Weiss, I. C., A. M. Domeney, C. A. Heidbreder, J. L. Moreau, and J. Feldon. 2001. Early Social Isolation, but Not Maternal Separation, Affects Behavioral Sensitization to Amphetamine in Male and Female Adult Rats. *Pharmacol. Biochem. Behav.* **70(2-3)**:397–409. [13]

Welch, J. M., J. Lu, R. M. Rodriguiz, et al. 2007. Cortico-Striatal Synaptic Defects and OCD-Like Behaviors in SAPAP3 Mutant Mice. *Nature* **448**:894–900. [8]

Wichmann, T., H. Bergman, and M. R. DeLong. 1994. The Primate Subthalamic Nucleus. III. Changes in Motor Behavior and Neuronal Activity in the Internal Pallidum Induced by Subthalamic Inactivation in the MPTP Model of Parkinsonism. *J. Neurophysiol.* **72**:521–531. [9]

Wichmann, T., H. Bergman, P. A. Starr, et al. 1999. Comparison of MPTP-Induced Changes in Spontaneous Neuronal Discharge in the Internal Pallidal Segment and in the Substantia Nigra Pars Reticulata in Primates. *Exp. Brain Res.* **125**:397–409. [9]

Wichterle, H., I. Lieberam, J. A. Porter, and T. M. Jessell. 2002. Directed Differentiation of Embryonic Stem Cells into Motor Neurons. *Cell* **110**:385–397. [11]

Wickersham, I. R., S. Finke, K. Conzelmann, and E. M. Callaway. 2007. Retrograde Neuronal Tracing with a Deletion-Mutant Rabies Virus. *Nat. Methods* **4**:47–49. [10]

Wicks, P., T. E. Vaughan, M. P. Massagli, and J. Heywood. 2011. Accelerated Clinical Discovery Using Self-Reported Patient Data Collected Online and a Patient-Matching Algorithm. *Nat. Biotechnol.* **29(5)**:411–414. [2]

Wietek, J., J. S. Wiegert, N. Adeishvili, et al. 2014. Conversion of Channelrhodopsin into a Light-Gated Chloride Channel. *Science* **344(6182)**:409–412. [10]

Williams, H. J., N. Craddock, G. Russo, et al. 2011. Most Genome-Wide Significant Susceptibility Loci for Schizophrenia and Bipolar Disorder Reported to Date Cross-Traditional Diagnostic Boundaries. *Hum. Mol. Genet.* **20(2)**:387–391. [3]

Willsey, A. J., S. J. Sanders, M. Li, et al. 2013. Coexpression Networks Implicate Human Midfetal Deep Cortical Projection Neurons in the Pathogenesis of Autism. *Cell* **155(5)**:997–1007. [2, 10]

Wing, J., and J. Nixon. 1975. Discriminating Symptoms in Schizophrenia: A Report from the International Pilot Study of Schizophrenia. *Arch. Gen. Psychiatry* **32(7)**:853–859. [4]

Winstanley, C. A., D. E. H. Theobald, R. N. Cardinal, and T. W. Robbins. 2004. Contrasting Roles of Basolateral Amygdala and Orbitofrontal Cortex in Impulsive Choice. *J. Neurosci.* **24**:4718–4722. [12]

Wise, R. 2004. Dopamine, Learning and Motivation. *Nat. Rev. Neurosci.* **5**:483–494. [12]

Wohr, M., F. I. Roullet, A. Y. Hung, M. Sheng, and J. N. Crawley. 2011. Communication Impairments in Mice Lacking Shank1: Reduced Levels of Ultrasonic Vocalizations and Scent Marking Behavior. *PLoS One* **6(6)**:e20631. [4]

Wolf, A. B., J. Valla, G. Bu, et al. 2013. Apolipoprotein E as a Beta-Amyloid-Independent Factor in Alzheimer's Disease. *Alzheimers Res. Ther.* **5(5)**:38. [7]

Wolters, A., F. Sandbrink, A. Schlottmann, et al. 2003. A Temporally Asymmetric Hebbian Rule Governing Plasticity in the Human Motor Cortex. *J. Neurophysiol.* **89**:2339–2345. [10]

Wolters, A., A. Schmidt, A. Schramm, et al. 2005. Timing-Dependent Plasticity in Human Primary Somatosensory Cortex. *J. Physiol.* **565**:1039–1052. [10]

Won, H., H. R. Lee, H. Y. Gee, et al. 2012. Autistic-Like Social Behaviour in Shank2-Mutant Mice Improved by Restoring NMDA Receptor Function. *Nature* **486(7402)**:261–265. [2, 4]

Wood, S. J., C. Pantelis, T. Proffitt, et al. 2003. Spatial Working Memory Ability Is a Marker of Risk-for-Psychosis. *Psychol. Med.* **33**:1239–1247. [12]

Wooding, S., A. Pollitt, S. Castle-Clarke, et al. 2013. Mental Health Retrosight. Understanding the Returns from Research (Lessons from Schizophrenia). RAND Corporation. [4]

Wooten, G. F., and R. C. Collins. 1981. Metabolic Effects of Unilateral Lesion of the Substantia Nigra. *J. Neurosci.* **1**:285–291. [9]

Worbe, Y., G. Savulich, V. Voon, E. Fernandez-Egea, and T. W. Robbins. 2014. Serotonin Depletion Induces "Waiting Impulsivity" on the Human Four Choice Serial Reaction Time Task: Cross-Species Translational Significance. *Neuropsychopharmacology* **39**:1519–1526. [12]

Wynn, J. K., C. Sugar, W. P. Horan, R. Kern, and M. F. Green. 2010. Mismatch Negativity, Social Cognition, and Functioning in Schizophrenia Patients. *Biol. Psychiatry* **67**:940–947. [10]

Xia, C. F., J. Arteaga, G. Chen, et al. 2013. [(18)F]T807, a Novel Tau Positron Emission Tomography Imaging Agent for Alzheimer's Disease. *Alzheimers Dement.* **9(6)**:666–676. [7]

Xu, W., W. Morishita, P. S. Buckmaster, et al. 2012. Distinct Neuronal Coding Schemes in Memory Revealed by Selective Erasure of Fast Synchronous Synaptic Transmission. *Neuron* **73**:990–1001. [10]

Yamanaka, S. 2007. Strategies and New Developments in the Generation of Patient-Specific Pluripotent Stem Cells. *Cell Stem Cell* **1**:39-49. [11]

Yanamandra, K., N. Kfoury, H. Jiang, et al. 2013. Anti-Tau Antibodies That Block Tau Aggregate Seeding *in Vitro* Markedly Decrease Pathology and Improve Cognition *in Vivo*. *Neuron* **80(2)**:402–414. [7]

Yang, J., B. Benyamin, B. P. McEvoy, et al. 2010. Common SNPs Explain a Large Proportion of the Heritability for Human Height. *Nat. Genet.* **42(7)**:565–569. [2]

Yang, L., D. Rieves, and C. Ganley. 2012a. Brain Amyloid Imaging: FDA Approval of Florbetapir F18 Injection. *N. Engl. J. Med.* **367(10)**:885–887. [13]

Yang, M., O. Bozdagi, M. L. Scattoni, et al. 2012b. Reduced Excitatory Neurotransmission and Mild Autism-Relevant Phenotypes in Adolescent Shank3 Null Mutant Mice. *J. Neurosci.* **32(19)**:6525–6541. [4]

Yang, Y. M., S. K. Gupta, K. J. Kim, et al. 2013. A Small Molecule Screen in Stem-Cell-Derived Motor Neurons Identifies a Kinase Inhibitor as a Candidate Therapeutic for ALS. *Cell Stem Cell* **12**:713–726. [11, 14]

Yizhar, O., L. E. Fenno, M. Prigge, et al. 2011. Neocortical Excitation/Inhibition Balance in Information Processing and Social Dysfunction. *Nature* **477(7363)**:171–178. [8]

Yu, D. X., M. C. Marchetto, and F. H. Gage. 2013a. Therapeutic Translation of iPSCs for Treating Neurological Disease. *Cell Stem Cell* **12**:678–688. [11]

Yu, T. W., M. H. Chahrour, M. E. Coulter, et al. 2013b. Using Whole-Exome Sequencing to Identify Inherited Causes of Autism. *Neuron* **77(2)**:259–273. [2]

Yue, Z., Q. J. Wang, and M. Komatsu. 2008. Neuronal Autophagy: Going the Distance to the Axon. *Autophagy* **4(1)**:94–96. [5]

Zaehle, T., W. C. Clapp, J. P. Hamm, M. Meyer, and I. J. Kirk. 2007. Induction of LTP-Like Changes in Human Auditory Cortex by Rapid Auditory Stimulation: An fMRI Study. *Restor. Neurol. Neurosci.* **25**:251–259. [10]

Zahs, K. R., and K. H. Ashe. 2010. Too Much Good News: Are Alzheimer Mouse Models Trying to Tell Us How to Prevent, Not Cure, Alzheimer's Disease? *Trends Neurosci.* **33(8)**:381–389. [6]

Zeeb, F. D., T. W. Robbins, and C. A. Winstanley. 2009. Serotonergic and Dopaminergic Modulation of Gambling Behavior as Assessed Using a Novel Rat Gambling Task. *Neuropsychopharmacology* **34**:2329–2343. [12]

Zhang, F., L. Cong, S. Lodato, et al. 2011. Efficient Construction of Sequence-Specific TAL Effectors for Modulating Mammalian Transcription. *Nat. Biotechnol.* **29**:149–153. [8]

Zhang, Y., C. Pak, Y. Han, et al. 2013. Rapid Single-Step Induction of Functional Neurons from Human Pluripotent Stem Cells. *Neuron* **78**:785–798. [1, 8, 11]

Zhao, S., J. T. Ting, H. E. Atallah, et al. 2011. Cell Type–Specific Channelrhodopsin-2 Transgenic Mice for Optogenetic Dissection of Neural Circuitry Function. *Nat. Methods* **8(9)**:745–752. [8]

Zhou, Z., H. Zhu, and L. Chen. 2013. Effect of Aripiprazole on Mismatch Negativity (MMN) in Schizophrenia. *PLoS One* **8**:e52186. [10]

Zhu, X., A. C. Need, S. Petrovski, and D. B. Goldstein. 2014. One Gene, Many Neuropsychiatric Disorders: Lessons from Mendelian Diseases. *Nat. Neurosci.* **17(6)**:773–781. [2]

Ziemann, U., T. V. Ilić, T. V. Iliać, et al. 2004. Learning Modifies Subsequent Induction of Long-Term Potentiation-Like and Long-Term Depression-Like Plasticity in Human Motor Cortex. *J. Neurosci.* **24**:1666–1672. [10]

Ziemann, U., W. Paulus, M. A. Nitsche, et al. 2008. Consensus: Motor Cortex Plasticity Protocols. *Brain Stimul.* **1**:164–182. [10]

Zimmermann, P., M. Gondan, and B. Fimm. 2004. KiTAP Testbatterie zur Aufmerksamkeitsprüfung für Kinder. Freiburg: Psytest. [10]

Zlokovic, B. V. 2008. The Blood-Brain Barrier in Health and Chronic Neurodegenerative Disorders. *Neuron* **57(2)**:178–201. [7]

Subject Index

Further Titles in the Strüngmann Forum Report Series[1]

Better Than Conscious?
Decision Making, the Human Mind, and Implications For Institutions
edited by Christoph Engel and Wolf Singer, ISBN 978-0-262-19580-5

Clouds in the Perturbed Climate System:
Their Relationship to Energy Balance, Atmospheric Dynamics, and Precipitation
edited by Jost Heintzenberg and Robert J. Charlson, ISBN 978-0-262-01287-4

Biological Foundations and Origin of Syntax
edited by Derek Bickerton and Eörs Szathmáry, ISBN 978-0-262-01356-7

Linkages of Sustainability
edited by Thomas E. Graedel and Ester van der Voet, ISBN 978-0-262-01358-1

Dynamic Coordination in the Brain: From Neurons to Mind
edited by Christoph von der Malsburg, William A. Phillips and Wolf Singer,
ISBN 978-0-262-01471-7

Disease Eradication in the 21st Century: Implications for Global Health
edited by Stephen L. Cochi and Walter R. Dowdle, ISBN 978-0-262-01673-5

Animal Thinking: Contemporary Issues in Comparative Cognition
edited by Randolf Menzel and Julia Fischer, ISBN 978-0-262-01663-6

Cognitive Search: Evolution, Algorithms, and the Brain
edited by Peter M. Todd, Thomas T. Hills and Trevor W. Robbins,
ISBN 978-0-262-01809-8

Evolution and the Mechanisms of Decision Making
edited by Peter Hammerstein and Jeffrey R. Stevens, ISBN 978-0-262-01808-1

Language, Music, and the Brain: A Mysterious Relationship
edited by Michael A. Arbib, ISBN 978-0-262-01962-0

Cultural Evolution: Society, Technology, Language, and Religion
edited by Peter J. Richerson and Morten H. Christiansen,
ISBN 978-0-262-01975-0

Schizophrenia: Evolution and Synthesis
edited by Steven M. Silverstein, Bita Moghaddam and Til Wykes,
ISBN 978-0-262-01962-0

Rethinking Global Land Use in an Urban Era
edited by Karen C. Seto and Anette Reenberg, ISBN 978-0-262-02690-1

Trace Metals and Infectious Diseases
edited by Jerome O. Nriagu and Eric P. Skaar, ISBN 978-0-262-02919-3

---

[1] available at https://mitpress.mit.edu/books/series/str%C3%BCngmann-forum-reports-0